PENG

BRITAIN'S

'Edgerton's book is a remarkable achievement. He re-envisions Britain's role in World War II and with it Britain's place in modernity. In place of a plucky island standing alone, he gives us a global empire of machines, not a welfare state, but a technocratic warfare state. The period will never look the same again' Adam Tooze, author of *The Wages of Destruction*

'An important corrective to the black-and-white portrait of the period that still prevails' James Owen, *Financial Times*

'Accessibly written and deserves a wide audience. Above all, Edgerton demonstrates that the war is a subject we haven't yet heard nearly enough about. *Britain's War Machine* is a considerable achievement' Graham Farmelo, *The Times Higher Education Supplement*

'It is a necessary and timely corrective to a great deal of conventional wisdom, and makes a real contribution to our understanding of the war' Richard Holmes, *Literary Review*

'This remarkable book shows that whatever the reasons for the length of time it took to bring Hitler to heel, the quantity and quality of British war material was not among them' Brendan Simms, *Sunday Telegraph*

'By coming at the subject at a new angle, it offers an unfamiliar picture of the war, one dominated not by bluff airmen, bayonet-wielding infantrymen, sailors on Arctic convoys, blitzed housewives, Bletchley codebreakers or Penguin-reading intellectuals, but by backroom boys – grey men with slide rules, workers toiling through the night shift in Orwellian arms factories, chemists in laboratories' Ben Shephard, *Observer*

'Required reading for all interested in 20th-century British history' *Library Journal*

David Edgerton is Hans Rausing Professor at Imperial College London. He is the author of a sequence of groundbreaking books on twentieth-century Britain: *England and the Aeroplane: An Essay on a Militant and Technological Nation*; *Science, Technology and the British Industrial 'Decline', 1870–1970*; and *Warfare State: Britain, 1920–1970*. He is also the author of the iconoclastic and brilliant *The Shock of the Old: Technology and Global History Since 1900*.

DAVID EDGERTON

Britain's War Machine

*Weapons, Resources and Experts
in the Second World War*

PENGUIN BOOKS

PENGUIN BOOKS

Published by the Penguin Group
Penguin Books Ltd, 80 Strand, London WC2R ORL, England
Penguin Group (USA), Inc., 375 Hudson Street, New York, New York 10014, USA
Penguin Group (Canada), 90 Eglinton Avenue East, Suite 700, Toronto, Ontario, Canada M4P 2Y3
(a division of Pearson Penguin Canada Inc.)
Penguin Ireland, 25 St Stephen's Green, Dublin 2, Ireland (a division of Penguin Books Ltd)
Penguin Group (Australia), 250 Camberwell Road, Camberwell, Victoria 3124, Australia
(a division of Pearson Australia Group Pty Ltd)
Penguin Books India Pvt Ltd, 11 Community Centre, Panchsheel Park, New Delhi – 110 017, India
Penguin Group (NZ), 67 Apollo Drive, Auckland, North Shore 0632, New Zealand
(a division of Pearson New Zealand Ltd)
Penguin Books (South Africa) (Pty) Ltd, Block D, Rosebank Office Park, 181 Jan Smuts Avenue,
Parktown North, Gauteng 2193, South Africa

Penguin Books Ltd, Registered Offices: 80 Strand, London WC2R ORL, England

www.penguin.com

First published by Allen Lane 2011
Published in Penguin Books 2012
001

Copyright © David Edgerton, 2011, 2012

The moral right of the author has been asserted

Printed in Great Britain by Clays Ltd, St Ives plc

A CIP catalogue record for this book is available from the British Library

978-0-1-410-2610-7

www.greenpenguin.co.uk

ALWAYS LEARNING   PEARSON

For Claire

Contents

Illustrations

Maps and Figures

Tables

Preface to the Paperback Edition

Since this book was first published evidence that the Second World War (and 1940 in particular) has become the foundation myth of modern Britain continues to accumulate. The Royal Wedding was celebrated with fly-pasts by a Hurricane, a Spitfire and a Lancaster bomber. The successful film *The King's Speech,* while nicely reminding us of the modernity of the British monarchy, traded on the theme of war as rebirth, brazenly falsifying royal and political history in the process. Particular accounts of the war continue to inspire radicals too. Caroline Lucas, Westminster's only Green MP, launched an initiative called *The New Home Front,* celebrating wartime Britain's fighting fascism abroad and waste at home; lauding its domestic austerity, economic nationalism, and commitment to a single cause. The war still matters profoundly.

This book presents a very different view of the war, and not surprisingly has caused some a certain shock and indeed offence. For one or two critics the mere fact of contradicting established authority and weight of opinion, especially about 1940, and about the weapons of the British army, was a measure of how wrong this book must be. Let me reassure the reader ready to find fault on these grounds that I write not in ignorance of the mass of literature which contradicts what I write, but in amazement at its continuing success.

I have also offended a few readers by suggesting that Britons were better fed in wartime than is commonly believed, and also by seemingly downplaying the human cost of the war to the British people. The offence caused is I think a measure of the extent to which British ideas about the nature of the war are still insulated from any appreciation of what was happening elsewhere in the world, a condition quickly cured by consulting Lizzie Collingham's marvellous book, *The Taste of War: World War Two and the Battle for Food* (Allen Lane, 2011). By comparison with all other belligerents excepting the Americans and Canadians, the British fought a rich people's war, dishing out far more suffering than was inflicted on them.

This book suggests that Britain was strong, particularly so in the early years of the war. This has caused surprise not just because it contradicts the dominant picture in the history books (if not necessarily in everyone's understanding), but also because Britain constantly lost battles in the early years of the war, took time to win, and was hardly the dominant power at the end of the war. If Britain was as industrially, technically and military strong in 1940–41 as I suggest, why did it not defeat Germany on its own and indeed quickly? The answers are found throughout the book, but it is I think worth highlighting some points. The first is that the argument that economic, technical and scientific strength leads to victory was a very British argument, much in evidence during the war itself, as well as later. It is there in the positive form: Britain will win the war because it is strong in these areas; and in negative form: Britain lost and therefore must have been weak in this regard. As analysts, however, we should not expect any such straightforward relationship. Might and mammon went together, ultimately, but not immediately, or always, and that is the assumption from which this book starts.

In any case not every battle was lost in the early years. The battle of France wasn't lost by the British army. In North Africa there were victories as well as defeats – at sea there were important victories over the Italian navy, and against the U-boat; also against the Luftwaffe. Yet in North Africa and the Far East the British were outfought by more experienced soldiers, who were usually less well-equipped than British troops. There is another answer which is perhaps even harder to accept – that the British placed emphasis on modern weapons which simply did not work very well. Britain had a fleet of great bombers, which in 1940–1941 could do little or no damage to Germany; on the other side, expensive night air defences were ineffective until 1941. In other words, the problem was not merely a lack of modern weapons, but overinvestment in them. They came to work only late in the war, and were even then less effective than hoped. The story of the war would have been very different had the bomber, and blockade, worked as British leaders hoped they would, for Britain would have emerged victorious earlier in the war, long before the USA or USSR could have achieved dominance. In the end the war was decided by the mass armies of relatively poor countries, locked in a titanic battle, of the sort neither Britain nor the USA participated in, nor fully understood. In other words, while the book is an account of a neglected British success, it is also a story of a peculiarly British failure, one which is incomprehensible from within standard British narratives about industry, machines and war.

Of those who have accepted my argument a number have asked me to say more about how the story I criticize was established, what its influence was, what interests it served, when and to what extent it was challenged. I say a little about this, particularly regarding the 'alone' myth. Yet there is much more to be said, and I have said some of it elsewhere. But it might perhaps be worth noting here that this book is a critique of many different traditions of writing about the Second World War, ranging from the left's Blitz to Beveridge story to the right's Battle of Britain, finest hour narrative. It is a critique too of the strange nationalist-declinist account of war that cut across the left–right divide in the late 1980s and early 1990s, as well as more recent accounts of British national identity in wartime which overplay 'alone' and the 'people's war' and national feeling more generally. I hope it is clear that, far from being a celebration of British power, this book is essentially a critique of many traditions of historical writing, of both left and right, which it seems to me profoundly misunderstand the nature and import of British power.

One friendly critic has taken me to task for not saying enough about Labour and labour. I have made some small changes to make clear my view that the political changes in May 1940 were of vital significance; he was right to remind me that there was a political convulsion, not merely a change of Tory prime minister. Yet I have also made clearer my view (also developed elsewhere) that there was a deep strategic continuity across that momentous change of political authority, and that Labour's role in production, on the home front generally, has been much exaggerated – the idea that Churchill and the Tories dealt with strategy and Labour covered the home front and production, deeply engrained since the war, seems to me to be quite wrong. I have made some other minor changes, some also in response to comments from readers, for which I am very grateful.

Preface

Hasn't everything interesting already been said about Britain in the Second World War? Do we really need another book on this subject? I hope to show that we do; that much remains unsaid; and that we need to reconsider, often drastically, many important arguments. In the course of researching and writing this book I myself have been astonished by what I have found, and I say this as someone who for thirty years has been criticizing received views about twentieth-century British history, particularly in relation to warfare, to the state, and to science and technology.

I hope that by the time you have read even a small part of this book, you will feel that the new evidence and interpretations it contains will make you rethink the history both of the war and of Britain's place in it. These are not topics to be dealt with lightly. What is known and what is believed about Britain in the Second World War have mattered to many people, for many reasons. Such views have profoundly affected British decisions to go to war, from Suez to Iraq. They shape, not surprisingly, the ways we understand recent British history, for the war is taken to be a pivotal event in that history. Indeed much writing about the war has been not just about the conflict itself, but about the entire history of twentieth-century Britain.

In these pages I shall describe an image of wartime Britain and its fighting power that will be unfamiliar. I shall describe the Britain that went into the war and the one that came out in terms that are quite different from most other accounts. I see the war not as something that tested and re-created the nation; which changed much, and yet did not change everything. Instead I see it as a distinctive period, characterized by sudden yet often temporary changes.

My account gives a central place to the armed services and Britain's method of waging war. Most distinctively it is based on a new material reckoning of the war effort which respects the particularities of the extraordinary global war in which Britain was a great player. I also reject the sentimental view, still widely encountered, that the war was, for Britain, a good thing. We now too readily associate the evils of war with war crimes; yet it is the fact of war itself which is the greatest crime against humanity. That the Second World War was perhaps necessary does not prevent its being a terrible disaster for humankind, and even being on the right side did not mean it was good for those who waged it.

Camden Town, December 2010

A NOTE ON QUANTITIES

This account has a lot to say about the material, weapons, resources and armed services so a simple guide to some of the measures used in describing them might be useful. This note may be skipped, but may be usefully referred back to when a quantity is not fully self-explanatory.

Conversion of past prices into modern equivalents is fraught with unavoidable difficulty. A very rough-and-ready way of converting early wartime prices would be to multiply by fifty to get the price of an article if the same one were to be made today; and by 250 to get the price of that article if it were to represent the same proportion of today's total national expenditure. Thus, a battleship cost around £10m in 1940. Building that battleship today might cost £500m, but to build a battleship representing the same proportion of current GDP would mean spending £2.5bn.

A convenient measure of mass for our purposes is the ton; the metric tonne and the short and long tons are close enough to be equivalent (1,000kg or roughly 2,200lb). Merchant ships were assessed according to various measures with different meanings: the gross registered tonnage (GRT), often called simply 'tonnage', is a special non-standard measure of *volume* of the ship, not mass, as is net tonnage, a measure of the cargo-carrying volume. The deadweight tonnage (DWT), by contrast, is the mass of cargo, plus fuel and consumables. The standard wartime Liberty ship had a GRT of around 7,000 'tons', a net tonnage of 4,300 'tons' and a deadweight tonnage of 10,000 tons. The US-built standard T2 tankers had a GRT around 10,000 'tons' and a deadweight tonnage around 16,000 tons. Warships were labelled by their total mass – called displacement, since it was the same as the water they displaced – measured in tons. The fully loaded HMS *Belfast* displaced 11,000 tons, roughly the same amount of water as could be carried in a wartime Liberty ship.

Some British weapons were known by the mass of the projectile they fired – like the 25-pounder. In this and other cases I give the standard metric measure of gun calibre to allow comparisons, as well as retaining the old name and the more familiar imperial measures. Where comparative measures exist in imperial measures I have retained them.

The size of armies was reckoned in roughly standardized units. In the Second World War all armies operated with divisions of around 10,000–20,000 men. A British infantry division had around three infantry brigades with supporting artillery, signals and engineers. Each infantry brigade usually had three battalions of very roughly 1,000 men. Usually around three divisions would make up a corps, which would have its own supporting troops; around three corps would make an army; and two to three armies would make an army group. The British Expeditionary Force of 1939–40 was equivalent to an army, the other major British/imperial armies being the 8th Army, the 2nd Army, the 14th Army, and the 1st Army (the numbers are arbitrary). The biggest, by far, British/imperial formation of the war was the Anglo-Canadian 21st Army Group (2nd Army plus 1st Canadian Army), which fought under Montgomery from Normandy to Germany. By contrast, Germany had three to four army groups in action throughout most of the war.

I

Introduction

Britain won the Second World War. Yet that victory is hardly commemorated. On the sixty-fifth anniversary of the conflict's end in Europe, 8 May 2010, a BBC radio news bulletin reported the extensive European and small British ceremonies in seventh place in its running order: after a murder, the sale of Harrods, the resignation of a German bishop and similarly vital matters. By contrast the early war years, a time of defeat, have become the subject of an almost obsessive fascination: evacuation, Dunkirk, the Battle of Britain, the Blitz, being 'alone' – all are at the centre of national narratives of the war in which a new inwardly focused nation was born. Memorializing the beginning and downplaying the final victory in Europe (let alone in Asia) meant Britain's role in the war as a whole could be comfortingly overstated. Britain's technical genius, exceptional mobilization and emergent welfare state could be celebrated ('the People's War'). This positive account was eroded by histories insisting on Britain's economic, industrial, technical and military weakness and its saving by the USA. With the end of the Cold War came the long-overdue full recognition of the Soviet role, which only accentuated the idea of a weak Britain which made a minor contribution to victory. The story became one of the defeat of a faltering power in 1940, which in one last heroic gesture bankrupted itself to save the world.

This summary can only hint at the extent to which our histories of the war are the product of wartime political debate and propaganda, of post-war nationalism, of the ideological exigencies of the Cold War and of backward projections of what post-war Britain was thought to have become: a welfare state, a craven satellite of the US, a decrepit industrial basket case, a national community whose identity centred to an unhealthy degree on nostalgia for the finest hour.[1]

The time is ripe for the revision of the core assumptions at the base of all these stories, the myths and the anti-myths of wartime Britain. This book gives a very different and novel account of Britain at war. It reveals a Britain that was a first-class power, confident, with good reason, in its capacity to wage a devastating war of machines. It had resources to spare, was wealthy enough to make mistakes, and could fight as it chose to rather than had to. It tells of plenitude and power rather than of scarcity and sacrifice. It tells of a great power which thought of war not in martial but in material terms, as befitted an industrial giant which remained at the heart of the world's trade. This great empire fought an internationalist war with other empires and states at its side. It could avoid putting the majority even of its servicemen and servicewomen in harm's way; it did not fight the sort of people's war contemporaries often (but not always) meant by the term.

Remarkable as it now seems, in 1939, and even in 1940 and 1941, British leaders were confident of victory. They believed that Britain's material strength, its unique place in the world, its industrial, technical and scientific capacity were warrants for victory. So much so, that even without a great ally, France, they could see the way not just to avoiding defeat but to victory. Empire gave Britain manpower, but even the British Empire, vast as it was, was not alone, and did not feel itself to be alone. It was supported economically and politically by much of the rest of world, from the United States to the states of the River Plate; from the African empire of a defeated Belgium, to the Dutch West and East Indies. Churchill's decision to fight on in 1940 was hardly irrational, reckless or merely heroic. The idea of a small island nation, standing 'alone', was far from being the central image of Britain it was to become after the war.

Political and intellectual opinion saw the war as one of machines and production and acted accordingly. On leaving office in May 1940 Chamberlain left Britain with the world's largest navy, the greatest aircraft production of any country, and a small but uniquely mechanized army. Britain planned to fight, and largely did fight, a war of machines, of warships, bombers and tanks, a war of science and invention, the next war, not the last. British leaders were premature military modernists. Churchill, for all his deep interest in history, and his presentation as a figure from the past, wanted a war of rockets, aeroplanes, radio and clever gadgets of all kinds, some of his own devising. Under him, more so than under Chamberlain, invention, research and development were central. So much so indeed that some senior scientists thought that the plethora of imperfectly tested and trialled new weapons were damaging the war effort.

Ironically, Britain's enemies, who did not place such confidence in machines, won battle after battle. Britain suffered very major defeats, against enemies less well equipped. In 1940 its air defence at night didn't work, and its ability to destroy Germany from the air turned out to be non-existent. The Empire suffered at the hands of inferior German (and Italian) forces in North Africa. Much more importantly, in 1941–2 much of the Empire was overrun by a Japanese army much inferior in numbers and most equipment.

These defeats were interpreted, influentially and wrongly so, as being the result of low arms production and of low arms quality. Such criticisms were central to the politics of the war in 1942, and powerfully affected how the war was to be thought about later, and by historians. Yet it had turned out that it took more than machines to win battles or wars. Defeat in 1942 masked British strength in machines – it was already the greatest producer of arms in the world. Contrary to the impressions then created not only did Britain have exceptionally large numbers of ships and aeroplanes, but from 1940 to the end of the war it had the most tank-intensive army of the period. The warfare state was one of plenty, of armed forces generously supplied with new equipment by new factories. In some key sectors, efficiency of production was the same as in the USA.

The build-up of this empire of machines made Britain an exceptionally mobilized society where millions of people were making and using modern armaments on huge scale. This warfare state was run by a wartime British government full of experts, of scientists and economists and businessmen. The technocrats were mostly men of the right. At the top were Chamberlain, who had himself studied science at what became Birmingham University, and Winston Churchill, a man whose closest wartime adviser was a scientist and who was himself a great enthusiast for machines. Wartime politics was to a considerable extent the politics of production, of machines and of experts, topics with which the Prime Minister himself was deeply concerned. Yet some of the government's greatest parliamentary critics were scientists and engineers.

During the war *laissez-faire* was done for, but not capitalism. Free trade was finished, but not the private ownership of means of production, or the flux of goods and labour. Britain's supplies of service personnel, of industrial capacity, of raw materials and food remained global rather than national, nor yet merely imperial. During the war Britain was not, by any reasonable standard, under siege or blockaded. Imports into Britain stayed at pre-war levels in value. Most food was not rationed, and was available in quantity. Meat and cheese imports actually increased. Oil products, fuel oil, petrol, aviation spirit and lubricating oils were all imported in unprecedented quantities. The context of British action was not, as so often suggested, one of weakness, isolation and austerity, but rather of abundance of key resources. The exceptional mobilization of armed forces and industry was, rather than an indication of national genius, a product of the support Britain got from elsewhere, and from the nature of its economy. Britain had many war fronts, and many home fronts. Appropriately enough, at the time the idea of the 'People's War' was often one where many different peoples were united in a common international struggle. The British war machine imposed hardships on and created opportunities for not just the peoples of the British Isles, but those of much of the world.

The defeat in the East was a devastating blow which weakened the Empire's capacity to fight Germany and Italy very considerably. The new necessity to fight on two fronts made the Empire dependent on the United States, which only now emerged as what Britain had until recently been, a great global power. In the immediate wake of these events and American entry into the war, British propaganda switched focus and made great play of the 'United Nations', especially but not only the 'Big Four', fighting to extirpate nationalist militarism from the world. Indeed, internationalism was an important and an underrated feature of wartime politics, propaganda and indeed strategy. From 1942, there were combined chiefs of staff, Allied supreme commanders in various theatres, and there was pooling and redistribution of all sorts of resources across national boundaries. For example, through Lend-Lease Britain now got nearly all its oil from refineries in the USA, whereas at the beginning of the war it had sought to avoid the neutral USA as a source at all. The great majority of tanks in British armoured formations came to be US-built Shermans. The RAF was a uniquely multinational force, seen by some as the progenitor of an international air police. The dependence on the US was not a straightforward matter of subservience to a greater power, but an attempt through a division of labour to maximize the exploitation of common resources.

Internationalism, superiority in wealth and in machines of war, and the command of the seas paid off in the end. For the richest belligerents, the USA and Britain, victory came at very low cost; for the poorest, the USSR and China, it came at a terrible cost in lives and wealth. Britain emerged as an extremely powerful warfare state, brimming with new war machines, whose use was now mastered as never before. It was not, as is routinely alleged, bankrupt, but merely under temporary financial pressure because of the sudden end of the war in the Far East and American decisions on Lend-Lease: Britain remained easily Europe's largest economy and, second only to the United States, a massive global trading presence. The unprecedented ruination of most of the European continent in 1945 was in appalling contrast to the many economic, military, social and political options available to post-war Britain.

This book deals with these questions by looking at the material: at machines and raw materials, industry and armaments, scientists and inventors, production and destruction, and at wartime thinking about all these matters. This is not as straightforward as it might seem. Henry Ford famously claimed that history was bunk; he was wrong of course, but his dismissive words are not so inappropriate when we consider our histories of the material and the technical not least in relation to war. Many of our accounts are rudimentary, seriously distorted by over-attention to some novelties when they were novel.[2] In this book I deploy a new approach I have developed for the study of the material which considers all the machines and structures, whether old or new, celebrated or ignored, whether they figure in theories of modernity or not. It is a story of tramp steamers as well as tanks, or refrigerated ships and stores as much as rockets, of the oil industry as well as atomic bombs. It is based too on much brilliant work from historians of many stripes on the material and economic history of the Second World War.[3]

Dealing with the material means dealing with experts. The stories experts told about particular issues have been hugely influential, in terms both of the internal history of a particular area, and of its relation to the rest of history. As we shall see, we need to beware analysing the significance of experts in terms of their own estimation of their role. Nor yet should we accept their descriptions of the material uncritically. The war was not a physicists' war; nor the war economy Keynesian; nor food merely vitamins, calories, protein and minerals as measured by nutritionists. Experts may or may not have made the modern world, but they have created the way we think about it, and we need to understand the often limited way we do this. The story told here of the relations of science, machines and war, of the economy at war, is different from earlier versions not only because of new evidence presented, but also because it is based on new ways of understanding these relations. For example, this book strongly qualifies the arguments, plainly already central to discussions in wartime, that modern war was overwhelmingly a matter of radically new machines and techniques derived from civilian and civilizing science, technology and industry.

This book is the culmination of a long process of rethinking crucial aspects of twentieth-century Britain, of putting the military and experts and machines into a narrative from which they have been studiously excluded. My argument has so far been carried out principally in academically focused books and articles and has not had much exposure to the general readers who I hope will enjoy this book. The roots of my argument lie in a dissatisfaction with many aspects of the standard histories of Britain, as they stood a generation ago, and are still often seen today. We have all grown up in the shadow of a relentless barrage of what I called *declinist* histories which indulged in inverted Whiggism, finding past failure to account for present decline. My concern was not merely to challenge particular theories of decline, but rather to note that viewing Britain's fate through this particular lens created a thoroughly distorted account.[4] Declinism sought to explain British relative decline – in fact caused by the growing success of other countries – by supposed national failures, whether it be in elite culture, education, industry, science and engineering, or the military.[5] Such faults were seemingly established by questionable comparisons with other countries and these were more often than not perversely misguided. Many British nationalists, militarists and technocrats imagined Germany as the template on which Britain should have built itself. Germany was seen as powerful because it lived up in their view to their ideal of modernity.[6] Yet it was clearly recognized by some, not least by Adolf Hitler, that there were striking asymmetries between Britain and Germany that were often to Britain's great advantage and made the defeat of Britain (as it indeed proved) insuperably difficult.[7] In the British post-war literature deeply imbedded but dubious assumptions about the relationships between nationalism, militarism, technocracy and modernity obscured the particularities and nature of British modernity, and indeed of British militarism. A further curious feature of declinist arguments was that, instead of contrasting past strength with more recent weakness, they insisted on presenting a picture of past weakness. For all the mountains of writing that implied otherwise, Britain has been one of a handful of great scientific, industrial and military powers of the twentieth century and its history needs to be written with that firmly in mind. In this book I do so without troubling the reader with the older declinist picture.

A second key assumption made (rather than argued for) in this book is the result of my ongoing challenge to the ingrained identification of the British state with the welfare state. The consolidation of the welfare state was one of the key narratives (particularly focusing on the Beveridge Report) even during the Second World War.[8] Yet it seems to me inescapable that the key characteristic of the British state has also been that it was also a warfare state – that is, a state which was organized around the successful prosecution of war, albeit of a very specific kind. The strategy typically pursued by the British warfare state I have described as 'liberal militarism' to distinguish it from the model of Prussian militarism. Its most striking feature has been, in contrast to the European model of mass armies, an obsession with masses of machines, specifically machines designed to destroy enemies both physically and economically.[9] This was based most obviously around the navy and air force, which sat at the heart of British strategy and need far more emphasis than they have traditionally had.

The strange assumption by many historians that welfare rather than warfare lay at the heart of the British experience has meant that the military were left out of key aspects of British history: the history of the state, the history of research, the history of expertise. This has created some spectacular distortions of our historical pictures of Britain. Even war, especially the Second World War, has itself been understood in ways which systematically downplayed the direct role of the military in the state, economy and society.[10] So powerful has this been that wartime investment in arms factories have been left out of investment statistics; wartime statistics left munitions out of imports; and the ample military rations distributed to millions of British service personnel disappeared from the story of wartime rationing. As this book shows, taking the British warfare state seriously, taking the military seriously, very substantially changes our picture of wartime industry, imports, rationing and the nature of the state machine.

The other major novel thread through this book is the recognition of the role of experts, especially since the war, in creating a broadly fallacious view of Britain. After the war, experts argued for more active industrial policies, more investment in research and development, more science in education. They fell into the habit of making their case by suggesting that Britain had been preternaturally dreadful in all these regards since the 1870s – that the British state and industry suffered from a fatalistic or incompetent failure to set priorities or nurture new inventions. This scarcely credible account was for a while widely believed. Similarly, military experts arguing in the decades after the war for increased spending on the armed forces would invariably point to the interwar period when, apparently, British war readiness had been crippled by an especially pacifistic and insular electorate and elite. These arguments came to be very widely believed, oddly, given that interwar Britain was a military superpower at sea and in the air, supporting the largest arms industry in the world.

We need to understand that Britain was a land of experts who as part of their lobbying had produced self-serving accounts which suggested appalling, systemic weakness and glamourized their own Cassandra-like outsider status, whereas in fact Britain had vast resources, with these experts standing in their midst. What were once, and sometimes still are, taken as truisms – for example, that Britain was good at inventing and bad at developing – often appear in this book as seriously misleading arguments. The experts created a sort of anti-history of British militarism and technocracy which this book leaves behind.[11]

Instead, it presents a realistic picture of experts of many sorts fighting each other, using devious and misleading arguments, dissembling and getting it plain wrong. It stresses their power and influence in wartime Britain and their particular political commitments. It shows indeed how the politics of expertise was central to wartime politics in ways which the standard party political and welfarist histories have not captured. Churchill was surrounded by experts of many stripes and ran the most expert government Britain has ever had. It shows how some experts successfully contrived, against all the odds, to present Churchill as a man who didn't understand modern war and his government as hostile to modern knowledge.

Finally, this book depends on a novel understanding of Britain's place in the world. Britain emerged, I suggest, more parochial than it had been during the war itself, both economically and ideologically. Post-war nationalism was much more significant and more novel than histories allow because it is not seen as problematic, and because the standard focus has been on a movement to the left.[12] Pre-war Britain had defined itself in part nationally (with 'England' very commonly used to subsume Wales and Scotland) but also imperially, as the 'mother country' with satellite dominions and colonies, and also as a liberal global trading and industrial power. During the war Britain, and its empire, was first one of the Allies and then a key member of the wartime 'United Nations', many of which were British dominions. This changed quite drastically in the circumstances of 1945 and its aftermath, when a more purely nationalist ideology won out, which profoundly affected our accounts of many different aspects of the war. Histories, commentaries, statistics focused on the United Kingdom or Great Britain, giving a distorted picture of the war. Empire and Commonwealth were written out as were numerous allies too, including the USA. It was in 1945 not 1940 that the idea of Britain 'alone' became central. The whole wartime emphasis on the United Nations disappeared. It was in this context that historians overstressed domestic social change and downplayed the crucial international aspects of the war.[13] The trend of recent years of putting Empire back into the story has helped but has resulted in an overemphasis on Empire in thinking about Britain's relations to the rest of the world. The role of the dominions in particular is receiving overdue attention given their economic centrality, and their vital importance in military terms too.[14] Yet it is necessary to insist, as this book does, on the importance of much of the rest of the world to the British economy and war effort in both peace and war. Indeed, foreign supply was critical and taking it into account in itself transforms our understanding of key aspects of wartime Britain hitherto recounted misleadingly in national or in imperial terms, or as a case of dependency on the US.

Britain's relationship with other countries during the war continues to be problematic, and a central issue in how the war is remembered. On his first visit to the USA as Prime Minister the recently elected David Cameron said in two separate television interviews that 'we were the junior partner in 1940'. In the context in which he spoke it is clear he was referring to Britain's relations with the USA, rather than the more plausible case of France.[15] In response to Cameron's appalling error, David Miliband, the Shadow Foreign Secretary and contender to lead the Labour Party, stated: '1940 was our finest hour', adding that 'We were not a junior partner. We stood alone against the Nazis.' As well as mouthing clichés he made a historical howler with the startling claim that 'Millions of Britons stood up and gave their lives to defeat fascism.'[16] The Prime Minister was forced, speaking from India, to acknowledge that he should have referred to 'the 1940s', and that 1940 'was the proudest year in British history bar none'.[17] These two heirs to Tony Blair also follow him in thinking in garbled historical clichés. Yet we can take some comfort in the fact that history writing on twentieth-century Britain and indeed the Second World War is getting ever richer, changing the established stories comprehensively, treating Britain with the critical distance and seriousness that British historians have long applied to the study of Britain's former enemies. Seventy years on from 1940, it is about time.

2

The Assurance of Victory

In 1939 and 1940 even critics of the government thought Britain was in a strong position to fight and win the war. Confidence in its economic power, its dominance of global resources, and increasing confidence in its armaments stand in contrast to later images of a nation that was economically weak and militarily unprepared. The optimistic view was not nearly as misplaced as it later seemed to be; more than that it is deeply revealing about British elite attitudes to the nature of war. These were themselves very different from what they were later supposed to be. The British elite believed Britain would win not because of its martial and militant qualities, but because of its industrial and economic strength.

In early 1939 Stephen King-Hall, former naval officer and the editor of an anti-appeasement newsletter, wrote of the 'world-wide economic struggle between the democracies and the totalitarian states', claiming that 'behind, as it were, a screen of military rearmament, the democracies are mobilizing their superior economic resources'.[1] King-Hall asserted in March 1939 that 'the Nazis are now beginning their death struggle': they would lash out and strike with 'long claws' but these were 'not so wonderful and fearsome as some think, as they are partly made of substitute materials'.[2] This was far from an idiosyncratic position. The *Economist*, a consistently anti-appeasing journal, noted in April 1939 that, within weeks, British aircraft output would exceed German, and that the Royal Navy was extending its already great lead over other navies.[3] In its issue of 2 September 1939, it was, from today's perspective, remarkably sanguine about the future. It noted that it was a 'commonplace nowadays ... that wars are won and lost in the workshop rather than on the battlefield' and that 'The Industrial

Revolution has now been fully applied to killing.' Given this, the out-look for Britain was good, since Britain was in a particularly strong position to fight such a war. According to the *Economist* the British and French empires, together with Poland and Turkey, had similar white populations to Germany and Italy, and similar levels of steel production, which suggested equality in the struggle to come. But other factors pointed to massive superiority for Britain and its allies: a very large colonial population, double the coal and motor car production, and more than three times the merchant shipping, and iron ore pro-duction, of Germany.[4] The advantages of the 'democratic Allies' in a 'totalitarian war in which full economic strength can be deployed is overwhelming,' it concluded.[5]

The *Economist* presented an important refinement of the argument which needs to be grasped in order to understand both British confi-dence and also the extent of British mobilization during the war. It noted the 'immense advantages' that a war of full mobilization 'confers on a wealthy nation'. The argument was that the wealth of a nation, its income per head, not merely its absolute size or economic output, mattered. In its estimate the total outputs of the British and German economies were about the same, which would suggest equal economic power to wage war. But Germany had twice the population, meaning it had about half the national income per head of Britain (This was, according to recent data, an overestimate – see Table 2.1). Germany would have to spend roughly twice as much as Britain just to keep its population going with basic supplies like food and other irreducible items. This would leave it with less, both relatively and absolutely, to spend on armaments and armed forces. What the 'wealthy democracies' had to do was fight off an initial attack, then reduce civilian consumption, using that capacity to give them an overwhelming advantage in armaments.[6]

The optimistic view of the underlying economic power of the Allies and indeed of Britain survived the beginning of the war, and more surprisingly still, as we shall see in the next chapter, the fall of France in 1940. In November 1939 the *Economist* wrote that Britain and France, even without their empires, had four times the iron ore pro-duction, three times the number of motor vehicles in use and twice the motor vehicle production as Germany and its 1938–9 conquests.[7] 'The immense staying power of democracy is the final guarantee of

Table 2.1 The British and German economies compared

		1938	1950
Germany's total output (GDP) per head as a percentage of the UK's		82.6	65.8
Germany's total manufacturing output per head as a percentage of the UK's		107.1	96.0
Percentage of workforce in manufacturing	UK	32.0*	35.0
	Germany	30.0**	31.0
Percentage of the workforce in agriculture	UK	8.0*	5.0
	Germany	30.0**	24.0

*1930 **1936

Source: S. N. Broadberry, *The Productivity Race: British Manufacturing in International Perspective 1850–1990* (Cambridge, 1997), Tables 5.1 and 5.4

Allied Triumph,' noted a government pamphlet called *The Assurance of Victory* of December 1939, focusing on the economic aspects.[8] Looking forward to 'the War in 1940' on New Year's Day the *Times* was supremely confident.[9] In March 1940 the *Engineer* asserted that 'The financial, industrial and commercial resources of the British Empire far exceed those of the enemy, and if they are fully mobilized, and remorselessly applied, they must become irresistible.'[10] Scientists too were confident, even in private. The physicist W. H. Bragg thought that 'Science has come on so fast in its applications and uses that the administrator in general does not know what he can make of it. Yet the development and use of science gives us a real advantage over the totalitarian country where originality is discouraged.'[11] 'There can be no doubt,' thought the *Engineer* in March 1940, 'that the relative positions of Great Britain and Germany as regards scientific preparedness show a complete reversal in our favour' compared with 1914. The progress that had been made during and after the war was such that 'to-day British industrial scientific research, official, and private, can well be claimed to be unsurpassed' excepting perhaps only that of the United States. Furthermore, the 'madness of racial discrimination' and 'interference of political propaganda' meant a decline in German science which 'is not unlikely to prove a decisive element in their defeat'.[12]

In 1939 and early 1940 there was confidence not only in Britain's general economic and scientific strength, but in its armaments too. In January 1940 the *Engineer* observed 'with the deepest satisfaction that our aircraft have, so far, proved beyond doubt superior to those of Germany. That was not expected by those who habitually deprecate the work of their own country, and accept the carefully fostered self-evaluation of Germany.' In its patriotic enthusiasm it proclaimed the 'incontestable superiority of British aircraft of all kinds', which was due to British engineers.[13] On the question of quantity too the professional view expressed in public was positive. *Aircraft Production* noted in its issue of April 1940 that the Air Minister had announced that 'allied output of aircraft now exceeded that of Germany'.[14] In fact Allied output had done so for quite a while.

Winston Churchill, who went into the Admiralty and the War Cabinet at the beginning of the war, admitted in his post-war memoirs to exaggerating British weakness before coming into office. He recalled 'being content' that his accounts of the dangers of air power 'should act as a spur'.[15] With regard to the navy, he recalled that 'my public speeches had naturally dwelt upon weaknesses and shortcomings and, taken by themselves, had by no means portrayed either the vast strength of the Royal Navy or my own confidence in it'. He claimed that he had felt that 'I had at my disposal what was undoubtedly the finest-tempered instrument of naval war in the world . . .'[16] Churchill's own warning as to propagandistic aims has not alas prevented many taking his exaggerations as descriptions of the real state of affairs.

THE WORLD ISLAND

Britain was a great economic power at the heart of a global sea-borne trading system; it was, in an apt phrase, a 'world island'.[17] It was the richest state in Europe, measured in income per head, certainly richer than Germany, though this was and is often forgotten by Britons too impressed with the idea of Teutonic efficiency. Although the heart of a great empire, Britain depended on global trade for essential materials;

its imports came from near and far, from Empire and from foreign sources. France traded comparatively much less with both the rest of the world and its empire. Global trading was not as difficult as might be supposed since even very long-distance sea transport was cheaper than short-distance land transport.[18] A military geographer of the 1930s hardly felt it necessary to tell his students that it was cheaper to send coal from Cardiff to Port Said by ship than to London by train; or that it was cheaper to bring wheat from the River Plate to London, than it was to send it on to Northampton.[19] The oceans united grain silos in Britain with those in Canada, Australia and Argentina, the cold stores of London with the meat packing works in Buenos Aires, Montevideo, Wellington and Brisbane, and the British motor car with oil refineries in the Caribbean, Iran, Palestine, Sumatra and Borneo, the steelworks of Sheffield with the iron ore fields of Sweden, pork butchers of London with the pig farms of Denmark, and the forests of the Baltic lands with the timber merchants of Britain.

Britain was seen as being fortunate in its relations with the wider world. Sir Alfred Zimmern, Professor of International Relations at Oxford, claimed in a lecture in 1934 that what he called the 'welfare states' – which included Britain and the USA – were gaining the ascendancy over the continental 'power states'. He claimed nothing less than that 'science has reduced them to a subordinate political position', for modern scientific armaments demanded key materials unevenly distributed throughout the world. This meant that only powers which controlled the seas had access to them.[20] The sea as a British-dominated highway which gave Britain power was still a powerful image at the centre of the thinking of Britain's elite.

It is important to distinguish this maritime and global orientation from an imperial one. Imperialism was central to elite thinking too, but it would be a serious mistake to understand Britain's relations with the world as primarily relations with the Empire. To be sure, Empire became more important to British trade in the 1930s. Imperial preference was introduced in the early 1930s, and the proportion of imperial trade increased, and was significantly greater than it had been in the Edwardian years.[21] Still, most of Britain's trade remained extra-imperial, as did its broader economic influence. One measure of that influence is that the sterling bloc extended well outside the Empire,

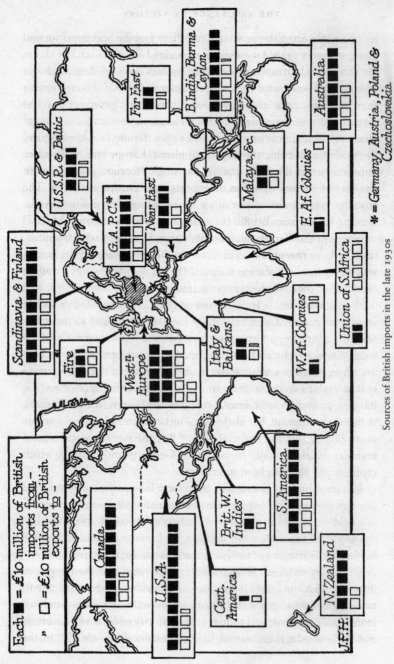

British imports and exports, late 1930s. Map by L. F. Horrabin

to include the Spanish-speaking River Plate and the Scandinavian and Baltic states. In 1938, for example, the United Kingdom imported raw materials, food, manufactures worth £919m (very crudely £45bn in today's prices corrected by a general price index). Of this just over a third (£371m) came from the Empire (including protectorates and mandated territories and territories under condominium); but Europe alone supplied almost as much with £308m (Denmark, Germany and the Netherlands being the largest suppliers). Europe was much more important than the USA, the largest single 'foreign' supplier, with £118m.[22] In terms of value of exports too, Europe mattered. The Scandinavian countries, for example, amounting to 17 million people, imported more from Britain (£40m) than did India or South Africa.[23]

In terms of monetary value Britain's greatest exports were manufactures, but in terms of bulk coal was overwhelmingly important. Britain was not merely self-sufficient in energy, but was the world's leading exporter of coal, the world's most important source of energy. Just before the war some 37 million tons of coal (excluding bunker coal in ships) was exported: this was about the same in weight as the entire annual wartime import bulk (divided as it was roughly equally between food, raw materials and tanker imports). It exported far more coal than it imported or consumed oil in all its forms. The one third carried in British ships went (in order of markets) to France, South America, Italy, Egypt and Canada. British coal went to Europe (mainly France), to the Mediterranean, but also to the southern hemisphere in what was overwhelmingly a non-imperial trade.[24] By contrast France was a great importer of coal (mainly from Britain, Germany and Poland), which made up half the weight of its imports.[25]

The great bulk of Britain's imports were biomass: timber and food (for humans and animals) being the main categories, though cotton and wool should not be forgotten. Oil was imported in similar quantities to food.[26] Britain was the largest importer in the world of all these bulk commodities, a point which needs particular emphasis.

The extent to which British food was imported, and the enormous distances which much of it travelled, confounds any idea that globalization is a recent phenomenon, and food-miles a measure only of interest today. In terms of tonnage, the majority of fruit, sugar, cereals and dairy products consumed in Britain came from abroad; in the case of meat the overseas sources were only just ahead of domestic supply. Among the standard foods local production dominated consumption of only potatoes, vegetables, liquid milk and fish.[27] Some estimates put imports at two thirds of the calories and one half of the protein consumed.[28] The cases of meat and wheat illustrate Britain's reliance on the whole world particularly well. Britain, the only major country which imported wheat free of tariffs, supplied domestically only 14 per cent of its wheat and flour up to 1932; after that a 10 per cent tariff for non-imperial wheat, and subsidies at home, increased the proportion to 20 per cent, rising to one third during the war.[29] In the 1930s the main suppliers of wheat were Canada, Australia, Argentina and Europe. Roughly half Britain's meat came from overseas. Beef and lamb came from the River Plate and Australasia; bacon (and eggs) came almost exclusively from continental Europe. The British were the great meat-eaters of Europe, eating 30 per cent more meat per capita than Germans. The kind of meat they ate was very different too. Per capita, the British ate half as much pork as the Germans but twice the beef and veal, and twenty times the lamb and mutton.[30] If, as many believed, having a small domestic agricultural sector was an index of modernity, then Britain was easily the most modern large nation in the world. Such a policy made sense: the policy of cheap imported food is 'one of the very real advantages Britain and the British worker hold over any other country in the world in times of peace'.[31] As we shall see, despite initial appearances and the arguments of nationalists, it was a crucial advantage in war too.

Britain's imports of raw materials were also enormous, though they typically travelled shorter distances than food. In terms of bulk the list was topped by timber, petrol and other petroleum-based fuels, and iron ore, in that order. Timber came overwhelmingly from the Baltic, petrol from the Caribbean and Middle East, and iron ore from Sweden, Spain and North Africa. Empire was again far from being the only or even the major import source.

Britain's enormous European trade was much less visible to contemporaries (and to historians) than its Empire and long-distance trade. Indeed the near trade with Europe was called 'home trade'.[32] Another reason is that it was too easily assumed that British trade was carried only on British ships. For example, a textbook on *Imperial*

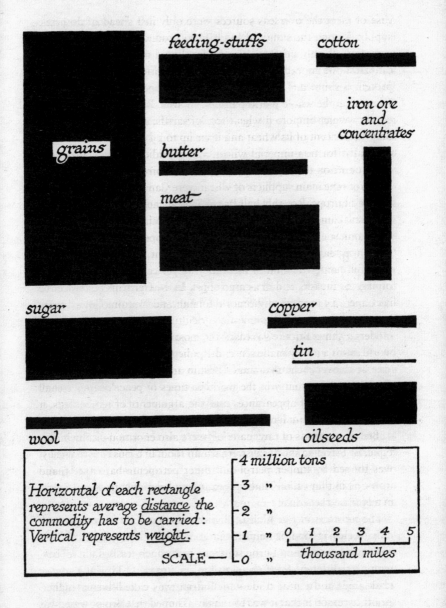

feeding-stuffs

cotton

grains

**iron ore
and
concentrates**

butter

meat

sugar

copper

tin

wool

oilseeds

*Horizontal of each rectangle
represents average __distance__ the
commodity has to be carried:
Vertical represents __weight__.*

SCALE —

4 million tons

3 "

2 "

1 " 0 1 2 3 4 5

0 " thousand miles

An imaginative representation of key British imports.
Two very large ones, timber, from nearby, and oil, from far away,
are not represented. Figure by L. F. Horrabin

Military Geography in its 1934 edition had a map showing 'the principal trade routes and the commodities they carry' nicely indicating the number of British ships over 3,000 tons operating on each route in 1923. It shows no trade whatever with Europe, though the non-imperial USA (108 ships) and River Plate (164 ships) figure prominently.[33] As much as 44 per cent of pre-war British imports, mostly timber and iron ore, by weight, came in foreign (largely Scandinavian) ships.[34] Indeed, the great majority of the important trade with Northern Europe was carried by foreign ships.[35] Two thirds of Britain's bulkiest export, coal, which went largely to Europe, was carried by foreign ships.[36]

By contrast most of the long-distance trade was carried in British ships. Britain's shipping fleet was the largest merchant marine in the world, making up over a third of the world's ocean-going gross registered tonnage. The overall size of the fleet (some 20 million gross 'tons') was about the same as it had been in 1914, giving the 1937 British Empire (overwhelmingly Britain) 32.5 per cent of the total world tonnage of tankers and dry-cargo ships. Its emergent rival, the USA, had a mere 15 per cent of the world tonnage, and Japan followed with 7 per cent.[37]

British shipping, like London's transport with its double-decker buses, trams and trolleybuses, and its distinctive taxis, had its own suite of rather special machines. Least distinguishable from the ships of others were its coal-fired freighters powered by reciprocating steam engines. These were slow ships, carrying all sorts of cargo, including the bulk dry cargoes like coal and wheat. They could typically carry around 10,000 tons of cargo, while their 'tonnage' was typically in the region of 6,000 'tons'. Many were tramp steamers, meaning that they plied the seas looking for cargoes, rather than operating to fixed timetables, as did liners, the name given to any kind of ship operating to an advertised schedule.

The most distinctive kind of British ship was the fast refrigerated liner ('reefer' in modern maritime jargon), carrying people and chilled and frozen produce from the southern hemisphere to Britain.[38] They were very different from the slow tramp steamers. Typically somewhat larger, and a good deal more modern, they were driven at higher speeds by much more powerful engines. They were often powered by diesel engines (the so-called 'motorships') or steam turbines, otherwise reserved for great passenger and naval ships. There were around one hundred of these ships in the British fleet. The Nelson Line (part of Royal Mail) had five passenger-reefer motorships built from the late 1920s at the Harland and Wolff shipyard: the *Highland Monarch*, *Highland Chieftain*, *Highland Brigade* and *Highland Princess*, and the *Highland Patriot* on the River Plate run.[39] The Blue Star Line (owned by the Vestey group) was the largest reefer operator. It had, for example, five large (13,000–15,000 'tons') oil-fired, turbine-powered ships built in the late 1920s for the South America run: the *Almeda Star*, *Avila Star*, *Andalucia Star*, *Arandora Star* and *Avelona Star*. Blue Star was also one of the main reefer lines to Oceania, and in the 1930s built ten motorships, all around 10,000 'tons', for this route. These along with a similar five for the Shaw Savill Line, which operated to Australasia, were called 'Empire Food Ships'; a further nine were built in 1939– 48.[40] The Shaw Savill Line operated the greatest of these passenger/ food ships, the *Dominion Monarch*, a huge motorship completed by Swan Hunter and Wigham Richardson in January 1939. She could carry over 500 first-class-only passengers and 12,800 tons of refrigerated meat and dairy products plus 3,600 tons of general cargo. She measured 27,000 'tons' and sailed at the high speed of 20 knots. The Port Line, New Zealand Shipping Co. and the Federal Line also operated new reefers to New Zealand. On such machines the British supply of meat (and long-distance butter and cheese) depended.

Reefers took around twenty-one days to travel to Britain from Argentina or Uruguay, and just over thirty days from Australasia (via South Africa, Suez or the Panama Canal). This difference in journey time was significant in that it put the Empire suppliers at a distinct disadvantage in the case of beef. While all lamb was frozen and was thus very long-lived, chilled beef, far superior to frozen, could not survive the long journey from Australasia. In South America low-quality animals were labelled 'freezers' and 'continentals', indicating their fate and destination. British companies and the British government invested a great deal of research effort into refrigeration in order to imperialize beef supply. Research at the Low Temperature Research Station, a government-funded institution at Cambridge, showed that chilling in an atmosphere containing 10 per cent carbon dioxide extended the life of chilled meat. The Department of Scientific and Industrial Research, the state patron of this imperial scientific venture, recorded the first shipment from New Zealand in June 1933, on the *Port Fairy*.[41] The trade grew, and the Vestey enterprise went into the Australian beef trade. There was a hope, hardly realized, that South Africa and Rhodesia would enter the market for both beef and lamb much more strongly.[42] The vast bulk of chilled beef still came from the River Plate.

Though far from uniquely British, the British oil tanker fleet was the largest in the world. The US fleet was a good second in size, though Norway was not far behind.[43] The US fleet transported fuel around the coasts of the US, which was not only a vast consumer of oil, but a net exporter too.[44] Britain's position was very different. It was the world's largest importer of oil. Its major oil companies had enormous fleets. Anglo-Saxon Petroleum, operating ships for Shell, had ninety-three large tankers in 1939, most of around 12,000 'tons'. These were all motorships of around 12 knots.[45] The British Tanker Company, owned by the state-owned Anglo-Iranian company, had ninety-one tankers of similar sizes, with a greater variety of engine types.[46] Created in the Great War, it was a patriotic enterprise and named all its ships the *British* something – for example, *British Courage*, *British Destiny*, *British Science*, *British Engineer*, *British Sincerity*, *British Zeal* ... There were smaller British-registered tanker lines owned by Standard Oil of New Jersey (Esso), the Anglo-American Line, and the Shell-owned Eagle Oil and Shipping Company, owners of the *San Demetrio*, the subject of a well-known 1943 film, *San Demetrio, London*.[47]

The world was criss-crossed with British-owned and British-registered ships, with British masters and engineers. Some two thirds of the crews were white British, the remainder being mostly what were called 'lascars', though they were used only on eastern routes.

It is a peculiarity of thinking about global history, and the history of international relations, that when machines are considered those that first come to mind are means of communication – ships, telegraphs, airlines. These machines can too easily obscure the international and global aspects of *production*. Similarly, Britain's foreign economic relations are primarily thought of in terms of trade, communication and finance, rather than production.[48] Yet British production was as global as British trade. Ships and telegraphs were hardly the only British machines operated by British managers and technicians and workers outside Britain. All around the world British technical experts were vital to the operation of British-owned facilities, many of which could boast of being the largest refinery, or meat-packing plant, or whatever, in the world. In the mid-1930s there was some £1bn in nominal capital in British enterprises operating overseas: £700m in foreign enterprises and maybe £400m in direct investments. Industrial subdivisions show the centrality of railways, mines, oil and plantations (rubber) in these investments.[49] These enterprises were British not only in ownership but often in management and in equipment. British-owned railways, which dominated in South America, usually had British managers and engineers, as did telephone companies, gasworks, water works, and rubber and tea plantations all over the world. The white British tea planters of Ceylon or India did not just supervise the complex process of growing and picking tea, but of manufacturing it in appropriately named tea factories. These buildings of four or five storeys, built of corrugated iron, were where tea was withered, rolled, fermented, fired, sorted and bagged. The Anglo-Iranian Oil Company had around 1,500 British staff (plus wives and children), comprising managers, chemists, engineers, geologists, etc., in Iran in the late 1930s.[50] The Burmah Oil Company refinery near Rangoon was 'well staffed by capable Scottish engineers'.[51] In meat production too British staff were numerous. The Vestey plant in Buenos Aires (the Anglo) had around one hundred British employees supervising the work of locals; the plant in Fray Bentos had a British community of between sixty and seventy adults and children.[52] There were British textile mills in Shanghai, like the China Printing and Finishing Company (which in fact also spun and wove cotton), a subsidiary of a British textile giant, CPA. China Printing was run by James Ballard, a product of Blackburn Grammar School and the University of London, from where he had a first-class degree in chemistry; Ballard was a 'much-travelled businessman, a lifelong admirer of the scientific world view and an enthusiast of all things American'.[53] The total numbers of British technicians involved are unknown but, for example, some 20 per cent of the members of the Institution of Mechanical Engineers were abroad. The branches in the Caribbean, China and South America each had fewer than one hundred members, whereas there were hundreds in India. This compared with around

1,000 in a branch in a highly industrialized British region. The proportion of chemists born and trained in Britain but working abroad was similar.[54]

The British dominions were in some senses part of the British diaspora, but partly too new nations. The dominions were vast settler territories which added 20 million white people or so to the UK population of under 50 million. The wealth of this world was astonishing. Some parts were, like the USA, richer than Britain in per capita terms, and by some measures of modernity more modern too. Although Britain was the most motorized nation in Europe, with twice the number of cars on the road per person than in Germany, car usage was considerably higher in some dominions. Empire, especially in the case of the dominions, was bound together with Imperial Airways, the BBC's Empire Service and Empire Food Ships. The modernity of Empire is clearly conveyed in the Empire Exhibition held in Glasgow in 1938, which was self-consciously modern, massive and imperial. The largest two buildings were the palaces of engineering (with an astonishing five acres of space) and of industry. Some individual industries had their own pavilions, as did some companies, such as Imperial Chemical Industries. The British and dominion governments had pavilions, but India abstained.[55] The tone of the Glasgow exhibition was more aggressively modernist than the 1924 Empire Exhibition held in Wembley and perhaps also than the smaller, though much better-known, little-Englander Festival of Britain of 1951.

THE WEALTH OF THE NATION

Nationalist critics of British capitalism focused on its international nature, the high export of capital and the high imports of goods. Their image was one of rentiers living in Britain drawing dividends from foreign-based enterprises which made their money from selling imported goods to British workers. The argument was that this internationalism weakened industrial activity as well as agriculture within the boundaries of the United Kingdom. In other words a powerful British capitalism did not imply a powerful productive base within the United Kingdom. In fact, by the 1930s Britain was already much more nationally focused as an economy than it had been before. By the mid-1930s sterling was off the gold standard, and was a managed currency kept low. The flow of capital was controlled and low, and was inward rather than outward. Where Britain had once been unique in its free trade, it became protectionist, giving preferences to Empire. It became, more than it had been for decades, an 'insular capitalism'.[56] British imports and exports were now the subject of tariffs and quotas, multilateral and bilateral deals. This made a very significant practical and ideological difference. As one economist put it: 'When I began teaching economics I used to tell my students that international trade was a misnomer as all trade was inter-individual; by 1938 international trade had become only too true a description.'[57]

The interwar national economy did not grow much, but it had changed a great deal. There is an image, dating from the 1930s, which shows Britain divided into those parts dominated by the hard-pressed export industries and those characterized by industries directed to the home market. While shipbuilding, cotton, iron and steel, and coal all did badly, construction, food industries and consumer durables did well. There was a great and obvious difference between a depressed South Wales, Tyneside or Clyde and the expanding octopus of Greater London – the former saw mass unemployment, the latter the expansion of new industries and firms.[58] With hindsight it is too easy to contrast a financial London with the industrial provinces, for it was precisely in Greater London that the nationally focused new industries grew, while the global industries of the northern industrial areas withered. Although the old industries like coal, and railways and cotton textiles were suffering in the interwar years, it is important not to neglect the sheer weight of their importance and competitiveness. Britain produced the cheapest ships largely by being the most efficient producer; it mined and exported vast quantities of coal, and still spun and wove huge quantities of quality cottons.

The great expanding centre of 'inner Britain', London, did not build ships but it built aeroplanes, it did not mine coal but it made electrical equipment, it did not grow food but it did process it – into beer, refined sugar, Horlicks and Mars bars. It made tyres, Hoovers, films. Take Hayes, on the western edge of London, home to the record company EMI with its research laboratory, where one of the two television systems in use in Britain was developed; to the Fairey Aviation company, one of the most important aircraft makers in the country; and to Nestlé, which launched the Milky bar and Nescafé in the 1930s. George Orwell worked here as a schoolteacher, and looked askance at its products.

There were many such places around London. In Elstree/Borehamwood one could find the 'British Hollywood'; Hatfield had the de Havilland aircraft works; Harrow had Kodak; Weybridge had Vickers Aviation; Brentford had the Firestone tyre factory; Wembley was home to the General Electric Company; Perivale had Hoover; and Dagenham the giant new Ford factory and the May and Baker drug company, which made the revolutionary sulphonamides, antibiotics of enormous power whose place in history has been overshadowed by penicillin. One very notable feature was that many of these enterprises were foreign owned, largely from the USA, as was the case for Mars, Hoover, Ford, Kodak and Firestone. Nestlé was Swiss, and May and Baker French-owned.

Not that modern industry was confined to Greater London. Nor should we distinguish old and new too clearly. A great example was a famous Manchester firm, the Metropolitan-Vickers company, one of the main electrical-equipment makers (which was controlled by US interests). It made heavy equipment mainly, from gigantic steam turbines to generating sets, electric motors of many sorts, and all the paraphernalia of electricity supply from transformers to rectifiers. It also made electronic equipment, including radar transmission gear. We have a material balance estimate for its plant which reveals how closely integrated the supposedly new and old in fact were. Each year 120,000 tons of material came into its Trafford Park works, and it left divided equally into finished product, refuse and flue gases. Those flue gases arose from burning 20,000 tons of coal for heating and process work. It was far from the only energy source: all the electricity came from the Manchester Corporation's Barton power station (which powered among other things 4,100 electric motors in the plant). The total quantity of metal coming in exceeded the coal – 16,000 tons of sheet steel, 2,000 tons of steel forgings, 4,500 tons of steel castings and 10,000 tons of pig iron, and 4,000 tons of copper. Other inputs were 5,500 tons of timber, 1,000 tons of slate and marble, 400 tons of insulating boards and 40,000 square yards of silk. The site employed 15,000 workers in 1939, double the figure in the early thirties, and 50 per cent more than before the Depression. Of these, 600 men worked in maintenance, 900 were trade apprentices and 400 were 'college apprentices', the majority of whom were university graduates.[59] Metropolitan-Vickers was one of Britain's main research firms, and was closely associated with physics and electrical engineering departments in important universities.

The research laboratory was, by the 1930s, a standard feature of the large manufacturing firm. The firm which did most research was Imperial Chemical Industries. The name of the company – as it said itself in 1929 – was 'deliberately chosen. The British Empire is the greatest single economic unit in the world, one in which every patriotic member of the great British Commonwealth has a personal interest.'[60] Imperial Chemical Industries' major project of the late 1920s was the imperial synthetic fertilizer plant based at Billingham. Its project of the early 1930s was synthetic petrol, also at Billingham. Probably the largest single research laboratory in Britain was that built in north Manchester for Imperial Chemical Industries' dyestuffs division. It was designed by the modernist architect Serge Chermayeff (also one of the architects of the De La Warr pavilion at Bexhill-on-Sea). Metropolitan-Vickers's Research Department at Trafford Park was one of many large ones in the electrical and electronic industries, including, for example, GEC at Wembley, British Thomson-Houston at Rugby and EMI at Hayes. Marconi also built a new laboratory, in Great Baddow in Essex, in the mid-1930s.[61] In fact, whatever the industry, there was hardly a significant manufacturing firm without research. The glass makers Pilkington employed members of the Pilkington family with science degrees, and in 1938 built a new Central Analytical and Research Laboratory.[62] Cadbury and Rowntree (in chocolate), Unilever (in margarine and infant foods), Glaxo (pharmaceuticals) and numerous others all had research laboratories, with many committed to researching and exploiting the new science of vitamins, very much the thing in food in the 1930s.[63] Even in the supposedly old industries like coal and steel and textiles, research also proceeded apace, in the laboratories of firms such as United Steel, Hadfields and Calico Printers Association (which produced terylene).

Comparing this research effort at the national level may not be fruitful, since firms competed with each other across national boundaries, and in this period especially collaborated too, but the USA was doing considerably more research, in total, and as a proportion of manufacturing output. Otherwise only Germany was comparable with Britain, though comparisons here are more uncertain.[64]

ARMS

In 1937 the writer E. M. Forster, in a contribution to a collection of essays called *Britain and the Beast*, noted that much had been said about the damage to the countryside, by 'private selfishness, too little about the destruction wrought by National Defence'. He was worried that 'the fighting services are bound to become serious enemies of what is left of England. Wherever they see a tract of wild, unspoiled country they naturally want it for camps, artillery practice, bomb-dropping, poison-gas tests.' He mentioned Salisbury Plain, Bere Heath, Lulworth Cove, Abbotsbury, Plymouth and Woolwich Arsenal, a 'park of death'.[65] Forster's image of a predatory modern British military hardly fits with accounts from 1940 and later which stressed the weakness of British preparations for war in the mid-1930s. But it helps us to recall a world in which many intellectuals and others lamented that military expenditures were increasing at the expense of social progress, that the applications of science to war were advancing faster than those to medicine.

It used to be argued with confidence, indeed vehemence, that by the early 1930s the British arms industry barely existed. It had been run down by naive governments beholden to a pacifist electorate, at least in the imaginary nation in which one of the largest arms industries in the world did not exist. For Britain had tens of thousands employed making armaments for the British and imperial governments and for export. It was the greatest arms exporter in the world. Some of Britain's largest firms were in the arms industry, headed by Vickers with its great subsidiaries Vickers-Armstrong, Vickers Aviation, Supermarine and the English Steel Company. The industry included great aircraft and aero-engine makers like Bristol, Rolls-Royce and many firms grouped under the umbrella of Hawker-Siddeley, for the aircraft industry was overwhelmingly an arms industry, for all that most literature on aviation suggested and suggests it is primarily a means of transportation. A good chunk of the shipbuilding industry, which was mainly civilian, made naval hulls, machinery and some other parts of warships. The arms industry included a state sector, designing and making guns, explosives and chemical weapons, and more besides. It would be tedious to list all the key firms but it is important to stress the depth and complexity of the specialized arms industry.[66] Arms production and its expansion depended on these firms, even more than was then expected. It was too readily assumed that after a few orders to give them experience of making arms, other firms, especially those experienced in mass production, would dominate the manufacture of weapons. In fact, more arms production was to mean an even more varied and much larger military-industrial complex.

We rarely think of the 1930s, unlike the years before 1914, as seeing an arms race. But arms race there was, involving all the major and many of the minor powers. Britain was a key participant in this collective militarization. It was moving along the path from welfare to warfare state. In 1965 A. J. P. Taylor made the perceptive comment that 'if one were to judge British policy solely from the story of armaments (as Hitler's policy is . . .), it would appear that Great Britain was marching consciously towards a war on which she was all along resolved'.[67] Taylor was right to point out that armaments production rose steadily according to established plans. Starting seriously in 1935, warlike expenditure was hugely increased, and at a rate much faster than in the Edwardian arms race. The 1939/40 planned expenditure (announced at the beginning of 1939) was for defence and civil defence spending of around £580m, just over half of all government expenditure, the equivalent of some 80 per cent of all government expenditure in the early 1930s. The peacetime air estimates for 1939/40, of over £200m, were twice the entire defence budget for 1933. Another measure of preparedness was that during the war forty-four overwhelmingly new ordnance factories were in operation; by March 1939 sixteen of these had been authorized and twenty-nine by December 1939; by December 1940 thirty-one were in operation.[68]

After the war there was a tendency to associate the policy of appeasement of dictators with disarmament or at least weak armaments. The

PEASHOOTERS

David Low's view of British arms production in 1940

policy of appeasement explains, it is suggested, weak rearmament. It is also proposed by defenders of Chamberlain that weak armaments explained appeasement. In fact rearmament and appeasement went hand in hand. For example, 1935 saw the Anglo-German naval treaty, double-dealing over sanctions on Italy and the start of rearmament. In that year a national government was re-elected with a huge majority on a platform which included a commitment to 'do what is necessary to repair the gaps in our defences', which would 'undoubtedly bring a considerable volume of work to the depressed areas'. The Labour and Liberal Oppositions both attacked what they saw as unilateral British rearmament.[69] From 1935 through to March 1939 appeasement and rearmament marched together: at the time of the Munich agreement British arms production was accelerating; the successful resolution to the crisis with the removal from Czechoslovakia of the Sudetenland gave Hitler's Germany another part of the former Austro-Hungarian empire. This seemingly successful appeasement of Hitler's ambitions did not slow down rearmament; rather, the contrary. The end of appeasement in March 1939, which came with Hitler's marching into the Czech lands, did not fundamentally alter the already huge scale of the rearmament programme, except that war became more likely.

Not only did the government rearm and appease, but a considerable body of opinion on the right wanted exactly this combination. Leo Amery, a leading Tory imperialist, too easily and misleadingly labelled an anti-appeaser, was deeply hostile to the League of Nations and in favour of national armaments. He was happy about giving more power to Germany in Europe, and rejected involvement in Europe other than at its fringes. He wanted to increase the navy and imperial shipping, and to be friendly to the two great potential threats to the Empire, Japan and Italy, including supporting their incursions into Manchuria and Abyssinia.[70] By contrast the left, initially at least, was hostile to both national rearmament and appeasement. Those who stood for appeasement and against rearmament were few and far between. Bertrand Russell might well be an example. He thought modern war would destroy civilization and was thus to be avoided even at heavy cost. While hostile to Nazism – as a nationalism – he was more hostile to communism, and thus even in mid-1939 he believed in appeasement. He became a militant warmonger only after the Nazi–Soviet Pact; like many he was more concerned with the Soviet threat than with the Nazi one.[71] Those who stood for rearmament and against appeasement were also a minority. Among them were Churchill on the right, and liberal and leftist figures who believed in a rearmament-backed anti-Nazi League of Nations. The latter were upholders of what were condemned as 'utopian' ideas by the appeasing 'realists' in the world of international relations.

In March 1939 appeasement was finished, and collective security against the Nazis was in. The government had adopted the policy of the Opposition, and there was now unprecedented political and press unity in Britain.[72] That unity was all the stronger in the wake of the Nazi–Soviet Pact of August 1939, which made anti-Nazis of anti-communists and anti-communists of anti-Nazis. Britain went to war in September 1939, somewhat reluctantly but deliberately, well-armed, united and confident of victory. It was under no immediate threat.

THE BRITISH FLEET

Britain's militarism was, by the standard of the continental nations, rather peculiar. Although Britain had successfully created a mass army in the Great War, this was felt by most advanced thinkers to be an aberration. In 1937 Winston Churchill pointed to a contrast between Britain and France:

> The British are good at paying taxes, but detest drill. The French do not mind drill, but avoid taxes. Both nations can still fight, if they are convinced there is no other way of surviving; but in such a case France would have a small surplus and Britain a small army.[73]

Governments of the 1930s agreed and it was not till 1939 that limited conscription and plans for an expeditionary force in Europe were made.[74] Britain was not like Germany or France, but this was not a measure of its backwardness. For British intellectuals of the time it was, on the contrary, evidence of greater wealth and modernity.[75] Even for the leaders of the British army, the navy was critical. Field Marshal Sir Philip Chetwode told the well-connected former Cabinet Office official Thomas Jones, in December 1939, that 'Ships may be dear, but to man them is far less costly than the upkeep of great armies'; he proposed a 'League of the Waters', made up of Britain, the US and Italy, who would between them control the seas.[76] He was confident that 'we shall beat the Germans because our money and credit will last longer'.[77] The reference to Italy illustrates nicely that in December 1939 the world was very different from what it would be in June 1940.

If we understand British strength not in continental military terms but in British terms, there is little doubt about relative British strength, bearing in mind that Britain never envisaged fighting a great power alone. The Royal Air Force of 1935 was without question one of the most powerful in the world – its rivals in size were the French and perhaps the American, with an emerging Luftwaffe threatening. The Royal Navy was the largest in the world, with the US navy just behind, followed by the Japanese navy. No European navy was anything like as powerful. On the other hand, all the great powers other than the USA had larger armies. Plans for expanding the army significantly came only in 1939. At the beginning of the year plans for a nineteen-division Field Force (a force to be deployed overseas other than for imperial defence) were announced; this was increased to thirty-two divisions in March 1939; the declaration of war saw it increased to fifty-five divisions (of which thirty-two were British). At the beginning of 1940 Britain had an army of 400,000 men on the Western Front (ten divisions), tiny by comparison with the French or German field armies.

The core of the Royal Navy, as of all the major interwar navies, was its fleet of battleships. These extraordinary things – there were about fifty in the world – were, apart from the greatest of passenger liners, the largest things afloat. Their engines – usually steam turbines fed by oil-fired boilers – were as powerful as those of entire power stations of the time. British battleships were driven by between 30 and 100 megawatts of power. The main guns could fire a broadside many times heavier than the bomb load of an interwar bomber, and do so again and again; the mountings could adjust the position of the gun with such accuracy that shells could hit targets many miles away; analogue computing devices (fire control gear) allowed the shells fired from one moving ship to hit another moving ship; the armour plating could protect against repeated hits from shells weighing hundreds of kilograms. In 1935 the British navy had, as a result of multilateral disarmament, fifteen battleships to the USA's fifteen and Japan's ten, the only powers remotely near Britain in comparative naval power. The *Nelson* and the *Rodney*, both British and both completed in 1927, were the newest in the world. The rest of the British fleet was not in fact much older than other fleets.[78]

Battleships took three to four years to build and cost many millions of pounds. If all the cost of building a battleship had been concentrated in one year it would have taken something approaching 10 per cent of the British defence budget of the early 1930s. They were mobile fortifications which like land installations could have a very long life and be radically overhauled and remodelled. In 1939 each of the major fleets had the same battleships it had in 1935, but some had by then been drastically changed. Every one of the fifteen British battleships of the 1930s had at least one large refit in the interwar years and five were 'reconstructed'. They were given new engines, had their gun-mountings modified to increase range, and were given new fire control gear – at total costs of £2m to £3m each. These costs approached half what it would have cost to build a battleship from scratch.

In 1939 the British fleet faced that of a minor naval power – Germany. Not only was the German fleet small, smaller indeed than the allied French fleet, but it was below the strength permitted under the 1935 naval agreement with Britain.[79] As Winston Churchill, the First Lord of the Admiralty, put it in September 1939: 'the Germans have only seven ships worth considering, to wit: the two 26,000 ton battle cruisers *Scharnhorst* and *Gneisenau*; the three so-called pocket battleships, which are really ill-constructed heavy cruisers, and two excellent 10,000 ton 8 inch cruisers'.[80] What a contrast to 1914! More threatening was the Italian fleet, which came into the war against Britain in 1940: it had seven battleships.

The Royal Navy was also strong in what turned out to be the new 'capital' ships, aircraft carriers. In 1935 the navy had six aircraft carriers to four each for Japan and the USA, while in 1939 it had seven and the others six each. Neither Germany nor Italy completed an aircraft carrier. By the end of 1941, had it not been for the losses of *Courageous*, *Glorious* and *Ark Royal*, Britain would have had eleven carriers to the Japanese eight (and the USA's seven). In terms of tonnage, the Royal Navy would have continued to have had a clear lead. Indeed, the only index by which the British fleet can be shown to be inferior is that it appears to have carried fewer aircraft.[81] In cruisers the British lead was even stronger than in battleships or aircraft carriers: in both 1930 and 1936 Britain had a cruiser tonnage some 50 per cent higher than the USA's; in 1939 Britain had twice the number of cruisers as did the US, and many more than any other country.[82]

In the interwar years the Royal Navy out-built all other navies in nearly all sub-periods and in nearly all classes of warship. Overall between 1928 and 1941 it completed 1 million tons (displacement) of warships while the United States managed 700,000 tons and the Japanese 600,000 tons.[83] Consider the battleship race of the 1930s. By the end of 1941 Germany had completed two battlecruisers with 11in guns and two battleships with 15in guns; and Italy two battleships with 15in guns, Japan one very large battleship (with 18in guns); and the USA two (with 16in guns). Britain, although the last to start, completed three battleships: *Duke of York*, *Prince of Wales* and *George V* (all with 14in guns), as did France (by 1940, two 13in and one 15in). Between 1931 and 1941 Britain completed five aircraft carriers, as did the USA and Japan, the only countries to complete any at all. In cruisers, the British performance was spectacular. In destroyers, too, the Royal Navy led the interwar construction effort. Only in submarines was it not first but second, but to Japan and not, as might be supposed, Germany.

BOMBERS

In retrospect, one of the most striking features of British attitudes to war in the 1930s was the overestimation of the power of the bomber. This was not the result of an exaggerated fear of the unknown by a gullible public and impressionable politicians; nor was it a straightforward error. Overestimates were made by interested parties committed to modern science and its key child, aviation. British air lobbies wanted to make others believe, and believed themselves, that their weapons had already transformed war. It was in their interests to claim that massive casualties would result from their use, whether by British or German forces: belief in the power of the German air forces was a necessary corollary to belief in Britain's strategic emphasis on air power. British defensive measures were as much a response to British belief in the power of the bomber as to the Luftwaffe itself. The air lobby succeeded. The British government acted quite self-consciously in preparing for the war of the future, in which aircraft would have a devastating impact, by building up bomber forces and defences against them.

The government went to war committed to the belief that the opening phase of the war would result in *millions* of British civilian casualties in a few weeks. Thus on 1 September 1939, before the declaration of war, there was a blackout and there were no longer public weather reports. There was mass evacuation of children from the major cities, which were left without schools or social services. Hospitals were emptied in readiness for mass casualties. Parts of government ministries were also evacuated: bits of the Admiralty went to Bath, Food to Colwyn Bay and Air Ministry production to Harrogate. For a fortnight east coast ports closed before a single bomb had fallen. All this happened a year before the Blitz.

The effects of bombing had been authoritatively estimated. The first estimate, which remained the standard one, was fifty casualties (not just killed, but wounded also) per ton of bombs dropped. This was based on estimates of bombs and casualties from sixteen night raids on London in 1917–18, which in the official figures amounted to roughly 1,000 casualties from 20 tons of bombs.[84] The Air Raid Precautions (ARP) department of the Home Office raised the estimate to seventy-two on the basis of information from air raids on Barcelona in March 1938.[85] Some evidence also from Barcelona suggested seventeen casualties per ton, but this was ignored.[86] The assessment of the impact of a ton of bombs, the province of the Home Office, was just one element in the calculation. The quantity of bombs to be dropped by the enemy had also to be estimated. This was done by the Air Ministry, which clearly had an interest in suggesting that the bomber would get through repeatedly, and it claimed that the Germans would be able to deliver 700 tons per day for an indefinite period.[87] That would lead to 2.5 million casualties in ten weeks. It was also expected that poison gas would be used, not only on civilians but on the military, and once again preparation for offence went hand in hand with preparation for defence. Offence got a higher priority.

The government was not willing to over-invest, as it saw it, in defence, so many measures were of more propagandistic than practical value.[88] The bulk of the population were to look after themselves, the poor would be issued with cheap shelters.[89] By February 1939 'Anderson shelters' were being issued to the working class, all before the war started and long before any bombing. The Anderson shelter, mass-produced from corrugated iron, was to be self-assembled by the householder, who would have to part-bury it and cover it in earth. The first tranche of orders from the steel industry in January 1939 (one fifth of the expected total), was for galvanized corrugated sheeting worth just over £2m, of 14 gauge thickness, weighing 3.2lb per square foot, the order amounting to 100,000 tons of sheeting, plus 18,000–20,000 tons of channels, plus a 'considerable tonnage of bolts and nuts', taking thirteen weeks to deliver.[90] The main protection from gas was the home-made gas-proofed room or shelter. As a supplement millions of very cheap civilian gas-masks (very different from those for the forces, including air raid wardens, and those for certain civilian workers who might have longer exposure) were issued at the time of Munich. Once more, the government had made ready long before.

The left argued that these defensive preparations were inadequate. It demanded *more* protection: scientists and architects of the left campaigned for better gas protection and serious deep shelters. The Communist Party launched a powerful campaign against the policies of Sir John Anderson as head of ARP and later Home Secretary. The party thought that its campaign was in part responsible for his replacement by Herbert Morrison in October 1940.[91]

When it came, the bombing was considerably less intensive, and less deadly, than was expected. This was true even of its most intense phase, the Blitz, which lasted from September 1940 to March 1941. Assuming that all the tonnage dropped in 1941 fell in the first three months (which is not unreasonable, as the Blitz ended in the spring) gives a daily rate average of around 200 tons for this period. This was thus well below pre-war estimates. The effect of the bombs was spectacularly less than the Home Office ratio predicted. There were some 150,000–200,000 casualties; two to three casualties per ton of bombs dropped.[92] Such numbers represent a great deal of human suffering, but a staggeringly smaller quantity than that predicted.

AIRCRAFT ENGINES

Investing in the capacity to build aircraft on a large scale was central to rearmament. While the Germans produced more in every year from 1934 to 1938, in 1939 the figures were about equal, with Britain moving into a clear lead in 1940. Turning out numbers was the easy part. There were huge uncertainties as to what the best types of aircraft and engines would be, and which particular models of each type would in fact be the most successful. Aircraft development was a very hit-and-miss affair. Although the increase in the capabilities of aircraft was astonishing, the precise form development would take was not known in advance. The two most successful four-engined bombers of the war, the Halifax and Lancaster, were initially designed as two-engine machines according to a 1936 specification; the bomber designed to a 1936 specification for a four-engined bomber, the Stirling, turned out to be the worst of the three. The Spitfire fighter was ordered as an interim design, to be quickly phased out once better machines were adopted, but went on to become the best and most adaptable British fighter of them all.[93] One of the most successful bombers of the war deviated from all standard notions of where aeronautical progress was headed: the Mosquito was made of wood. By the same token we should not condemn out of hand the aircraft of 1939–40 which came to look like dead ends, such as the two-man fighter the Boulton Paul Defiant, where the observer sat in a powered turret with machine guns, which was the third Merlin-powered modern British fighter of the era.

Engine development illustrates the issue of uncertainties especially well. Engines are particularly significant because the costs of engine development were much greater than those for airframes, and because in many respects developments in engine power and efficiency drove aeronautical change more generally. The Bristol company was one of the most successful aero-engine firms of the 1920s, and in the 1930s was developing the use of the sleeve-valve for aero-engines. More efficient at high speed, sleeve-valves did away with the complex and heavy machinery to operate poppet valves. They seemed to open the way to engines of much higher power. The Bristolian engineer Roy Fedden, educated at public school but not university, led a team who developed a series of such engines by 1938: the Perseus, the Hercules (57,000 built) and the Centaurus (8,000 built). The Hercules powered many wartime aircraft (for example, the Halifax bomber), while the Centaurus was a much-troubled, very powerful engine which only came into use at the end of the war. In fact nearly all the new large engines developed in Britain from the 1930s had sleeve-valves and all were either cancelled or long delayed. The Napier Sabre, a very powerful inline engine, of the same size as the Centaurus, was successfully tested in 1939. The engine only became successful when it adopted the Bristol sleeve-valve, and towards the end of the war.[94] Rolls-Royce developed the sleeve-valve two-stroke Crecy, the Exe, the Eagle and the Pennine.[95] The Eagle was the only one which went into even small production, the last piston engine aero-engine from Rolls-Royce. The Crecy, which promised so much in the mid-1930s, was by 1944 producing the same power as the refined Merlin.[96] The Merlin was an inline engine using poppet valves which was developed from 1932. Troubled at first, it was to prove very adaptable, becoming the most made British engine of the war, defying the expectations of the 1930s. Despite the huge effort that went into the sleeve-valve it turned out to be no better than the developed sodium-cooled poppet valve even for the most powerful engine.[97] At the end of the war Britain had three sleeve-valved engines in use, compared to none in Germany and the USA, whose largest engines with the power of the Centaurus and Sabre, such as the Wright Duplex Cyclone (R-3350) of the B-29 bomber and the Pratt & Whitney Twin Wasp, retained the poppet valve.

THE PROBLEMS OF AIR DEFENCE

With the expansion of the bomber force, and rapid development in aero-engine and aircraft design, came the expansion of air defence capacities. This was to be a very complex multifaceted business, involving not just radar and fighters, but acoustical devices, observers, bombers, aerial minefields, barrage balloons, guns, rockets and more besides.

The particular problem of detecting attacking aircraft was dealt with in many ways. In the early 1930s acoustical mirrors made of concrete were thought promising, and a huge array was about to be built when a new kind of detector appeared.[98] The feasibility of radar (or rather radio-location as it was then known) for aircraft detection was very quickly established in 1935 in experiments using a large radio transmitter. The reaction was dramatic. A development laboratory was set up in the same year, working mainly for the air force, but also for the army, and naval radar got going very quickly too. By 1940 Britain had a network of radar stations to detect incoming aircraft; some radar sets aboard aircraft; and radar-controlled guns for anti-aircraft gun batteries and for ships' guns. The electrical and radio industries were closely involved. Both television and radar depended on high-power, high-frequency radio transmission and reception, and on cathode-ray tubes, as well as other products of an emergent electronics industry. For example, Metropolitan-Vickers made the Chain Home transmitters and Cossor the Chain Home receiver sets. Pye adapted a new television set to make the first successful airborne radar. In only five years therefore a completely new set of machines were deployed which radically changed Britain's defence capability. The Chain Home and Chain Home Low radars for fighter defence cost about £5m to December 1940, out of a total cost of RAF radar for research, equipment and operation of £10m. The army radar programme cost another £10m over the same dates, most of it concerned with air defence; indeed, more expenditure was involved with gunnery radar (some £6m) than with the first systems associated with fighter defence. By December 1940 very little was spent on naval radar.[99] To put it another way, the total radar equipment installed by the end of 1940 cost about the same as a battleship (£10m) and was divided between the equipment to be used with fighters and that to be used with guns.

Detection and establishing the position of attacking aircraft were only part of the story: destroying or misdirecting attacking aircraft was the central aim and this was extremely difficult. It was certainly envisaged that fighters would intercept bombers (though fighter-on-fighter dog-fights were seen as features of the Great War, made impossible by the high speeds of modern aircraft), but this was just one element. By the 1930s specialized heavy guns were available which could fire shells huge distances into the sky. By May 1940 there were over 1,000 heavy AA guns defending Britain (of 3in or over, most were in fact new 3.7in and 4.5in guns); by the peak in 1944 there were over 2,500, with the 3in guns all retired and a new 5.25in gun also deployed.[100] The first new gun emplacements were built by early 1939, a process which would continue for many years. They were everywhere; for example, in London there was a battery on the peak of Primrose Hill (Regent's Park); another on Hampstead Heath; another in Finsbury Park. In 1937 it had been hoped that Britain's AA would depend on mobile 3.7in guns, and fixed 4.5in guns, but in fact the need for speed meant that cheaper fixed 3.7in guns were to become very important, giving Britain a much larger number of fixed gunsites than ever envisaged. These guns were to be radar-controlled, giving them in principle enormously enhanced capabilities. By 1937 there were prototype gun-laying radars of great accuracy, and by autumn 1939 the GL Mk1 gun-laying radar was in service.[101] It was a crucial development, allowing gunners to aim at aircraft hidden by cloud or at night: it could pick up targets at seventeen miles, and provide accurate and continuous range information from eight miles.[102] However, in practice, it worked poorly at first.

By contrast fighters could not (yet) be controlled by radar and could not be used at night. The early radars could locate approaching aircraft looking out to sea, but not over land. Once over Britain aircraft had to be followed by a developing Observer Corps, only possible by day. In any case, fighters could find invading aircraft only by day.

One issue which was particularly troubling therefore was how to deal with aircraft attacking at night. The committee of scientists dealing with air defence were divided on the issue. In 1936 A. V. Hill's proposal for 'making hostile aircraft visible against illuminated cloud' was downgraded in favour of what he called 'foolish experiments with bombs tied to parachutes', the idea of F. A. Lindemann.[103] In fact, both systems would be developed to the point of very large-scale testing and beyond, though both are now hardly known at all. It has recently been discovered that Hill's system, named Silhouette, was tested with the establishment of fifty high-powered floodlights covering 200 square miles in Essex. The system was built from early 1939 and killed off in early 1940.[104] The bombs attached to parachutes, also called 'aerial mines', were to be developed in many different forms, as we shall see in Chapter 4.

In the mid-1930s, because the speed of bombers was not so different from that of fighters, it was proposed to drop bombs over bombers, not with the hope of hitting them, but of getting the bomb close enough to cause damage. The difficulty was that this needed a fuze which could detect the bomber. Three types of 'proximity fuze' were invented and developed before the war started. The first detected the sound of the aircraft, with a microphone, and was developed in the Air Defence Experimental Establishment. The second, promoted by Patrick Blackett and developed by the Royal Aircraft Establishment (RAE) out station in Exeter, was the photoelectric (PE) fuze. Five hundred were ordered for tests before the war, from Metropolitan-Vickers, and were regarded by the outbreak of war as workable devices. A radio proximity fuze was developed at the Air Defence Experimental Establishment.[105] Although developed for anti-aircraft bombs, proximity fuzes came to be associated very strongly with the extraordinary British rocket programme of the late 1930s, yet another aspect of the air defence story. Both PE and radio fuzes were developed for 3in anti-aircraft rockets. Both types of fuze were taken to the USA by the Tizard Mission in late 1940.[106] The radio fuze would become a famous US development.

Intelligence reports from Germany led to work starting on rockets at Woolwich in 1935, and by July 1936 a large programme started under Dr Alwyn Crow and coordinated by the Air Defence Research Sub-Committee of the Committee of Imperial Defence (CID). The priority was an anti-aircraft rocket, but the second priority was a large rocket with ranges of between 500 and 900 miles (much longer than that of the V2). Soon attention focused only on the first, with the aim of producing alternatives to first the 3in and then the 3.7in AA guns. Two-inch rockets were tested in 1937, and in 1938–9 a 3in rocket on a new range in Jamaica. Before the war a 5in short-range land-to-land rocket was also developed.[107] In 1939 a new establishment was built for rocket development under the nondescript title of Projectile Development Establishment, at Fort Halstead near Sevenoaks in Kent. One estimate put the cost at over half a million pounds. The buildings were vacated during the air raids of 1940.[108] And, as in so many cases, it would turn out that 'in practically no case did the finished article follow the original specification for use'.[109] Indeed, as we shall see, the AA rocket was a costly failure, but the rockets were to find many uses at the end of the war.

TO WAR

Britain was not preparing to fight a war alone. While it abjured collective security, rearmed and appeased, an alliance with France of one sort or another was, hardly surprisingly, always in the picture. After all, far from being eternal enemies, France was Britain's most important ally. Anglo-French armies had defeated the Germans on the Western

Front in the Great War, and in the late 1930s they came close again in an alliance against Germany. Indeed it was inconceivable that Britain would have gone to war against Germany without France.

The relationship was strengthened after Munich. There were staff talks, and discussions on joint economic control, raw materials and other matters relating to a possible war.[110] In April 1939 the British government introduced limited peacetime conscription as a gesture to French as well as British opinion.[111] With the outbreak of war, the British Expeditionary Force went straight into northern France and extensive collaborative programmes were started. In November 1939 the Supreme War Council announced a joint economic effort, with Anglo-French Executive Committees looking at Air, Munitions and Raw Materials, Oil, Food, Shipping and Economic Warfare.[112] Purchases in the USA were coordinated. In December 1939 Anglo-French financial collaboration was started, with pooling of foreign assets, and mutual consultation on external financial matters, and a close linking of sterling and the franc.[113] In January 1940 the Minister for Coordination of Defence, Lord Chatfield, told the Lords that 'A mutual exchange of scientific knowledge over a wide field was concerted and already in progress between the two Governments before the outbreak of the war'; we have had, he said, 'members of the scientific organisations of the two countries working in British and French scientific establishments. We have, I may say, a complete exchange of scientific information.'[114] One journal reported that by April 1940 Anglo-French collaboration had 'reached an advanced stage'; there was to be a 'complete pooling of resources. Scientists, engineers and manufacturers, etc., will work together in the closest possible association.'[115] Indeed, the French Ministry of Armaments told the British nuclear scientists of French nuclear work and they had the Norwegian stocks of heavy water.[116] Given this closeness, it was perhaps not surprising that in March 1940 a joint postage stamp was announced to celebrate the Anglo-French alliance, which in an early attempt at Franglais was dubbed the *stambre* or the *timp*. It was due to be issued in November 1940, but events intervened.[117]

In 1939 the British and French governments believed they were capable of defeating Germany in the medium term. They decided to stick a powerful spoke in the Nazi wheel, first by making a set of commitments to the independence of Eastern European nations, and then, following the invasion of Poland, to declaring war *on* Germany without even a hint that they were themselves under immediate threat. Both were in what they took to be a temporary position of strength and acted so powerfully and decisively against Germany for that reason: 'the Polish issue gave a moral gloss to what was in fact a decision about when was the best time to fight for Britain and Germany, not for Poland', according to Richard Overy.[118] Hitler, for his part, once he had recovered from his surprise at the fact that Britain and France had declared war, took this as a good time to fight a broader war. For, in Adam Tooze's powerful argument, he too recognized the overwhelming strength of Britain and France with the USA in the background. For Hitler it was now or never.[119]

Economics and politics mattered in how the war was understood. The *Engineer* suggested in January 1940 that:

> There is no industrial people in the world that does not now admit the need of international trade. Before the present war opened, the first little steps towards a better understanding had been taken. Possibly it was because the leaders of Germany foresaw that the universal extension of such an understanding would put a limit to their will for supremacy that war was launched.[120]

In other words, this was a war for freer trade. It was significant that the war was not just against the right. The stunning Nazi–Soviet Pact of August 1939 thoroughly discomfited the left and the right in Britain and France, making opposition to war from the right much weaker.

It is misleading to think of 1939–40 as the phoney war or twilight war, or the Sitz-krieg, or bore-war, with all the sense of lack of authenticity and reality these terms now convey. This was war as Britain and France wanted to wage it and win it. The dominant propaganda line in Britain in the early months of the war was 'assurance of victory', the policy was economic offence and military defence.[121] Britain was fighting a war 'for King, for Empire, and for Freedom', as British propaganda put it.[122] One reason we still think of the early war as phoney is that the air force was not used for bombing and that Germany was not attacked on land either. But the navy went on to the offensive. German shipping and trade were swept from the seas. Imports into Germany were blocked, and from November 1939 its exports too, both extraordinary changes to a world economic system in which Germany was a more significant trader than France. Given the huge inferiority of the German navy, engagements were few, so much so that the victory over the *Graf Spee* in December 1939 was much over-rated. The U-boat threat to British shipping was weak at this time and the magnetic mine was defeated early on.

There were Anglo-French moves towards taking the offensive based on concern that Germany was strengthening faster than they anticipated.[123] The French pushed for action in the Balkans (rejected by the British), for military help for the Finns against the Soviet Union, and for the interruption of Swedish supplies of iron ore to Germany. There were also plans, which while taken seriously were never agreed, to bomb (with an Anglo-French force) the Soviet oil fields in Baku. Had these plans worked, they would have made Britain and France have to fight both Germany and the Soviet Union.[124] That might perhaps be another indicator of Anglo-French confidence as well as their anti-communism. One plan was put into action. The British and French would take offensive action against the Soviet Union, in defence of Finland. The attacking forces would reach Finland via Narvik in Norway and take over the Swedish iron ore mines on the way. The decision to proceed was made in March 1940, despite the opposition of Norway and Sweden. But Finland came to terms with the USSR on 12 March, making the whole project redundant. A plan was now activated to mine Norwegian waters to stop German imports of ore. But the Germans invaded Norway in April and the British (with French and other forces) then botched an invasion of Narvik in May 1940. But in the meantime the ten divisions of the British army stationed in France were defeated alongside a French army ten times that size. With that the war changed dramatically. The fall of France, and indeed of Western Europe, dramatically changed Britain's position.

The failures of 1940 cast a powerful shadow backwards condemning governments of the 1930s for lack of foresight, vigour and preparedness for war. The work of conscious rearmament was to get short shrift. William Beveridge proclaimed in June 1940 that 'Early misfortunes in war cannot be avoided by pacific democracies, since it is not in the character of a pacific democracy or its chosen leaders to believe in war before it happens and prepare for it'.[125] Churchill, looking back to September 1939 in his post-war memoirs, wrote of 'The glory of Old England, peace-loving and ill-prepared as she was, but instant and fearless at the call of honour.'[126] And yet the problem was not lack of preparation for war. Britain rearmed on a scale unprecedented in peacetime. Furthermore Britain did not go to war to defend its borders or those of close allies. It went to war, allied with France, in pursuit of great interests, by choice. It went to war believing in victory, and did so because it believed in its economic power and the power of its new scientific weapons.

3
Never Alone

Britain was not 'alone' between June 1940 and June 1941; nor did it believe itself to be. If anything was alone it was the mighty British Empire. But the Empire had allies and trading partners which meant that especially materially it was very far from isolated. The Empire was strong; indeed, in some respects, relative to other powers, it was at its strongest in this part of the war. So much so indeed, that it could realistically expect to defeat Germany. Yet the belief that Britain in this period had been alone, and was weak, came to be very important, especially after 1945.[1]

For the first weeks of Churchill's premiership Britain was still fighting with France, a point powerfully reflected in his speeches. In his very first broadcast as Prime Minister, on 19 May 1940, Churchill spoke of living through 'beyond doubt the most sublime' period in the history of Britain *and France*, and went on to say that they were 'Side by side, unaided except by their kith and kin in the great Dominions and by the wide empires which rest beneath their shield – side by side the British and French peoples have advanced to rescue not only Europe but mankind.'[2] In his Dunkirk speech to the House of Commons on 4 June 1940 he also invoked both the Empire and France: 'The British empire and the French republic, linked together in their cause and in their need, will defend to the death their native soil, aiding each other like good comrades to the outmost of their strength.' In this speech Churchill made the famous statement: 'we shall defend our island, whatever the cost may be, we shall fight on the beaches, we shall fight on the landing grounds, we shall fight in the fields and in the streets, we shall fight in the hills; we shall never surrender . . .' But this comment was *preceded* by one often ignored but of equal significance: 'We shall go on to the end, we shall fight in France, we shall fight on the seas and oceans, we shall fight with growing confidence and growing strength in the air.' Empire and the new world were very much part of the story too, for even if the British Isles or part of them fell, 'our empire beyond the seas, armed and guarded by the British fleet, would carry on the struggle' until the 'new world' came to the 'rescue and liberation of the old'.

In the Dunkirk speech Churchill spoke of fighting 'if necessary alone'. Following the capitulation of France at the end of June, Churchill often used the word 'alone'. He was almost always referring to the British Empire. For example in his 'Few' speech to the House of Commons on 20 August 1940 he referred to 'The British nation and the British Empire finding themselves alone'.[3] On 8 May 1945 in his official declaration of the end of the war in the House of Commons and a broadcast he recounted: 'After gallant France had been struck down we, from this Island and from our united Empire, maintained the struggle single-handed for a whole year until we were joined by the military might of Soviet Russia.'[4] But on the same day he made a brief speech to the crowds in Whitehall, in which 'alone' played a crucial role but now referred clearly to the 'British nation' and the 'ancient island'.[5] This was no slip. At the 1945 general election 'Mr Churchill's declaration of policy to the electors' gave much attention to Empire and imperial trade, but used a very national rather than imperial 'we': noting of the Empire and Commonwealth that '*We* shall never forget *their* love and steadfastness when *we* stood alone against the German Terror [emphasis added]', and that 'During a whole year of this great war Britain bore the burden of the struggle alone.' Such terms did not reflect general usage in 1945. The free-trading Liberal manifesto was much more generous (and accurate) in noting the 'sacrifice and steadfastness of the people of these Islands, the British Commonwealth and Empire' who in 'standing alone for a whole year against the insolent might of Germany and her Allies' had 'saved the world'.[6] It is sobering to note that the Liberal manifesto spoke out against imperial preference, while Churchill was in favour. Churchill had referred to the nation being alone at least once before. In the broadcast he made on the second anniversary of his taking office, in May 1942, it was the island nation not the Empire which had been alone.[7] During the war

"VERY WELL, ALONE"

"So our poor old Empire is alone in the world."
"Aye, we are—the whole five hundred million of us."

Two views of 'alone' from June and July 1940:
that of the New Zealander Low, working for a Canadian newspaper baron,
and that of the British illustrator Fougasse (Kenneth Bird), working for *Punch*.
Low's cartoon, much reproduced, is one of the most misleading images in
British history. He recognised, in a garbled caption to a reproduction of
the cartoon later in 1940, that the empire was helping.

this was exceptional, but after it, in his history of the war, he was clear: 'we were all alone. None of the British Dominions or India, or the Colonies could send decisive aid . . .'[8]

In 1940–41 the image of the British nation being alone, while present, was not dominant. An example is the J. B. Priestley-narrated documentary *Britain at Bay* (1940), which does refer to Britain 'alone' and 'at bay', stressing that it was an 'island fortress' and making no reference to Empire, though at the end troops from the dominions are shown arriving in Britain.[9] The Empire alone was an important theme, but so too was the idea that the Empire was fighting with others. Going it 'Alone', or ceasing to be alone, either in the national or in the imperial sense, does not appear in the Pathé newsreel *Review of the Year* for 1940 and 1941. Instead, the reviews stress the link with France at Dunkirk, and the defence of Greece in autumn 1940.[10] In the newsreels of the 'alone' period there are odd references to being alone, but they are outnumbered by references not only to Empire but to allies. Thus a February 1941 newsreel shows Churchill with the Free French leader General de Gaulle and the Polish leader General Sikorski watching a demonstration of the power of British tanks.[11] The British Gaumont newsreel report of the Inter-Allied Conference of 12 June 1941 spoke 'of all the nations that stand beside Britain in the fight against Nazism' and reported a vigorous speech by Churchill denouncing the 'vile race of quislings' who were collaborating with the Nazis.[12] The conference brought together the British Empire leaders, with the heads of the governments in exile of Poland, Czechoslovakia, Belgium, the Free French, Norway, the Netherlands, Luxembourg, Yugoslavia and Greece. Another indication is the BBC's weekly broadcast on Sunday before the nine o'clock radio news of a compilation of the national anthems of Allied nations. On 12 May 1940 the anthems of the previously neutral Netherlands and Belgium were added; many others were to follow until the programme was forced off the air in June 1941. The reason was that it was deemed unacceptable to hear the 'Internationale' (then still the Soviet 'national' anthem) played on the BBC.[13]

'Alone' hardly figures in the documentaries and writings of the time. The documentary *London Can Take It!* (1940), by Humphrey Jennings, made for the USA with a voice-over from an American correspondent, Quentin Reynolds, had no 'alone'; nor did Reynolds's effort *Christmas Under Fire* (1941), about Christmas 1940 in Britain. European unity was a theme of the Boulting Brothers' film *Dawn Guard* of 1941, which has a Home Guard soldier noting that Europeans have united to fight Hitler, a scene set after Dunkirk.[14] George Orwell's patriotic polemic *The Lion and the Unicorn*, written in late 1940 and published in 1941, had Britain fighting fascism with China, and from October 1940 with Greece.[15] However, a 1942 production called *Battle for Freedom* about the contribution to the war of the British Commonwealth and Empire, made for the USA, made the remarkable claim in an inter-title that 'For two years Britain fought alone.'[16]

BLOCKADED BY CONQUEST

Materially Britain could never have been alone, given its dependence on overseas supply. Defeat in Europe meant more than the loss of a great allied army, air force and navy, it also meant the loss of crucial trading partners. The Germans imposed a very important and effective blockade which cut off major supplies to Britain. This was not done only, or even mainly, by submarines, but by the takeover, directly or indirectly, of major suppliers of food and raw materials to Britain. The loss of these 'near supplies' was very serious.[17] Britain got the bulk of its bacon and eggs from Denmark and the Netherlands, as well as large quantities of butter. The loss of raw materials was even more significant. Germany was able to blockade Britain's Baltic timber imports, which were reduced to a trickle.[18] In 1938 the main bulk of imported timber came in three categories indicating its main uses – construction, mining and railways: 1.5 million standards of sawn softwood; 821,000 standards of pit props; and 117,000 standards of railway sleepers. Imports soon fell to 300,000 standards of sawn wood, 11,400 standards of pit props and 13,100 standards of railway sleepers.[19] Such was the cataclysmic decline in Britain's biggest import, central to its production.

Britain's second bulkiest raw material import was iron ore. Most came from Sweden (via Norway) and French North Africa, and 'The

IRON ORE

Total Imports 1938:

5.1 million tons

	million tons
TUNIS	1.3
ALGERIA	1.3
FRANCE	0.3
SWEDEN	1.3
NORWAY	0.6

IRON & STEEL MANUFACTURES

Total Imports 1938:

1,341 thousand tons

thousand tons	
257	FRANCE
273	BELGIUM
89	SWEDEN
34	NORWAY

PAPER MAKING MATERIALS

Total Imports 1938:

2 million tons

	million tons
TUNIS	0.1
ALGERIA	0.2
SWEDEN	0.6
NORWAY	0.3
FINLAND	0.6

PAPER & CARDBOARD

Total Imports 1938:

1.1 million tons

million tons	
0.2	HOLLAND
0.2	SWEDEN
0.1	NORWAY
0.2	FRANCE

BUTTER – Total Imports, 1938: *476 thousand tons*

HOLLAND	DENMARK	Other SCAND.N & BALTIC STATES
36	118	64

thousand tons

BACON – Total Imports. 1938: *343 thousand tons*

DENMARK	Other S. & B. STATES	HOLLAND	POLAND
169	28	26	23

thousand tons

EGGS – Total Imports, 1938: *27.7 million great hundreds*

DENMARK	S. & B. STATES	HOLLAND	POLAND
9.5	1.8	5.9	2.6

million great hundreds

FRESH, PRESERVED and TINNED FRUIT
Total Imports, 1938: *1,969 thousand tons*

ITALY	FRANCE
64	47

thousand tons

FRESH, PRESERVED and TINNED VEGETABLES
Total Imports, 1938: *653 thousand tons*

HOLLAND	ITALY	FRANCE
178	40	16

thousand tons

Sources of British imports now shut off by Nazi domination of Europe, late 1940. The black section of each bar represents sources *not* impeded by Germany. Figure by L. F. Horrabin. Horrabin makes the point that Britain's loss is not necessarily Germany's gain, e.g. Dutch intensive agriculture quickly collapses as it is reliant on imported feed, which is now in turn shut off by British blockade.

German push to the West had cost Britain at a minimum six million tons of iron ore a year.'[20] William Beveridge thought that the loss of iron ore from Europe and North Africa was a decisive blow such that 'economically, nothing but the new world can save us'. Beveridge envisaged a mass evacuation of women, children and munitions workers to the new world, since with no iron ore Britain could not produce munitions.[21] There were similar stories for other commodities, among them linen – the cut-off in supplies meant that, uniquely in the United Kingdom, unemployment in Northern Ireland increased at the beginning of the war as a result of the collapse of this largely feminine industry.[22]

Britain had to shift its supply of key materials from Europe 'to a degree and in a manner which could not have been dreamed of by the pre-war planners'.[23] But this could be done. In the case of timber, home production came to dominate. The Forestry Commission had planted a lot of trees during and after the Great War and now these and others were cut down. In the war up to 98 per cent of pitwood was home-produced; without it coal could not be mined. Here Britain did manage alone. In other cases it was control of the seas and continuing support from overseas which allowed Britain to cope.

Britain did not sever all links with Europe. On the contrary, as the propaganda suggested, new links were forged, and these helped Britain shift its source of supplies to more distant shores. At the beginning of the war, German attacks on neutral shipping, and political pressure on neutrals, meant that Baltic ships, other than Swedish, no longer sailed to Britain.[24] It proved impossible to charter ships from the Dutch and Swedes, and very difficult to do so from the Norwegians and Greeks.[25] By February 1940 foreign ships were only bringing in 25 per cent of British imports.[26] However, after the conquest of Europe, many of the ships came over to the British cause. The bulk of the Norwegian fleet (including its very large tanker fleet, third in the world after those of Britain and the USA; it had nearly 20 per cent of world tanker tonnage) came over, as did about half the Danish fleet.[27]

In the summer of 1940, confident of their power to do so, the British got tough on the high seas. A ship-warrant scheme ensured that 'for the ill-disposed there were to be no bunkers, or stores, or insurance, or water or credit, no access to dry-docks, no Admiralty charts, no help or guidance or supplies of any sort'.[28] The seas were free only to allies, and to friendly neutrals like the USA. A new British maritime order was established at sea. Speaking in July 1940 Churchill was bullish:

> Is it not remarkable that after ten months of unlimited U-boat and air attack upon our commerce, our food reserves are higher than they have ever been, and we have a substantially larger tonnage under our own flag, apart from great numbers of foreign ships in our control, than we had at the beginning of the war? Why do I dwell on all this? . . . I dwell on them because it is right to show that the good cause can command the means of survival; and that while we toil through the dark valley we can see the sunlight on the uplands beyond.[29]

Britain was expelled from the European economy, and became an island reliant on long-distance trade as never before. Yet it was at the centre of a new global order. It was closely allied to governments in exile which controlled great overseas empires – the Dutch and the Belgians. It could draw on resources from all over the world. A Britain truly alone, one forced to become self-sufficient, would certainly 'neither have made effective war nor even maintained [its] civil population'.[30] Even in war Britain depended on an 'international economic order' in which it had a privileged place. This was a great strength. Britain could import food rather than growing it, it could import oil rather than go to the costly trouble of making it from coal, and it could, if necessary, import manufactures, from tanks to tractors, on a vast scale. Even if trade was far from free, and not very safe, it continued in wartime and allowed Britain to wage war in efficient ways.

So powerful was Britain in the world economy that it could in effect force many people around the world to supply it with goods for credit. Many suppliers had no one else to sell to. So the supplies kept coming in even as British exports fell. British exports dropped in late 1940, with the prospect of future US help; they would fall again with Lend-Lease.[31]

The US position was special. From the beginning of the war Britain aimed to save dollars, using available shipping to get supplies from further away. At the end of 1940 and through 1941 Britain was forced to get supplies from the USA to replace lost European trade. These supplies either were not available elsewhere or were so far away that they consumed too much shipping capacity. For these reasons, after Dunkirk, supplies from the USA increased in proportion quite radically. Saving dollars was no longer a priority, and indeed as a result there came iron and steel rather than iron ore.[32] Food from the USA, especially bacon and eggs, appeared in Britain, replacing imports from Europe.[33]

In terms of fighting units the British Empire was largely but not entirely alone.[34] When the British government decided, in September 1939, to equip a field army of 55 divisions it was planning for only 32 of these to be British. These formations excluded imperial garrisons and internal security troops. In the revised plan of late 1940, there were still 55 divisions to be raised by 1942 but now 34 were likely to be British, 9 Indian, 4 from African colonies, 3 each from Canada and Australia, and one each from New Zealand and South Africa.[35] The expectation was that 21 would be needed mostly for the Middle East, leaving 34 for both active home defence in Britain and other overseas operations. It was estimated that Britain might have 20–30 divisions for offensive overseas operations by spring 1942.[36]

Imperial divisions were deployed in Europe from the very beginning of the war. A Canadian infantry division landed in Britain in December 1939 and another in the summer and autumn of 1940, with more units following. An Australian Imperial Force, also of volunteers, was being formed from September 1939 for action in Europe; events meant that instead of posting to Britain or France, it was to fight in the Middle East from the end of 1940. A New Zealand Expeditionary Force of one division went to the Middle East in 1940, with some units diverted temporarily to Britain. An Indian Army division was formed in Egypt in 1939.

In the case of the air force the imperial dimension was also of significance from the beginning. At the end of 1939 an Empire Air Training scheme was agreed whereby Canada would become the base for advanced training for all the imperial air forces, including the RAF. This extraordinary programme turned out 24,000 pilots a year, plus all the other aircrew (all volunteers, even the British ones) and would require a huge infrastructure of aerodromes and associated facilities, as well as aircraft. By the end of the war over 40 per cent of

Bomber Command, based only in Britain, came from the dominions, with additional numbers from allies.[37]

Later in the war the British Empire forces had a higher proportion of non-British personnel than in the great imperialist war of 1914–1919. In fact, the fighting British Empire of 1945 had around 8 million in uniform, out of 12 million mobilized, only just over half from the British Isles. Apart from the conscript British army, its largest component was the largest-ever volunteer army, the Indian Army of 2.5 million.[38]

Non-imperial forces should not be forgotten. In 1940–41 governments in exile were not merely collections of refugee politicians. Some came over with armed forces, and from 1940 many units, some of very substantial scale, from Poland, Czechoslovakia and other countries, would fight with British equipment and under British command.

The idea of Britain 'alone' clearly conveys weakness. Indeed, the idea that Britain had been militarily weak in 1940 is a central part of the 'alone' story – it is not merely that it was weak after the fall of France, but that it was weak before then also. There is another story to be told. Contrary to all reasonable expectations, including those of senior German soldiers, in 1940 the German army defeated superior Allied and neutral forces decisively and quickly.[39] The Germans put their own victory in 1940 down to their race and their Führer, not to technical superiority.[40] It is well-known indeed that they were right not to ascribe their success to mechanical plenty: in the first half of 1940 the combined Anglo-French production of tanks was 1,412 and the German 558; between January and May 1940 Anglo-French aircraft production was twice the German.[41] The British, however, invoked technical and material failures to explain German victory and British defeat. Even though the British army was not at the centre of the German thrust, and was not large enough to have a key role in the campaign, the failure of the whole was interpreted as a failure of the part. Furthermore, the failure of this minor part of the British war effort was made to stand for the British war effort as a whole. From the first, and for decades later, the defeat in France was associated with supposed British technical inferiority and general lack of preparedness for war. Invoking technical and industrial failure was a standard wartime response to military failure, as we shall see, and is still found in many histories.

LESSON *of* RECENT DEFEATS

Low's analysis of defeat, 1940

The great shocking victories of the German armies in Western Europe in April, May and June 1940 required explanation. The *Engineer*, a high-quality professional paper, produced an editorial in June 1940 called 'War against a Machine' in which it was claimed that:

> Germany has at length opened our eyes to what an engineers' war really is. It is not a war against flesh and blood, but a war against a colossal machine; a machine in which men are accounted no more than the cogs and ratchets of the mechanism. It is ruthless and senselessly destructive because, being a machine, it is without a soul, with no moral sense, and no mind. The whole army of Germany is but a single vast 'tank', driven forward at the will of its commanders but careless of the blood it sheds either of its people or of others.

This was a direct echo of much other propaganda in those days, including J. B. Priestley's Dunkirk broadcasts, with their evocation of German colossi and British little ships.[42] It is there too in the famous polemic against British leaders of the interwar years *Guilty Men*, a highly successful character assassination published in July 1940. As we shall see, the *Engineer* and Priestley and many others besides misread the character of the Wehrmacht.

The *Engineer* was very clear about what the British Empire had to do to fight back: 'the Moloch we are opposing is, in fact, a colossal Robot, which can be destroyed, not by appeals to any of the emotions which move humanity, but only by a machine more powerful than itself.' 'The Empire is awake!' it exclaimed. This was, it said, 'an engineers' war against a machine'.[43] Indeed the idea that Britain should respond with ever greater machine power was a common one. In June 1940 William Beveridge was certain that 'The surrender of France does not entail defeat for Britain', only that it left it with no land bases to roll back 'Germany's mechanical invasion by the mechanical superiority which ultimately can be obtained from the new world'. Victory now depended 'on domination in the air, achieved both by mechanical superiority and by the petrol starvation of Germany'. Britain, threatened by siege rather than invasion, could nevertheless build up its 'munition machine', but should avoid creating a large defensive army and 'hasten by all possible means the date of our new mechanical offensive'.[44] That confidence in the mechanical strength of Britain and Empire was not as misplaced as it now seems.

THE GREAT RIFLE CRISIS

Ignoring for the moment the air force and the navy, how strong was the British army following the fall of France? Is it the case that Britain had virtually nothing in the way of arms or was virtually disarmed in 1940?[45] In one respect it was: the army was short of rifles. However, this was the product of a hubristic belief in the power of the machine rather than indifference to the necessities of modern war. Britain was temporarily short of rifles not because of industrial and military weakness, but because it was not in the business of creating in peacetime an old-fashioned mass army. Under 200,000 rifles were produced between 1939 and 1941.[46] In 1939–40 there was one major rifle plant in Britain, the BSA Small Heath works in Birmingham, turning out the No. 1 rifle, standard issue for the forces. Small Heath production fell as a result of air raids from around 8,000–10,000 per month in May–September 1940 to less than half this for a long time afterwards.[47] At one point no rifle barrels were being produced at all.[48] At the time German rifle production was at least ten times larger (see Table 3.1). The problem was not quite as bad as these figures suggest – there were a total of 1.5 million standard .303 rifles available in 1940, and over half a million had been reconditioned between September 1939 and May 1940.[49] But there arose a serious deficit of rifles for home defence following the disaster in France. Churchill managed to secure 750,000 rifles from the United States and ensured they were loaded on to more than one ship: he said he would 'not breathe until they arrived'.[50] They helped equip the Home Guard, created in 1940, which was some 1.7 million strong.[51] These rifles were not a gift from the US, but were bought for cash.[52] The need for them has helped support the view that Britain was weak in 1940, linked as it is with images of the Home Guard drilling with pitchforks.

Table 3.1 Output of principal army weapons, British and German, September 1939–May 1940

	Rifles, 000s		Machine guns, 000s		Field and medium artillery		Medium anti-aircraft artillery		Tanks (all types)	
	Ger.	UK	Ger.	UK	Ger.	UK	Ger.	UK	Ger.	UK
Last four months of 1939	279.0	18.7	12.7	6.9	773	—	192	224	247	314
First four months of 1940	310.4	26.8	14.7	7.4	675	51	317	234	283	287
May 1940	101.6	11.1	5.2	2.9	217	63	86	94	116	138

Source: M. M. Postan, *British War Production* (London, 1952), p. 109.

Rifle production represented a major temporary problem, but it would increase very rapidly. Indeed, the story of subsequent British rifle production gives us a clear picture of the global way in which Britain produced arms. First, there were long-established plans to make a new mass-producible version of the standard rifle, the Lee-Enfield No. 4, should a mass army be needed. This rifle was turned out in large numbers from new ordnance factories at Maltby and Fazakerly starting in very late 1941, and from a new BSA factory at Shirley. In 1942 production reached nearly 600,000 per annum. Wartime domestic rifle production amounted to only around 2 million, less than the size of the British army, and much less than the size of the imperial armies as a whole. Domestic production was only part of the story. Overseas supplies of British rifles were to be essential for both British and imperial forces. Indeed, more (3 million) Lee-Enfield rifles were made abroad than in Britain, a combined total of over 5 million for forces of the British Empire.

Production of the No. 4 started overseas, as in Britain, at the end of 1941. The Long Branch arsenal near Toronto in Canada would make nearly one million, and the Savage Arms company in Massachusetts in the USA over one million.[53] These rifles were interchangeable with British No. 4s. Production of the No. 1 rifle continued at the BSA Small Heath plant until 1943, but very large numbers were made overseas. Just over 400,000 were made at the long-established Lithgow Small Arms Factory in Australia and especially at a new factory at Orange,[54] while the long-established Ishapore rifle factory outside Calcutta built and repaired nearly 700,000.[55] Thus was the gigantic imperial army supplied.

TANKS

If there was a mechanical Moloch on the Western Front it was the British Expeditionary Force. The commanding officer of German Army Group B, which defeated the British Expeditionary Force in May 1940, noted in his war diary that 'the scene on the roads used by the English retreat was indescribable. Huge quantities of motor vehicles, guns, combat vehicles, and army equipment were crammed into

a very small area . . . There lies the materiel of an army whose complete-
ness of equipment we poor wretches can only gaze at with envy.'[56] This
comparative description of the two forces is not well-known, even
though it is well-attested that the German army faced, and knew it faced,
stronger combined French, British, Belgian and Dutch forces than it had
itself, whether in terms of troops, artillery, tanks or aircraft.[57]

The British army of the 1930s was small, but was indeed well-
equipped.[58] Among the new weapons were the Bren gun (the name
derived from *Br*no, site of the Czech arsenal which developed the
original, and *En*field). The Bren was a light machine gun which was
to see long service in the British army. Brens were first made in Enfield,
and were the main cause of employment increasing from 900 in January 1937 to 3,400 in January 1939.[59] Like the Lee-Enfield rifle it was
produced in imperial arsenals. Contracts were placed in Canada in
1938, with a private firm and the Long Branch arsenal; they were also
to be made at Lithgow and Ishapore. The Bren was an infantry
weapon, but it was sometimes carried in a lightly armoured tracked
vehicle called first a Bren-gun carrier and later the Universal carrier.
Designed by Vickers, it went into production in 1936 and by the end
of the war over 100,000 had been made, an unsurpassed record for an
armoured vehicle. Again it was also made outside Britain. Around
half the total was made in Britain; the rest came from Canada and the
United States. The Ford Motor Company was the main producer in
all three countries. The carrier was the smallest of a whole range of
tracked armoured vehicles which were made for the British army.
They ranged through light and medium tanks to heavy tanks as well
as many specialized types.

The small British Expeditionary Force was well-supplied with
modern light and medium tanks. Of the latter it had 100 infantry
tanks (around seventy-five were the 11-ton Matilda I armed with a
machine gun, the remaining twenty-five or so the very different 29-ton
Matilda II armed with the 40mm 2-pounder gun). In addition the 1st
Armoured Division which went to France in May had around 150 A9
and A10 cruiser tanks, giving Britain a total of 250 medium tanks in
France.[60] Yet the German armies had a total medium tank force
(meaning tanks larger than the Panzer II) of 1,456.[61] According to
other figures, 961 German medium tanks advanced into France of

which only 683 were armed with anti-tank guns.[62] This compares with 175 comparable British medium tanks in France (omitting the seventy-five or so Matilda Is with a machine gun only).[63] Yet the impression given then and since was and has been very different. Oliver Lyttelton, the British Minister of Production, painted an image in 1942 which remains influential: 'The only British tanks mounting more than a machine gun which fought in France in those battles, in which the Germans deployed between four and five thousand tanks, were 23 Mark II infantry tanks and 158 Cruisers.'[64] This does not compare like with like – it compares the medium British tanks with all German tanks, including light ones (with which the British were themselves liberally supplied), many of which were armed with machine guns. Furthermore it compares the tanks of the equivalent of one British army with those of the many German armies which invaded France.

The British forces in France were more tank-intensive than the German, on a like-for-like basis. The BEF had nine infantry divisions, supplemented by one infantry and one armoured division. The 175 British medium tanks were the equivalent of more than two panzer divisions of 1940, which had seventy-six medium tanks each.[65] On the basis of the German standard, the British forces therefore had roughly two armoured divisions to ten infantry. The German 6th Army (part of Army Group B) had sixteen infantry divisions, two panzer divisions and one motorized division.[66]

Any suggestion that the British tanks were inferior in terms of fire-power, or indeed armour, would be misplaced. The Panzer IV carried a 75mm artillery piece, very different from an anti-tank gun. The Panzer III tanks were, at this time, armed with a 37mm anti-tank gun and the 38t with a 37.2mm gun; both were less effective than the British 40mm anti-tank gun. This was mounted on 175 British medium tanks, the Matilda II and all the cruisers, as well as equipping the artillery anti-tank units. The 40mm 2-pounder gun was better than any German anti-tank gun deployed until 1942, when the long 50mm gun came in.

Just as remarkable is that on a like-for-like basis British medium tank production was about one third to one half that of Germany in 1939, and two thirds of it in 1940. In 1939 Britain produced 235 cruiser and infantry tanks (ranging between 11 and 29 tons in weight), and 418 in the first half of 1940.[67] German production (of the Czech 38t, Panzer III and Panzer IV – all in the 10–20 ton range) was around 500–600 in 1939, and around 1,500 for the whole of 1940. In 1940 Britain produced over 1,000 medium tanks.[68]

Table 3.2 German and British tank and armoured fighting vehicle production, 1939–41

| | | Sept–Dec | | | |
	Pre-war	1939	1939	1940	1941
Germany					
Total armoured vehicles*	3,503	370		1,788	3,623
Medium tanks*	387	355	500**	1,679	3,358
UK					
Total armoured vehicles†		314††	969	1,399	4,841
Medium tanks‡	146	104	235‡‡	1,209	4,890

Medium tank is defined as all tanks heavier than the Panzer II, thus including for this period the Panzer III, Panzer IV, Panzer 35t and Panzer 38t, and the British Matilda I, Matilda II, Valentine, Churchill and various types of cruiser tank.

* Includes self-propelled guns, assault guns, etc.: http://en.wikipedia.org/wiki/German_armored_fighting_vehicle_production_during_World_War_II. 1940 figures from tables for individual tanks checked against**.

** Calculated from http://sturmvogel.orbat.com/GermAFVProd.html. Excludes captured.

† CSO, *Statistical Digest of the War*, Table 126. Includes special-purpose and self-propelled guns on tank chassis.

†† Postan, *British War Production*, p. 109.

‡ All calculated from data kindly supplied by David Boyd. See his site http://www.wwiiequipment.com.

‡‡ Postan, *British War Production*, p. 103.

If the number of British tanks in France has been underestimated, the consequence of their loss is exaggerated. This was already the case during the war. Viscount Cranborne told the House of Lords in 1942 that the number of tanks in Britain was disastrously low:

> Our position in June, 1940, after the French collapse, was that we were standing absolutely alone against the greatest mechanized Army the world has ever seen. Germany had ten armoured Divisions; we had 200 light tanks armed with machine guns and 50 infantry tanks – that was all that remained to us for the defence of our country.[69]

By contrast Churchill later gave figures for 103 cruisers, 132 infantry tanks and 252 light tanks left in Britain after the fall of France.[70] These quantities, the equivalent in medium tank numbers of three to four panzer divisions, tally with the production figures and the numbers lost in France. British medium tank losses in France, say 250, match production up to December 1939, leaving over 400 medium tanks made in the first half of 1940 available.[71] Some of these had gone abroad.

That Britain had hundreds of medium tanks in mid-1940 ready to repel an invasion helps explain why the government could have made the decision in August 1940 to send fifty-two cruisers and fifty infantry tanks to Egypt immediately, together with light tanks and other equipment, to supplement armoured forces already there. Harold Macmillan later claimed that it was 'to the everlasting credit of the Prime Minister and his colleagues that, even when facing this supreme test, they sent out convoys of tanks and munitions to save the Middle East'.[72] This decision was, according to Churchill, 'awful but right'.[73] This was not heroic folly, but calculation. The decision was a measure of British tank strength, and of confidence in the ability to repel an invasion. The tanks were put in four fast ships, which sailed around the Cape, a journey of about one month.[74]

WARS IN THE AIR

The very visible French failure on the Western Front was followed by the glories of the Battle of Britain. From the summer into the early autumn, RAF fighters based in southern England destroyed 50 per cent more enemy bombers and fighters than they lost. The resulting defeat of the Luftwaffe by the RAF in the summer of 1940 was in many respects the culmination of steady planning in air defence over many years. One crucial aspect was completely unexpected and un-prepared for – the famous dog-fights between Spitfires, Hurricanes, Messerschmitt 109s and 110s. Fighters were expected to intercept bombers, not deal with other fighters. Dog-fights were regarded as things of the past, specifically of the Western Front in the Great War.[75] Otherwise it was a triumph of system and organization, of radar, of observers, of command and control systems. British victory was not, however, the result of what are usually taken as British values tri-umphing over what are taken to be German or Nazi values. If any air force conformed to the usual image of how British fighters operated – a matter of improvisation and individualism – it was not the RAF but the Luftwaffe. If one of the forces was organized with Teutonic efficiency and regimentation, it was the RAF, not the Luftwaffe.[76]

The pacifist writer Vera Brittain noted in 1940 that 'The bombers have a heavy, massive hum, quite different from the lighter, more casual-sounding British machines. All the difference between the Teutonic and Anglo-Saxon temperaments seemed to lie in those two familiar noises.'[77] For all the propagandistic image of the Germans destroying Warsaw and then Rotterdam from the air, the commitment to the bombing of cities was a British rather than a Nazi phenomenon. British bombing of Germany was not in retaliation for the Blitz, a case of the Germans reaping the whirlwind they had sown. It predated not only the Blitz, but also the Battle of Britain. Bomber Command launched the first general bombing offensive against cities in the war on 11 May 1940. The Luftwaffe was prohibited from bombing cities not in the front line. It was not till September 1940 that Hitler allowed the Luftwaffe to start British-style bombing of Britain, following the bombing of Berlin.[78]

There was no shortage of new aircraft in Britain in 1940. Modern types had been in production for years, and just as importantly gigan-tic new factories were ready to increase production. In 1940 Britain out-produced Germany in aircraft, just as the propagandists stated. Even the high level of production before May 1940 was not deemed enough. One of Churchill's very first acts on becoming Prime Minister was to create a new Ministry of Aircraft Production, under Lord

Beaverbrook. Beaverbrook wanted a rapid increase in production, and issued appeals to workers and managers. More importantly he decided to give special priority to five types already in quantity production. On 15 May representatives of the Ministry of Aircraft Production agreed that at least until the end of September 1940 all efforts were to be concentrated on the production of Wellingtons, Whitley Vs, Blenheims, Hurricanes and Spitfires.[79] We may note that three of these types were bombers, instruments of offence. The priority was over by October 1940.[80] Indeed, we have the testimony of the Director of Engine Production that ministry officials and people from industry were pushing Beaverbrook hard to rescind the order. At a meeting on the matter in late June an unmovable Beaverbrook was called away to Downing Street to be told of the French capitulation. He returned to the ministry late at night and agreed to the plan to reintroduce production of new types, such was the continuing confidence in victory.[81]

One of the main aims of the rearmament programmes was to build up a powerful air force which could bomb Germany. Big twin-engined bombers like the Wellington, Whitley and Hampden were built, aeroplanes at least as powerful as the Heinkels and Junkers of the Germans. Despite enormous efforts and expenditures these programmes were, as of 1940–41, failures. The British bombers soon discovered they had to fly at night because air defence was more effective than had been envisaged. Flying at night meant they rarely found their targets. Far from being capable of delivering a knock-out blow, they caused minimal damage to Germany. It is difficult to find a similar example of such a catastrophic failure of a new technical system on this scale. For the enthusiast for counterfactuals this raises the question: what if all the effort that had been devoted to the bombers had gone into tanks or rifles? A prescient prime minister might have done exactly that in the late 1930s. It is sometimes in effect suggested that the British government of 1938 was indeed prescient in anticipating 1940, but not in this sense. The idea is that Chamberlain wisely appeased and delayed war until Britain had enough Spitfires and radar gear to win the Battle of Britain. But using Spitfires and radar in the summer of 1940 stemmed from the military disaster which might not have happened if other policies had been followed.

ANOTHER FAILURE: BLITZ

Late 1940 and early 1941 saw a second British failure. London and many other port and industrial cities were to be bombed with near impunity by the Germans. One reason the Germans could bomb successfully while the British could not was that as well as using electronic navigation aids they were operating from bases not in Germany, but in France and the Low Countries. The other reason was that air defences hardly worked at all. The means of defence were many and varied, including guns, barrage balloons, rockets, aerial mines and more. One of the least known was the extensive system of decoys to mislead bombers, a very large programme in which major towns and cities and industrial and military installations were shadowed by various types of decoy, the most common of which simulated burning towns. They drew off 2,000 tons of bombs, perhaps some 5 per cent of those dropped.[82] The Stockwood decoy, a small set-up which blazed away during raids on Bristol, hoping to confuse German bombers as they approached from the south-east, is credited as one of the more successful ones.[83] Indeed, it is not impossible that the decoys were the most successful form of air defence during the Blitz.

Despite huge investments in fighters and lesser investment in radar, as well as anti-aircraft artillery, barrage balloons and other measures like proximity fuzes, illumination of the skies and rockets, the German bomber got through. The simple tactic of bombing at night rendered the great British air defence system essentially inoperative. This was a matter of some import. More British civilians died during the Blitz than British soldiers in the Battle of France. Yet the German Blitz, while more successful than the British bomber attacks on Germany of 1940–41, did not live up to the horror stories painted by many (including Churchill) in the 1930s; nor indeed to later accounts of the Blitz's destructiveness.

The reasons for this British failure are difficult to grasp since radar could see in the dark. The problem was that during the Blitz radars could not yet be used to direct fighters sufficiently closely to bombers to attack them; nor could they yet be used to direct gunfire accurately enough.

What has proved difficult to grasp, then and since, is that in 1940 a great modern army, a great bomber force and a great air defence system were defeated or inoperative. Britain suffered great losses in 1940 not because it did not have modern technical means, but because these means did not work anything like as effectively as had been confidently expected by a forward-looking British leadership.

Huge efforts were to be made to make the systems work. In the early years radar-controlled guns were very ineffective: the number of 'rounds per bird', shells fired per aircraft brought down, was 20,000. By the spring of 1941 it was down to 4,000.[84] The situation was improved by the redeployment of guns, and the building of very large flat mats, just above ground level, around radar sets to reduce variability in signals caused by different landscapes. These octagonal mats of wire and wire mesh or netting for a time absorbed all the output of galvanized wire and caused shortages of netting ('chicken wire') for poultry pens.[85] There was soon an effective system of ground control radars to direct fighters, together with airborne radar systems. By early 1941 the Mk II GL radar was coming in (it could detect at thirty miles, follow at seventeen miles, and control AA fire from eight miles to an accuracy in range of fifty yards.[86] Towards the end of the war the Mk III made a further great jump. In 1944 aircraft attempting a 1940-type Blitz would have been blown out of the sky by a multiplicity of weapons, just as were the great majority of V-1 flying bombs. Just as dramatic was the change in capability of the British bombers: by 1944 they laid waste to German cities. In fact, right across the board, British forces would learn to use their material superiority to very considerable effect.

CONFIDENCE

In May 1940 the Cabinet and the War Cabinet gave consideration to a possible offer of terms from Hitler. The choice as it presented itself was not between surrender and the continuation of the war with the possibility of heroic defeat. It was one between, on the one hand, an independent Britain ceasing to fight and reaching an accommodation with Hitler and, on the other, a belligerent Britain fighting not for national independence but for the defeat of Germany. The latter position won, and did so remarkably easily and quickly.[87] Indeed, the historian of the British blockade wrote of June 1940 that 'the German victories produced no mood of defeatism; the mood was far rather one of fresh confidence and release with the opportunity of total economic war opened up by the new reality of total danger', a claim endorsed by the then (Labour) Minister of Economic Warfare, Hugh Dalton.[88] Harold Nicolson, a refined National Labour MP, in July 1940 exuded confidence in victory: 'We shall win,' he wrote to his wife.[89] That optimism did not disappear in the following months, rather the contrary. In September 1940 the British Chiefs of Staff were very confident that the British Empire would be victorious, possibly in 1942. By then, they reckoned, Germany would be severely weakened by lack of oil, general blockade, bombing, and by active British military operations against it. An invasion attempt by Germany was a distinct possibility; but a German victory was not in the frame: 'Time is on our side, provided we can continue to draw on the resources of the world and build up our armaments with the minimum of interference, whilst continuing to subject the enemy to the utmost rigours of blockade,' they stated.[90]

Such a belief rested not only on what turned out to be an underestimate of Germany's capacity to extract work from European peoples but on what would seem in later years an implausible belief in Britain's material strength.

It seems extraordinary that Britain was confident, around the time of Dunkirk, which was when the potential overtures from Hitler were rejected, that it could not merely avoid defeat but win the war. So extraordinary, indeed, that the government have been seen as coming to the right decision, but for the wrong reasons: the Cabinet, it is rightly suggested, was swayed by what turned out to be overestimates of the weakness of the German economy, and its vulnerability to blockade, and of the potential power of British bombing.[91] But there were other reasons to believe that Britain would win which seem implausible if we assume that Britain was economically and militarily weak in 1940, and that it depended for survival on the United States. Yet this was not the case to anything like the extent which subsequently seemed obvious.

In 1940 Britain's economic strength was a matter not of faith but of calculation. Figures supported the view that Britain was strong and had a greater capacity to wage modern war than Germany. Geoffrey Crowther, the editor of *The Economist*, estimated that the British Empire had a higher total income than newly Nazified Europe, that Britain and the dominions alone came in at 40 units (arbitrary ones, used by Crowther for convenience), while the whole of Nazi Europe managed 55. This meant, given the higher population of Nazi Europe, that Britain and the dominions had a much higher income per head. Britain and the dominions could therefore afford to put larger and better-equipped forces into the field than Nazi Europe could. Imperial car production, taken as an index of modernity and war potential, was 700,000 per annum (overwhelmingly in Britain), while Germany and France combined managed only 500,000. The upshot of this economic analysis was that the British Empire could win the war against Nazi Europe.[92] Crowther was commenting on war potential, but actual arms production told, in some respects, a similar story. In 1940 aircraft production was greater in Britain than in Germany. British tank production was catching up with German (see Table 3.2).

There was another important reason for British confidence. Britain was safe from a successful German invasion. It was believed, certainly by Churchill, that the British Isles were secure before, as well as after, the Battle of Britain. In his memoirs he was adamant that even if the Germans had had the training and tools for amphibious warfare (which he was clear they did not) their task 'would still have been a forlorn hope'. He asked himself 'how we had the nerve to strip ourselves of the remaining effective military formations we possessed' by sending additional forces to France after the defeat of the BEF: the answer was that 'we understood the difficulties of the Channel crossing without the command of the sea or the air, or the necessary landing craft'.[93] The second British expedition to France had, like the first, to be evacuated: over 136,000 British and Canadian troops were brought home from Cherbourg. After this and the French surrender, Churchill still insisted that if the Germans even managed to reach England's beaches, they would be destroyed. Churchill explicitly told his ministers in a directive on 4 July 1940 that 'there are no grounds for supposing that more German troops can be landed in this country, either from the air or across the sea, than can be destroyed or captured by the strong forces at present under arms'.[94] In support of his argument his memoirs compared the British defensive land forces of 1940 to the smaller German forces defending Normandy in 1944, and pointed to the vast invasion force the allies had needed to defeat it.[95] The chiefs of staff took a different view[96] but Churchill's view prevailed. What this confidence implied is worth spelling out. It meant not only sending more troops to France in May 1940, but also that there was no great move to gather all Britain's forces at home to repel an invasion. For example, very large naval forces remained in Gibraltar[97] and substantial British forces in Egypt. Indeed, these units were to be reinforced from Britain in late 1940, and with tanks, as we have seen. Still, a significant programme of building pill-boxes, anti-tank barriers, gun emplacements and much else was put in place.[98]

MODERN WAR

In the famous 'The Few' speech of 20 August 1940 to the House of Commons, Churchill declared that this war was turning out to be quite unlike 1914–18. In that war 'millions of men fought by hurling enormous masses of steel at one another. "Men and shells" was the cry, and prodigious slaughter was the consequence'. But this new war was a 'conflict of strategy, of organization, of technical apparatus, of science, mechanics, and morale'. The great French mass army was 'beaten into complete and total submission with less than the casualties which they suffered in any one of half a dozen of the battles of 1914–18', as he graphically put it. However, this gave cause for hope since this kind of war was 'well suited to the genius and the resources of the British nation and the British Empire'. This new kind of war was 'more favourable to us than the sombre mass slaughters of the Somme and Passchendaele'. A free, united, committed Britain, 'nurtured in freedom and individual responsibility', devoted to developing the 'arts of war' would 'show the enemy quite a lot of things that they have not thought of yet'. He even commented that since 'the Germans drove the Jews out and lowered their technical standards, our science is definitely ahead of theirs'. And, he insisted, 'Our geographical position, the command of the sea, and the friendship of the United States enable us to draw resources from the whole world and to manufacture weapons of war of every kind, but especially of the superfine kinds, on a scale hitherto practised only by Nazi Germany.'[99] This was no mere propaganda for a distressed and defeated empire. In September 1940 he argued privately to colleagues that 'It is by devising new weapons, and above all by scientific leadership, that we shall best cope with the enemy's superior strength.'[100] The examples he gave were new inventions to find and hit aircraft, the only named one being the UP, the 'un-rotated projectile', the codename for rockets.[101] As we shall see in the following chapter and others, all sorts of new weapons of war were being actively developed in the summer and autumn of 1940, many with personal encouragement from Churchill himself.

These sentiments were shared by Frederick Lindemann, his personal adviser, who in August 1941 minuted Churchill that 'we shall beat the enemy not with large masses but with comparatively small numbers of men armed, equipped and trained to the highest degree (airmen, tank troops, navy)'. For Lindemann this priority had a useful domestic policy consequence. Defending a decision to maintain beer supplies, Lindemann argued for leaving workers in industry to provide the population with 'reasonable amenities even at the expense of, say, the infantry'.[102] Churchill strongly agreed with this position. For both of them an inflated infantry depriving the nation of workers to brew beer and provide other amenities was not the way to fight the war. Instead happy, well-supplied workers would create a powerful scientific and mechanical armed force. Churchill and Lindemann, far from rejecting the strategy of the 1930s, were endorsing it. Even in war they held to the desire to concentrate on machine-intensive elite armed forces in order in part to limit the militarization of society, what I have called liberal militarism. It had, as the discussion of beer supplies suggests, a dash of good old-fashioned Edwardian Tory socialism too. It was a way of thinking far removed from that of advocates of a people's army or those who endorsed hair-shirted wartime austerity.

In March 1941 Churchill insisted to colleagues that the defeat of the enemy would be brought about 'by the staying power of the Navy, and above all by the effect of air predominance'. The army would be used for 'operations of a secondary order, and it is for these special operations that its organization and character should be adapted'.[103] In September 1940 there were plans made for an even more extraordinarily tank-dominated army than already existed. Anthony Eden, the Secretary of State for War (that is, army minister), was aiming at five armoured divisions, plus ten army tank brigades, reckoned at the time to total ten (large) armoured divisions.[104] Another way of putting this is that Britain would have a tank force of twenty armoured brigades. This would indeed have been the army with 'an exceptional proportion of armoured fighting vehicles' that Churchill wanted.[105] Some idea of the centrality of the tank to Churchill's thinking is revealed by his thoughts on attacking continental Europe. Travelling to the USA in December 1941, once the US was in the war, Churchill looked to joint Anglo-American multiple attacks on continental Europe in 1943, which would provoke uprisings in their support.[106] The idea was to have 600,000 men in the armoured units and a million in others. The attacks would involve forty armoured divisions/armoured brigades, half of them British.[107] Forty armoured divisions/brigades was an extraordinary number, many more than the entire German army had, and many more than the Allied forces would ever use in North-West Europe in 1944–5. At the end of the war the US had sixteen armoured divisions in Europe; the British had six armoured divisions, and at least six additional armoured brigades. As we shall see, even this was a much larger force than they faced.

THE EMPIRE AS SUPERPOWER

Given the sheer scale of its total forces, could the British Empire, with its associated exiled forces, and its access to world resources, have defeated Nazi Germany and Italy? The assumption made retrospectively is that this was impossible, but was it? Churchill spoke as if Britain could: 'Give us the tools and we will finish the job' is how he ended a broadcast in February 1941, during the Blitz, in a comment addressed directly at the USA. The idea that '*we* will finish' the job now strikes us as fanciful propaganda. Yet Churchill's line was consistent. In April 1941 he was confident of victory with US help. The

US and the British Empire have 'unchallengeable command of the oceans' he claimed and would soon have 'decisive superiority in the air'. They have 'more wealth, more technical resources, and they make more steel, than the whole of the rest of the world put together'.[108] An internal 'Review of Future Strategy' completed in June 1941 claimed that active belligerency of the US was necessary to victory, but did not envisage large US forces in Europe. It suggested that the build-up of the blockade, bombing and subversion meant an opportune time to strike Germany was autumn 1942, long before the US could be fully engaged.[109]

From June 1941 Britain's position was potentially greatly strengthened. It now had a great ally, the Soviet Union. Britain offered help with supplies of arms and equipment, via the Arctic route. And in the key period of late 1941 and early 1942 these British supplies were significant. By the end of 1941 Britain had delivered 249 Valentines and 187 Matilda II tanks, of which the majority were by then in the hands of the Red Army, and these accounted for no less than 25 per cent of the medium/heavy tanks, and 30–40 per cent of the medium/heavy tanks defending Moscow at this critical point. On 1 January 1942 there were ninety-nine Hurricanes in service with the Red Air Force; 16 per cent of the fighters defending Moscow in December 1941 were Hurricanes or US-built Tomahawks, supplied by Britain. Britain also supplied radar sets, machine tools, anti-submarine sonar and raw materials.[110]

There arises an intriguing counterfactual question: could the British Empire, in alliance with the USSR, have defeated Nazi Germany and Italy? Given that this is how the war was being fought, with the British anticipating victory, the answer could be yes. Indeed, if the war had been one of tank and aircraft production Britain alone would have stood a good chance of beating Germany. The British economy, though smaller than that of Germany, out-produced it in arms as the economists predicted: it was ahead in aircraft from 1940 and in tanks in 1941–2.[111] In field, medium and heavy artillery it was only slightly behind Germany, despite its much smaller army, in both 1941 and 1942.[112] Churchill clearly expressed confidence in such a victory. In September 1941 Churchill told the Commons: 'we are masters of our fate', and in this speech and others celebrated great British advances, including the building up of a huge Middle East army. In his memoirs Churchill called the chapter dealing with autumn 1941 'The Mounting Strength of Britain'. He noted that:

> we had made formidable increases in our military power and were still steadily advancing in actual strength and in the mastery of our many problems. We felt ourselves strong to defend our Island, and able to send troops abroad to the utmost limit of our shipping. We wondered about the future, but, after all we had surmounted, could not fear it.[113]

This is by no means a fanciful picture. By the end of 1941 Britain had achieved significant victories. In the Middle East it had destroyed two Italian armies and conquered most of the Italian empire in North Africa. It had also taken control of Iraq and Vichy Syria and Lebanon. At sea the U-boat was under control, not least because British cryptographers could read the promiscuous signalling between U-boats and headquarters. At the end of 1941 Churchill planned a British invasion of Sicily for 1942, having envisaged winning in all North Africa.[114] An early British victory was within the realm of possibility.

DEFEAT OF THE EMPIRE

At the end of 1941 and early 1942, however, the position of the British Empire was decisively weakened. In December 1941 Japan attacked not only US forces in Hawaii, but that other US imperial territory, the Philippines. It also attacked, simultaneously, the British Empire in the East, starting with Malaya and Hong Kong, and the Dutch East Indies, closely allied to Britain. The Empire was to suffer a much more significant defeat than in 1940.[115] The surrender of Singapore, with losses of over 130,000 troops to an inferior enemy, was a greater humiliation than the capitulation of Kut of 1916, with losses much greater than those in France in 1940. That was just the start: there were other surrenders – in Hong Kong, for example. Large numbers of expatriate men, women and children, tens of thousands, were also incarcerated by the Japanese conquerors. Britain lost its richest colony, Malaya, with its supplies of tin and rubber, as well as the resources of the Dutch East Indies. The loss of rubber and quinine supplies could be made up by expensive synthetic programmes, but the loss of British and Anglo-Dutch supplies of oil and refineries was serious, even though they were denied to the Japanese. As the Japanese advanced in Burma the Rangoon oil storage farms and refinery were sabotaged by the departing British, causing what was said to be the world's largest ever man-made fire.[116] The war in the East now required vast supplies of oil from elsewhere.

Even more important was that large numbers of troops and military supplies which might have been used against Germany now had to be used against Japan. As Churchill told Parliament in January 1942:

> There never has been a moment, there never could have been a moment, when Great Britain or the British Empire, single-handed, could fight Germany and Italy, could wage the Battle of Britain, the Battle of the Atlantic and the Battle of the Middle East and at the same time stand thoroughly prepared in Burma, the Malay Peninsula, and generally in the Far East against the impact of a vast military Empire like Japan, with more than 70 mobile divisions, the third navy in the world, a great air force and the thrust [sic] of 80 or 90 millions of hardy, warlike Asiatics.[117]

In Churchill's image Japan was seemingly of comparable strength to Britain. Yet Churchill insisted it was neither troops nor equipment which were the problem, but the capacity to move them by ship.[118] Behind this claim was the assumption that the British Empire could indeed fight Germany and Italy single-handedly, for Churchill does not invoke the Soviet Union here.

Had there been no defeat in the East, the Empire could have put more troops and equipment into the war in Europe and the Middle East. Some Australian and New Zealand troops which were withdrawn from Europe could have served there, as could the African Divisions which were sent to Burma. Indeed, it is not too fanciful to imagine the Imperial 14th Army, which fought in Burma, fighting instead in the Mediterranean and Europe.

The significance of the British defeat in the East is underplayed. Indeed, starting from Churchill's memoirs the late 1941 and early 1942 period is seen positively, as the consummation of a great Anglo-American alliance that would lead to certain victory. For Churchill,

77

the attack on Pearl Harbor was a critical and welcome event for Britain.[119] For A. J. P. Taylor writing in 1965, 'No greater service than Pearl Harbor was ever performed for the British cause', and he went on to claim – astonishingly – that 'the private war between Britain and Germany was ended'.[120] The Anglo-Soviet alliance is written out of this story, as is, more surprisingly, the terrible loss to the Empire and thus to British strength. The defeats of the Empire are not seen as causes of weakness but as evidence of a profound existing weakness, from which Britain had to be saved by the USA.

SAVED BY THE USA?

The view that the entry of the USA transformed the war against Germany, essentially making victory possible, was a commonplace into the 1960s. Yet in literature from the 1980s and 1990s one finds a much more exaggerated story – one that argues that Britain and the British Empire were saved by the US in 1940. For Britain in 1940 the 'prospects were bleak . . . for survival, let alone victory, US assistance on an unprecedented scale was clearly vital'.[121] According to the author of *The Myth of the Blitz*, 'the greatest single fact suppressed by the Myth of the Blitz is this: in 1940, because Churchill refused to give in, world power passed decisively away from Britain to the USA'.[122] Another recent historian could write: 'Between June 1940 and June 1941 Britain could stand alone only with American help', despite noting that Britain paid for this help.[123] Perhaps most categorically of all, two leading students of British imperialism note that Britain's 'gentlemanly capitalists . . . like the empire they controlled, were saved from liquidation in 1940 by American aid'.[124] One particular example is the claim that the US supplied 100-octane fuel to British fighters in the Battle of Britain, giving them a crucial edge: this is a double myth in that the impact of the new fuel was not as great as implied and because Britain had huge stocks of 100-octane, and supplies from many sources, mainly British.[125]

The relationship formed with the US in 1940 was to be very important, especially as the seeds were sown for a very different kind of war and relationship in 1942–5, one in which US support is underestimated. In late 1940 Britain was rapidly and deliberately running down its dollar reserves; it was buying from the USA without heed to its longer-term economic needs. It was doing this because it knew from the end of 1940 that US-financed help was likely to be available for the future. The Lend-Lease Act was signed in March 1941. This meant that future deliveries, in practice those that arrived once the US was in the war, would be financed by Lend-Lease; indeed, it was not till the US entered the war that terms were agreed.[126] We need to distinguish the levels of American help in the later period, and a different war, from the help received earlier.[127]

The early relationship represented not British weakness but British strength. For Britain went to the USA with dollars, in order to supplement its supplies of arms, a sign not of dependence but of wealth. Britain would continue to pay for the bulk of increasing supplies from the US in hard cash till the end of 1941. British cash launched the USA on its role as the 'arsenal of democracy'. Before Lend-Lease, some emergency US-financed supplies did come over, like a few old destroyers in 1940 in exchange for bases, but nothing major. The important supplies from the USA in this period came from large British (and French) orders placed before the war for aircraft and for machine tools. With the fall of France, Britain took over the French contracts, a further sign of confidence. In fact the Americans reneged on some of these contracts – in July 1940 it became clear in what 'amounted almost to a breach of faith' that the US would not comply with its obligations. The British acceded to a reduction in the ex-French order in September 1940, but there was 'considerable shock' when this was reneged on too, later in the month. The British were forced to accept an Anglo-American joint committee to allocate these orders. As the British put it: 'It is we who are releasing our contractual right not asking for a favour.' The reason the US did not comply was that, as it started to rearm, machine tools became short. It claimed also that if sent to Britain machine tools would be destroyed in bombing, or not be used to full capacity. There were dangers, some officials thought, in giving the impression to Americans that Britain was being pounded harder than it was – the destruction of machines by bombing, it was said, 'has, in fact, been negligible'.[128]

In 1940 new orders to be paid for in dollars were being placed by the British on a huge scale: for Liberty ships, rifles, tanks, aero-engines, and very large quantities of propellants and explosives.[129] These were meant as important supplements to strong British and imperial production. In 1940 the United States was a great industrial power, universally regarded as having an unrivalled capacity to generate new things, as well as to produce them more efficiently than anyone else. Yet its military machines and capabilities were neither better, nor cheaper, nor produced on a large scale compared with the belligerent Europeans. The scientist Ralph Fowler, looking into the matter from Canada in September 1940, claimed that 'the Americans have damned little to offer' and that they were 'too apt to talk of paper projects as if they were already in the ironmongery stage'.[130] And, indeed, when the British team led by Sir Henry Tizard went to the US with new British developments in October 1940, there was again a clear sense that the British had more to offer the Americans than vice versa.[131]

The usual story is that the British arrogantly believed that British arms were better than American ones: they thought the Stirling bomber superior to then available variants of the B-24; that the 87.6mm 25-pounder field gun was better than any US model such as the 105mm; that the 94mm (3.7in) AA gun was better than the US 90mm; that British tanks were better too. The Americans, this story goes, did not want to produce what it would not itself use when it came into the war, and they didn't like what they saw. Thus the Americans refused to build the 25-pounder and the 3.7in AA gun; both were built in Canada.[132] The British could not get their own tanks built in the USA, and in June 1940 were forced to order the Grant/Lee tanks that would see action in North Africa in 1942. But while the official historians note some important exceptions like the Oerlikon and Bofors guns (neither in fact of British design), they rather surprisingly miss some very important cases which tell a very different story.[133]

The US was to produce British designs on a very large scale. We have already mentioned the Lee-Enfield Rifle No. 4. The Rolls-Royce Merlin engine was produced in large numbers by the Packard company (the agreement was made in September 1940) in Detroit. This engine was made not only for Britain but also for US aircraft, perhaps most famously in the P-51 Mustang fighter, which ended up being by far the most successful and most produced US fighter of the war. The British Mark XIV bombsight was built in very large numbers in the USA by the Sperry company and its subcontractor, the GM subsidiary A. C. Sparkplug, in a huge plant in Flint, Michigan.[134] Named the T-1 Bombsight, it was apparently not used by the USAAF but was fitted to US aircraft used by the RAF, as well as British-built aircraft.[135] The No. 19 tank radio, designed by Pye Radio, was manufactured in large numbers in the USA and Canada as well as in Britain.[136] The Universal carrier was made by Ford from 1942, in the USA (and from earlier by Ford in Canada). The 57mm 6-pounder anti-tank gun, also used in tanks, first made for British use in the USA from 1941, was to become the standard US army anti-tank gun too. In 1941 Britain arranged for the manufacture of British jet engines in the USA, and British scientists wanted to build the British bomb in Canada, with equipment from the USA.[137]

EMPIRE LIBERTY

One of the most important British designs to be produced on a large scale in the USA was that of a standard tramp steamer, a humble machine, but a vital one. August 1941 saw the launch of the *Empire Liberty* in the north-eastern British shipbuilding town of Sunderland. If to modern ears the name may be oxymoronic, to many British contemporaries it expressed what they were fighting for, Empire and liberty. But there was a prosaic reason why this ship got its resonant name. *Empire* was the standard element in the names of over 1,300 ships acquired by the British state during the war. They included most of the newly built British merchant ships, old ships bought from America and ships requisitioned from the enemy, like the *Empire Windrush*.[138] But new wartime British merchant ships built abroad did not carry the name. Among them were sixty or so ships built in the USA with names starting with *Ocean,* for example *Ocean Vanguard*; twenty-six Canadian-built ships with names starting with *Fort*; and 182 ships, most with names starting with *Sam*, as in *Samothrace*, built in the USA.[139] These were all of the type that came to be known as Liberty ships. *Empire Liberty* was so named because it was from her drawings that the Liberties were derived, as well as the Oceans, Forts (and Parks – ships of the same type built for Canada), as well as British-built ships.[140] They added up to over 3,000 ships – the greatest-ever production of a single design of ship. *Ocean Vanguard* beat the *Empire Liberty* into the seas by one week.[141] The first Liberty, the SS *Patrick Henry*, was launched a month later; the first Fort, the *Fort St James*, a month after that.

British orders for these ships were crucial in expanding the US and Canadian shipbuilding industries. In September 1940 a British Ship-building Mission, headed by Robert Thompson of the Sunderland shipbuilding firm and including Harry Hunter, the technical director of North-Eastern Marine Engineering, went to America.[142] They were charged with ordering sixty tramp steamers of around 10,000 dead-weight 'tons' each.[143] They visited many shipyards and marine engineering works, as well possible sites for new yards, in both the USA and Canada. The only business that could be done was with a great West Coast civil engineering contractor, Henry Kaiser, who happened to have a small shipbuilding firm in Seattle. By November 1940 the outline of the deal was agreed: $96m (£24m) to include $9m for two new shipyards.[144] The ship to be built was to be Thompson's ship 607, whose plans he had brought with him, engined with Harry Hunter's efficient reciprocating triple expansion machinery. However, the Admiralty decided suddenly that the specifications should be those of a later Thompson ship, the 611, later called *Empire Liberty*, whose drawings were just being finished in Sunderland.[145] This British deci-sion to build the Oceans (and the contemporaneous deal for twenty-six Forts from Canada) 'was without doubt one of the most momentous supply decisions of the whole war' according to the official historian of North American supply.[146] Thompson had to go back to Britain in December 1940 to get the much more expensive deal approved; on the way home his ship was torpedoed, though he was rescued.[147]

In January 1941 the US decided to start its own emergency ship-building programme. To save time it was based on the same British design converted to oil rather than coal-burning, to the tune of 200 ships. The first ones were laid down in April 1941, at the same time as the British Oceans. The programme was to be expanded hugely, with

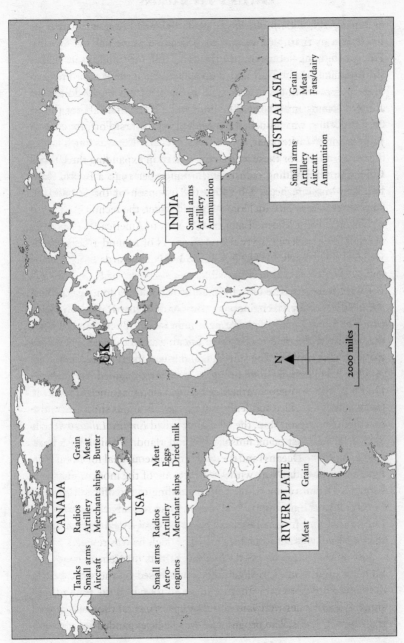

Overseas production of food and British-designed equipment

CANADA

Tanks   Radios   Grain
Small arms   Artillery   Meat
Aircraft   Merchant ships Butter

USA

Small arms   Radios   Meat
Aero-   Artillery   Eggs
engines   Merchant ships Dried milk

RIVER PLATE

Meat   Grain

UK

INDIA

Small arms
Artillery
Ammunition

AUSTRALASIA

Small arms   Grain
Artillery   Meat
Aircraft   Fats/dairy
Ammunition

N

2000 miles

2,700 Liberty ships built. As a result the world fleet was larger in 1946 than in 1939, despite war losses equivalent to half the world's pre-war shipping fleet, some 30m 'tons'.[148] The Liberty ships were also the main reason the post-war world fleet was more steam-intensive than the pre-war one. In 1930, 30 per cent of the world fleet were oil-fired steamers; by 1945 the figure was 52 per cent.[149] They raised the proportion of world steam-powered vessels from 77 per cent to 81 per cent.[150] In 1940–41 the British merchant fleet was much larger than that of the US; by the end of the war the US fleet was three times larger than the British. From having around 15 per cent of world shipping before the war, the US ended the war with well over half, the largest national merchant fleet the world had ever seen.

CONCLUSION

The aggressive, defiant and powerful British Empire of 1940–41 had not been saved by the USA but rather had been spectacularly overtaken by it, not in 1940 but in 1942, in the wake of defeat in the East. Before that the United States was critical, as were other parts of the world, to Britain's fighting strength, and especially so after the loss of supplies from Europe. From 1942 the US became much more important both as a source of critical supplies and as a co-belligerent. It became, in 1942–3, what Britain had been at the end of 1941, the richest, greatest naval and air power, the greatest mercantile power on the seas, the greatest producer of tanks and aeroplanes. The USA took on many of the strategies pioneered by Britain, supporting allies, pursuing machine-intensive ways of warfare involving low casualties. US support to Britain allowed Britain to mobilize to the extent that it did, to continue with its own exceptionally machine-based, low-casualty way of war, and to maintain its citizens in what by European standards were enviable conditions.

Just as the defeat of 1940 obscured British strength in 1940 and earlier, so did a greater defeat eighteen months later obscure the increased strength of the Empire in 1941. The Empire, not entirely alone militarily, and certainly not alone at all materially, was a very great power – at sea, in the air, and to some extent on land too. But defeat at the hands of the Japanese in the East, followed by defeats in North Africa in 1942, again seemed to suggest a fundamental weakness in the forces and capacities of the Empire. Yet while the defeats of 1940 became central to histories of Britain, the import of the imperial disasters of 1941–2 has been obscured. They became marginal to English or British history, defined narrowly as what mattered to the people of the mother country, the home islands. There was to be no imperial Dunkirk moment, or imperial Blitz, no surge of imperial unity. There was no pride to be derived from defeat – only embarrassment, humiliation, a tendency to downplay Empire. Yet these defeats were at the very least as significant in terms of the loss of relative British power as those of 1940, and probably much more so.

4

Cronies and Technocrats

Winston Churchill came into office following the botched April 1940 British operation in Norway, for which he was partly responsible. He led a transformed coalition government now including the Labour Party, ready to prosecute a more offensive war. But on the same day, the Germans invaded the Low Countries and France. Churchill was soon faced with military defeat in France and the French surrender, which changed the nature of the war drastically. Yet this moment of defeat was a moment of glory. He was, for the moment, Hitler's biggest antagonist, a great player on the world stage. Soon enough, however, the great imperialist would preside over the greatest and most humiliating defeat the Empire ever suffered. A great patriot, he put great British forces under foreign command; and under him Britain saw its most dramatic relative decline.

Some aspects of this astonishing personality are not sufficiently understood because they contradict a certain image of Englishness which Churchill is taken to embody. For example, he was an intellectual in a country, England, which many, especially in his own party, supposed was fortunate in having none.[1] He made his living by the pen, mainly as a historian, and wrote at a level which did not disgrace the Nobel Prize for Literature awarded to him in 1953. Secondly, in a nation of supposed amateurs, he considered himself an expert; indeed, specifically an expert in military matters, in a nation held to be pacific.[2] In an unpublished draft of his memoir/history of the Second World War, he noted that as the Chancellor who had returned Britain to the Gold Standard at the $4.86 rate in 1925, he had been misled: 'I had no special comprehension of the currency problem, and therefore fell into the hands of the experts, as I never did later when military matters were concerned.'[3] On becoming Prime Minister he took charge of the military, giving himself the title Minister of Defence and making himself the direct superior of the service chiefs. Thirdly, Churchill was a great enthusiast for science and machines, particularly in relation to war, in a country where the elite, and especially the old aristocratic elite from which Churchill came, were thought to be either above such matters or sunk in rural idiocy. According to the wartime Minister of Production Oliver Lyttelton, one of Churchill's most important qualities as war leader was 'his eager readiness to listen to new, sometimes fantastic, ideas thrown up by scientists, engineers and academic figures. This receptivity in a man over sixty, with small claim to be regarded as a good listener, or so it would seem, had long and deep-seated origins'. His contribution in these matters, Lyttelton judged, ranked 'very high amongst his contributions to victory'.[4] According to Lyttelton, the country owed Churchill a 'very great debt for the way in which he has inspired scientific bodies to experiment with new weapons and to invent new explosives and new means of waging war'.[5]

Lyttelton's was not, however, a common view. Churchill's opponents, and they were many on both the right and the left, were to label him as hostile or at best indifferent to science and the necessities of modern warfare, especially during 1942. Thus the Labour peer (then a very rare beast) Lord Strabolgi told the House of Lords in 1942 that 'It is said that the Prime Minister has a great regard for scientists and employs scientists. I am very glad to hear it because there could be no greater antithesis than between the brilliant mind of the Prime Minister and the scientific mind; they are completely opposed to each other.'[6] Wartime critics of Churchill were to write most of the history of British science in the war, trading on clichés about British elite attitudes to science, and playing up their own contribution. This is one area where Churchill the historian's account was not influential.[7] Yet he provided in many respects a richer vision of science at war than his critics would, as well as more than enough evidence for the falsity of the accusation that Churchill was an antediluvian romantic.

Churchill's account, in his history of the Second World War, of the technical and material aspects is much broader in scope than most subsequent accounts.[8] One reason was that it is centred on instruments of war. His history is replete with references to Nellie, PLUTO, Mulberry, Window, UP (all terms to be explained below) as well as radar, atomic bombs and, most surprising of all, chemical warfare, taboo for the scientific intellectuals during and after the war. His volumes have a very particular cast of technical characters. The scientist most regularly referred to was Churchill's 'trusted friend and confidant' (as he described him), Frederick Lindemann.[9] The young and junior official R. V. Jones, who had been Lindemann's student and had a hand in writing the history, appears in connection with German navigation beams and rockets.[10] Most of the other references to scientists are slight. A little detail is given in the case of Dr Jackson, 'one of our leading spectroscopists', who worked on Window, a radar-jamming device used against Germany from 1943.[11] There are passing, often *ex officio*, references to Sir Henry Tizard, Edward Appleton, Robert Watson-Watt and George Thomson. On the other hand there are very many references to inventors and engineers: to Geoffrey Pyke and his ice-ship, to Charles Hopkins, a senior naval constructor, to Dr Alwyn Crow, the rocketeer, and Stewart Blacker, inventor of the Blacker Bombard.[12] However, by far the most referred-to inventor is Major, then Colonel (later Major-General Sir), Millis R. Jefferis, the man who ran an organization which came to be known as Churchill's 'Toyshop'. Churchill wrote most about people he knew, liked, worked with and had garlanded with honours, yet are now practically unknown. He was often personally associated with their machines.

Many of the stories which follow are about machines which seem crazy; others are about machines whose power and ingenuity seem obvious. Yet as the stories unfold it will become clear that such judgements are products not of any careful assessment but represent a rough and more often than not probably unsafe and often unjust verdict of history. Enthusiasm for machines and their possibilities was no guarantee of support for sensible assessment of the power of a machine; equally, those who urged the application of cold reason could often be caught out by the limits of their own reasoning as much as changes in circumstances. The stories might be read as parables, perhaps just instructive fables, for a mechanical age. One moral is that technical enthusiasms can create great waste. Another is that waste is unavoidable in technical development.

CHURCHILL AS INVENTOR

As we have seen, Churchill showed a very clear preference for mechanical forms of warfare, for machines over manpower. This went back to the Great War, where he served not only as First Lord of the Admiralty, but also Minister of Munitions.[13] Churchill took a personal interest in many war machines of many types. Perhaps least surprising, given he was an Edwardian and had twice been First Lord of the Admiralty, was his interest in battleships. As Prime Minister he kept design work going on the cancelled Lion class battleships and as late as 1941 interested himself in the technical details. He wanted to ensure that they would carry 16in guns and that a central citadel not be compromised by providing a space for stowing aircraft. He was anxious too about the differences between the King George V class battleships then being completed and the new American battleships, corresponding with the First Sea Lord on this question.[14] But his interest and concern stretched to every sort of device, from rockets to anti-tank 'puff-balls', from atomic bombs to jet engines. Churchill was a regular attendee at demonstrations of new gadgets and one who clearly believed that Britain had great reserves of ingenuity which would be critical for victory.

His involvement went further than interest, support or even enthusiasm. According to a post-war reflection by Charles Goodeve, the senior naval scientist of the Second World War, Churchill was 'an inventor of no mean repute'.[15] That surprising encomium was justified. When Churchill was at the Admiralty in the First World War he was the key figure behind the invention of the tank, which he entrusted to the Director of Naval Construction, then the greatest technical officer of the state, and a 'Landships Committee'. The first tanks were built by a Lincoln firm, William Foster, makers of steam traction engines and other agricultural equipment. In the Second World War, and in the same job, Churchill again turned to invention. He came up with an idea, a sketch, for a gigantic earthmoving mole that would cut huge trenches. Once again he was trespassing on army business, and once again his machine was overseen by naval constructors and built in Lincoln.

Churchill's invention was first called 'White Rabbit No. 6', then 'Cultivator No. 6' and then the 'NLE Tractor', hence the nickname 'Nellie'. It was a machine which moving forward very slowly cut a trench along which soldiers, and in a later planned version vehicles, could advance on the enemy's trenches. His tactical notion was that two to three hundred of these 100-ton monsters would be used along a front of twenty to twenty-five miles moving through the night from one front line to the other. Churchill wanted a means of 'breaking a deadlock on the French front without repetition of the slaughter of the previous war'.[16]

The project was handed over to the Directorate of Naval Construction in November 1939, where an assistant constructor, Charles Hopkins, ran the project with the very large sum of £100,000 for experiments. Within six weeks a model had been made, and was soon demonstrated to the Prime Minister, Neville Chamberlain and the French military leaders. By February 1940 the government had approved the making of 200 narrow types, and forty wide ones. In March 1940 the project was handed to the Ministry of Supply, which set up a section called Naval Land Equipment to run it under Charles Hopkins, who was given the gloriously oxymoronic title of Director of Naval Land Equipment. The moles were designed and key parts built by the earth-moving equipment manufacturers Ruston-Bucyrus of Lincoln, a company owned by the old Lincoln firm of Ruston and Hornsby (also a maker of traction engines), and the US excavator firm Bucyrus-Erie.

The project was overseen by a very high-level committee chaired by Churchill as First Lord, which included the Minister of Supply, the Director of Naval Construction and the Secretary of the Admiralty; the fourth and last meeting was on 1 May 1940.[17] But after he became Prime Minister, Churchill still received regular progress reports from Charles Hopkins into 1941 and was involved in all key decisions on the mole until the end of the war.

The unorthodox arrangements and the air of eccentricity around the mole should not disguise the huge ambition and scale of the project. The minutes of the first meeting record that Churchill 'asked for the completion of 240 machines by May 1941'.[18] The machine had not yet been made, nor its main components tested. Costing an estimated

£8m it was a project comparable to a very large and complex arms factory or a battleship.[19] The 240 machines, each many times heavier than the heaviest tanks, would require over 20,000 tons of steel, and 1/18th of national production of steel castings.[20] On one estimate each machine involved 36,000 components, 71 sub-assemblies and 250 (or 350 – sources vary) suppliers.[21] They were to be powered by a Rolls-Royce Merlin aero-engine, precious things in early 1940. Hopkins visited Rolls-Royce in March 1940 and told them he would need 200 marine versions of the Merlin engines by the end of the year, and twenty to forty spares by June 1941. Rolls-Royce were making 148 marine Merlins for export to the Dutch and French to use in torpedo boats, but they had no spare capacity and pointed out that 200 Merlins for the mole meant 200 fewer fighter aircraft.[22] Churchill was unabashed and wrote a strong letter demanding 'immediate, personal attention' of the Secretary of State for Air to his request for the 200 Merlins; he got only one engine for experiments.[23] The Merlin was replaced with two Paxman-Ricardo diesels in each mole. This was no trivial change in what was still a plan: the mole needed redesigning and, as Hopkins put it, 'production of 500 diesels of this size within a year far exceeds anything attempted before in this country'.[24]

All the above arguments were about a machine not yet made. It was not till 17 April 1940 that a test rig of the business end of the mole was tried out in the presence of the Minister of Supply, Leslie Burgin. He wrote a seven-page report on the trial himself, and in a covering letter to Churchill breathlessly reported that 'Whilst I was on board the machine, 20 yards of trench some 3 feet odd wide and 3 feet 6 inches deep were cut ... in a manner absolutely novel and indeed revolutionary.'[25] Yet within a few weeks of this glowing assessment of its possibilities the machine was essentially redundant, as there was no longer a Western Front for it to operate on. Churchill, now in Downing Street, reduced the order to thirty-three machines, its priority reduced from the highest to 1(b). In the autumn of 1940 the order was further reduced to four 'infantry' types (earlier called 'private' types) and four of the larger 'officer' types, for research purposes. As a security measure, the name was changed to the 'NLE Tractor', given that 'full particulars' of the 'Cultivator' had been given to General Gamelin, Admiral Darlan and other French officers.[26] The pilot was still being assembled in early 1941, prompting Churchill to complain: 'We are four months late. I am most anxious for a field trial.'[27] This did not happen until the summer and Churchill attended a demonstration on 6 November 1941.[28] As Churchill recalled of the trial: 'This mammoth mole could cut in loam a trench five feet deep and seven and a half feet wide at half-a-mile an hour.'[29] By April 1943 the four infantry models had been made, and the first 'officer' type was due to appear. However, the Secretary of State for War, contending that the only possible use for the machines was earth moving, recommended the aborting of the officer type and the transfer of the staff to tank development work. Churchill agreed, but instructed that the four 'infantry' types already made should be kept 'in good order', adding: 'Their turn may come.'[30] Towards the very end of the war, the Secretary of State for War proposed to have the four infantry moles broken up. Churchill ordered that one be kept, as indeed happened, but only until the 1950s, when it was scrapped.[31] He was 'responsible but impenitent' he wrote in a two-page appendix on his invention in his history of the war.[32]

It is tempting to look at Churchill's mole and to assume, since no one else made one and none was needed, that the whole effort was thoroughly misplaced. Yet, in a different scenario, the mole could have become a famous and decisive weapon, one which might have led, say, to an Anglo-French march on Berlin in 1941. As it happens, the context it was designed for disappeared. What it testifies to is the extraordinary commitment of the Chamberlain government to the development of machines of war – so deep and wide was the British effort that it included not just vast numbers of universal carriers and tanks but these moles too. They were the product not just of an elite who wished to fight a mechanical war, but of a nation which had the capacity to produce the means to fight it.

LEO AMERY'S FLYING ARMOURED CAR

Churchill was not the only inventor in his Cabinet. His imperialist Secretary of State for India and Burma, Leo Amery, like Churchill an intellectual, was the inventor of an idea whose silliness strikes even the non-engineer. It was a 'light armed, and possibly slightly armoured, vehicle, propelled over desert sand by air screw and possibly able to lift into the air for short periods' using its small wings.[33] His proposal went to the Admiralty, in July 1940; the Director of Scientific Research, Charles Wright, described it as an aeroplane that could move on land.[34] He was drawn into correspondence with Amery, but passed the suggestion on to the Ministry of Supply's Director of Scientific Research, Herbert Gough. Amery now put up a further suggestion of a 'light air-driven armoured sleigh on ski runners' for use in countries like Norway.[35] The doubtless very busy Gough was forced to reply in tones of barely concealed contempt: the performance of the originally suggested vehicle would be worse than a light armoured car since this one was encumbered with wings. Wings, he pointed out, 'would be a nuisance to a land service operation in restricted spaces, and because of a lack of ground clearance when operating over uneven surfaces, or banking on bends'.[36]

A year later Amery was to come up with a slightly more practical suggestion for a purpose close to his heart – imperial unity. Regretting that it was not possible to hold an Imperial War Cabinet 'in the flesh', he looked forward to doing so via television in the future. In the meantime, he enquired of Churchill's adviser Frederick Lindemann whether 'the development of wireless telephony' made it possible to do so during the war. He imagined people listening and chipping in by radio from around the world, providing a 'demonstration of effective Empire unity'.[37] In a rather droll reply Lindemann wondered how good reception would be and pointed to the difficulties of preventing the 'enemy eavesdropping'.[38]

MORRISON SHELTER

The Blitz hit Britain in the autumn and winter nights of 1940–41. This was not the time to be in a trench or an Anderson shelter in one's garden. There was a demand for indoor shelters. The choice of a new shelter was made in No. 10 Downing Street, not in a committee meeting, but at a demonstration. On New Year's Day 1941 two prototypes were on show there: one, with a flat top, was by Professor John Baker;

the other, with a curved top, by a Dr Merriman. The scene was rendered as follows in an official account: 'The PM entering the room found a convenient seat on the flat-topped shelter and hailed "curved-top" as "just the thing". Professor Baker suggested that the convenience of "flat-top" as an article of furniture was perhaps causing its qualities as a shelter to be hidden.' The upshot was that both would be produced and the public would have a choice. But tests on production models in March 1941 showed the curved top to be 'markedly inferior' and it was abandoned.[39]

Baker's own account was much more dramatic. He claimed that Churchill had himself come up with the curved-top idea, and had sketched an arched shelter for Herbert Morrison, the (Labour) Minister of Home Security. Morrison got the idea worked up into a design and mock-up in his ministry. Baker, who worked in the ministry, was asked to look at sketches, but instead designed his own shelter, based on his novel principles for construction in steel. This involved using the buckling of steel to absorb energy. Baker had a mock-up of his shelter made and, to the surprise of the chief engineer of the Ministry, took it to Downing Street as a last-minute challenger to the official shelter inspired by Churchill. He recorded:

> The Prime Minister presumably taking it for a piece of furniture, sat on the edge of the table shelter and looked at the arch type. 'That's the kind of thing to give them' he said and the various points of the shelter were discussed. Mr Morrison then drew the Prime Minister's attention to the fact that he was sitting on another shelter. He got off it and the Minister of Home Security described its virtues. This impressed me very much since he brought out all the points I had retailed to him a quarter of an hour before about an engineering product he had heard of for the first time. The Prime Minister addressed some question about the shelter to the Chief Engineer, he passed it on to me and I answered it . . . The Prime Minister then concentrated on me and plied me with questions.

Churchill, he continued, 'was eventually satisfied, thumped the top of the table shelter, said, "that's the one, make 500,000 in the next three months and give them to the people. Show them that it is safe, blow a house up on one, put a pig in it, put the inventor in it", said he poking me in the ribs.' Not surprisingly, Baker recorded that 'It was most probably the most memorable three quarters of an hour of my life and I am never likely to forget it.' In this version, Morrison asks Baker to redesign the arched version as well.[40]

Baker's novel design came to be produced on a large scale. The 'steel table shelter' was issued free for those earning under £350 per annum and was on sale for £7. It was better known as the Morrison shelter, though the Baker shelter, even the Churchill shelter, would have been much more appropriate. Had the bombing continued, it would perhaps have become even more famous than the Anderson shelter. Yet, given that the bombing was coming to an end, until a brief resumption in 1944, it was to save few lives. Britain's cities would have been better off without it, in that the steel could have been used for something else. Of course, in this case, as in many others, it had to be assumed that Britain was likely to be bombed in the near future.

CHURCHILL'S TECHNOCRATS

Churchill's wartime administrations were made up of large numbers of men whose primary expertise was not political, but rather technical or administrative. We might take a deep breath and call it a government of technocrats. Such a description jars for important and interesting reasons. An expert, or technocratic government, we are invited to think, is one appointed on merit and experience untainted by politics or personal connection. Furthermore, British enthusiasm for expertise in this period is strongly associated with the left. Yet most of the expert ministers in Churchill's government were clearly figures of the right. Furthermore, while they were not party-political figures, they were often personal cronies of Churchill's.

Churchill was not the first wartime prime minister to bring non-politicians into ministerial positions. Lloyd George had done so in the Great War. In the Second World War, in this respect as in so many others, Chamberlain led where Churchill followed. Chamberlain brought in six non-politicians to his Cabinet; some were appointed even before the war started. The one who would have the most illustrious government wartime career was Sir John Anderson. Anderson had been a famous civil servant, serving in Ireland and at the Home Office. He retired from the home civil service to become Governor of Bengal, one of the great political-administrative jobs in the Empire. He was called back into the British government in 1938 as Lord Privy Seal to run the Air Raid Protection department of the Home Office. Anderson had a science degree and in the brief interlude between Bengal and his return to the Home Office was a director of Imperial Chemical Industries. At the first meeting of the Civil Defence Research Committee in May 1939, he gave an address to the assembled scientists and engineers in which he recalled he could have become a scientist and that he 'always had a very soft corner in my heart for scientific work. I have always envied a scientist because he can make contact with the outside world and can make himself master of subjects in a way an administrator about his business can never hope to do.'[41]

Two of Chamberlain's six expert ministers had naval backgrounds. Lord Chatfield, a gunnery officer and former head of the navy, was made Minister for Coordination of Defence, standing above, until Churchill's arrival in September 1939, the three service ministers. Lord Hankey, 'the man of secrets' who had for years been the secretary to the Cabinet and the Committee of Imperial Defence, had been a Royal Marine artilleryman. He was appointed Minister without Portfolio, taking up many sensitive tasks, many concerned with science and engineering. The fourth expert was a lawyer and administrator in the steel industry, Sir Andrew Duncan, who was made President of the Board of Trade in January 1940. Frederick Marquis, soon to become Lord Woolton, a businessman with a science degree who was made Minister of Food in April 1940, was to become the most familiar to the public. Sir John Reith, former engineer and former head of the BBC, became Minister of Information. All except Chatfield were in office when Churchill became Prime Minister and many stayed till the end of the war. Anderson thrived, serving in the War Cabinet, as both Lord President and Chancellor, and charged with vitally important responsibilities, like chairing the Manpower Committee and leading the British atomic bomb programme, an appropriate role for someone who had in his youth been an expert in the chemistry of uranium.[42]

Lord Woolton continued as Minister of Food and was promoted into the War Cabinet as Minister of Reconstruction in 1943. Andrew Duncan ran the Ministry of Supply for most of the war. Hankey had important responsibilities even after he left the Cabinet in 1942. Reith, who loathed Churchill, went on to briefly become Minister of Transport, and then Minister of Works, a position he left in 1942.

Churchill brought personal cronies into key ministries and into the War Cabinet.[43] The first outsider to join Churchill's small War Cabinet was the Canadian newspaper baron Lord Beaverbrook (August 1940), owner of the *Daily Express* and *Evening Standard*. He was an unscrupulous buccaneer, described by a minister who came to like him as having a 'streak of vindictiveness and even cruelty'; he brought with him into the new Ministry of Aircraft Production his own cronies, 'personal supporters' as well as 'noted industrialists'.[44] He later became Minister of Supply, during which time he also conducted crucial negotiations on supply in the USA and Moscow. In February 1942 he took charge of all the supply issues as Minister of War Production but resigned after two weeks, whether for reasons of health or politics is not clear. He went on to agitate for a second front. He was succeeded in the crucial role of Minister of Production, the only production minister in the War Cabinet, by another businessman crony, Oliver Lyttelton. Lyttelton had been head of the British Metal Corporation, a major player in global tin, zinc and copper industries.[45] He entered the Cabinet in 1940 as President of the Board of Trade and then had various important jobs in the Middle East. He was an old friend of Churchill's who regularly dined in Chequers while a mere Controller of Non-Ferrous Metals in the Ministry of Supply, a job he took up at the beginning of the war.[46]

Churchill appointed many other industrial and scientific cronies to his Cabinet. Brendan Bracken, chairman of the *Financial News* and founding editor of *The Banker*, entered the Cabinet in 1941 as Minister of Information. Frederick Lindemann (Lord Cherwell) entered in 1942 as Paymaster General, having been at Churchill's side for the whole war. F. J. Leathers (Lord Leathers), 'a great shipping expert, known personally to me',[47] became Minister of War Transport in 1941. The work of this ministry, a civil servant recorded, was 'aided by the fact that our Minister Lord Leathers was one of those with the most intimate access to Winston Churchill'.[48] Leathers, of working-class background, was in the coal and oil lighter business in London (that is to say, in a company which owned 'lighters' – large barges which transferred goods from ship to shore) and became a major figure in shipping. Churchill met him on the board of a shipping line, Churchill's only business interest.[49] In 1940 Leathers was brought in to advise on coal shipping and was able to sort out the problems caused by ships laden with coal for a now defeated France.[50] Not all the experts appointed to Cabinet by Churchill were cronies. Among the others were Baron Portal, who ran the family banknote business, and Sir John Grigg, a unique case of a permanent secretary of a ministry becoming its minister, in this case Secretary of State for War, in 1942.

It was not merely experts who were well represented in Cabinet, but men with science degrees. As the liberal politician Viscount Samuel put it in 1943:

> This present Administration is different, I think, from any that have preceded it, in that there are four Ministers holding important posts in the Government who are themselves trained scientists. Sir John Anderson, Sir Stafford Cripps and the noble Lords, Lord Woolton and Lord Cherwell, have all been trained in science in their youth and have scientific degrees, and the last is, of course, a scientist *de carrière*.[51]

(He might have added Grigg, who was a mathematician.) Of these only Cripps could claim a connection to the left, though by 1942 he was an independent.[52]

The running of the economic, industrial and technical war was an immensely complicated bureaucratic business. The idea that Churchill was unconcerned with such matters and left this domestic side of the war to Labour ministers is, though a commonplace, highly misleading. The presence of Labour in the Coalition was of momentous political rather than administrative significance. Firstly, Churchill put his cronies in key production ministries: Beaverbrook from 1940 to 1942, then Lyttelton, 1942–5, were both in the War Cabinet as the key production ministers. In addition, as well as Leathers at War Transport, most ministers of Supply and Aircraft Production like Andrew Duncan, Lord Brabazon and John Llewellin were Conservatives. The only Labour ministers concerned with war production apart from Ernest Bevin as Minister of Labour and National Service were A. V. Alexander at the Admiralty and Herbert Morrison very briefly at Supply in 1940. The independent Sir Stafford Cripps was Minister of Aircraft Production between 1942 and 1945. Bevin, a member of the War Cabinet, was an important figure in production, but not the final arbiter, much less the directing force. In the early years Beaverbrook was at least as important, and in the later years Oliver Lyttelton and Sir John Anderson were each almost certainly more significant than Bevin in production matters. However, the most important figure was Churchill himself. He took a personal and direct interest in this vital matter, not least through Lord Lindemann. Lindemann was Churchill's trusted adviser on all the material aspects of the war effort.[53] Lindemann's almost daily minutes to Churchill give an interconnected story of things and experts, little of which is reflected in the vast literature on both.

THE ENIGMATIC LORD LINDEMANN

In February 1942 Sir Henry Tizard, adviser to the Ministry of Aircraft Production and member of the Air Council, asked at a lunch of the Parliamentary and Scientific Committee: 'what previous Prime Minister of England ever had a scientific adviser continually at his elbow?'[54] It was a rhetorical question, but one which significantly underplayed the significance of the role played by Frederick Lindemann, Lord Cherwell. When Churchill became First Lord of the Admiralty and a member of the War Cabinet, Lindemann became his personal scientific adviser. But within a month he had an additional and vital duty as an adviser on production and the war economy. Lindemann installed a staff of 'half-a-dozen statisticians and economists whom we could trust to give no attention to anything but realities'. At a time when each department collected statistics on its own basis, Lindemann's team presented Churchill with 'tables and diagrams' and analyses of departmental plans; Churchill had from the beginning his 'own sure, steady source of information, every part of which was integrally related to all the rest'.[55] Lindemann's duties were described by Churchill in Parliament in 1941 in such a way that makes clear he was a great deal more than a scientific adviser: 'Lord Cherwell is one of my personal assistants, and advises me ... upon the scientific and statistical aspects of our national defence. He is ... now specially charged with the duty of warning me of short fallings in any part of our war supply.'[56] At the Potsdam conference in 1945 Churchill introduced Cherwell to Stalin as:

> the man who had been advising him about a balanced use of our resources for the war effort, drawing his attention to shortfalls, and generally keeping his eye on the whole scene of government, to inform him if anything was amiss; in fact he had acted as a kind of Gestapo for him.

Stalin, showing a sharp wit and knowledge of the recent British election campaign, replied immediately: 'I thought it was only Mr. Attlee who had a Gestapo.'[57]

Lindemann was a brilliant, contradictory figure – a bachelor don who lived frugally in his college, yet frequented the salons of power; a deeply impolitic man who was a significant political figure; a vegetarian who wanted the wartime civilians to have more meat; a scientist too aloofly aristocratic even for the upper-middle-class academics of Oxford. No scientist ever had more influence in British history; and probably no academic either. Yet far from being celebrated by the scientific community as an exemplary figure, he was and continues to be abused. As 'Churchill's scientific adviser' he is described as a 'pet scientist' or 'courtier' who did his master's bidding and prevented the voice of reason being heard in government. Despite his being written about more sensitively and intelligently than most wartime scientists (though principally by economists and politicians), the conventional view of his many detractors (many but not all from the left) is still influential.[58] The style and content of a wartime observation by the head of Coastal Command – 'I do not really dislike old Lindypops – I find him rather a pathetic figure, the enemy of so many and cursed by ill-health' – are exceptional in their warmth.[59]

Lindemann was Professor of Physics in Oxford and was the head of the Clarendon Laboratory. Oxford scientific research, with the important exception of chemistry, did not loom as large as that done in Cambridge and had a different tone. Lindemann was at the heart of a world of well-connected and wealthy Oxford scientists, often like

Lindemann himself with connections to industry. Among his Clarendon Laboratory colleagues were Thomas Merton, an Old Etonian, who marked his elevation in 1937 to the chair of Spectroscopy at Oxford by giving up academic research. He increasingly lived the life of a country gentleman and art connoisseur but also 'changed from a scientist into an inventor . . . he joined the small but illustrious band of wealthy inventors'.[60] Merton was Treasurer of the Royal Society from 1939, and a vice-president from 1941. Another was Alfred 'Jack' Egerton (another Old Etonian), who went to Imperial College in 1936. Egerton was Secretary of the Royal Society from 1938 through the war and was a life-long friend of his brother-in-law and fellow chemist, Stafford Cripps. The spectroscopist Derek Jackson, a Lindemann protégé, owned a share of the *News of the World*. Married to one of the Mitford sisters, he was bisexual, a national hunt rider, a pro-Nazi Moseley-ite, and an anti-Semite. He spent the war in the RAF, doing important work on the radar countermeasure method codenamed Window.[61] He was about as far from the standard historical image of the wartime British scientist as it is possible to imagine.

In the 1930s and 1940s there were a number of hereditary nobles in the scientific and engineering communities, many of whom were active in public and industrial life. Dudley Gladstone Gordon, Lord Aberdeen, second son of the 1st Marquis of Aberdeen and Temair, was a refrigeration engineer who ran one of the leading firms, J. and E. Hall Ltd of Dartford, supplying machinery for ships and cold stores. He was president of the Federation of British Industries during the war and became chairman of Hadfields, the steel and arms firm, at its end.

Charles Howard, the 20th Earl of Suffolk, who was also the 13th Earl of Berkshire, an Oxford-trained chemist, was a British liaison officer with the armaments ministry in Paris who returned to Britain in May 1940 by destroyer carrying industrial diamonds removed from Antwerp and Amsterdam, the French stock of heavy water and nuclear scientists, and a consignment of American machine tools, picked up en route in Bordeaux. Harold Macmillan, then a junior minister in the Ministry of Supply, described him as 'a mixture between Sir Francis Drake and the Scarlet Pimpernel'. In Britain he turned to defusing bombs, together with his secretary and his driver; all three were killed

LORD · DUDLEY · GORDON · D·S·O· LL·D· M·I·Mech·E ·
PRESIDENT · F · B · I · 1940 — 43 ·

The aristocratic refrigeration engineer and industrialist
Lord Dudley Gordon, 1943, by Malcolm Osborne

dealing with their thirty-fifth bomb, in 1941.[62] He was awarded a posthumous George Cross. John Grimston, 4th Earl Verulam, was an electrical engineer and businessman, the founder of Enfield Cables, and the head of the British Bedaux company.[63] While social ascent into the world of industrial research was common, social descent into such a career was rare: John ('Tony') Giffard became the 3rd Earl of Halsbury in 1943 (the 1st Earl was a reactionary Lord Chancellor, the original editor and compiler of the continuing *Halsbury's Laws of England*) while working on creep-resisting steels for the jet engine programme at the Brown-Firth Laboratories in Sheffield.[64] Lack of family money forced Giffard to train as an accountant. He took an external London degree in chemistry and maths as a private student. In 1935 he went to work as a chemist in the Lever Bros combine, working on the processing of whale oil among other things, before joining Firth Brown in 1942.[65] By contrast, Evelyn Boscawen, 8th Viscount Falmouth, was immensely rich. He was a Cambridge-trained engineer who spent most of the war as head of the Fire Division of the Research and Experiments Division of the Ministry of Home Security. Victor Rothschild, 3rd Baron Rothschild, also rich, was a Cambridge biologist. He became assistant to Sir Harold Hartley, working on chemical warfare, but in 1940 transferred to MI5 to run counter-sabotage operations, where as Colonel Lord Rothschild he (among other things) tested gifts of drink and cigars for the Prime Minister, a personal friend.[66] In 1944 he defused a bomb which had been found in a crate of onions imported from Spain, for which he might have expected an MBE or a commendation. Churchill got him the George Medal.[67]

What makes Lindemann stand out from this aristocratic and wealthy scientific world was his intimate and long-standing friendship with a major political figure, Winston Churchill, and his own political concerns and ambitions. He had stood unsuccessfully for a vacant Oxford University seat in Parliament in 1936, basing his candidature on his expertise in air defence. He and Churchill were both strenuous critics of the air defence effort yet at the same time were insiders, sitting on key government committees. Thus Lindemann was a member of the Air Defence Committee of the Air Ministry in 1935–6 and from 1938 (together with Churchill) of the senior Air Defence

Research Committee of the Committee of Imperial Defence.[68] Although a scientific adviser to the Baldwin and Chamberlain governments, he, like Churchill, was more of an activist than an adviser, a critic of the state scientific apparatus and of its most trusted advisers.

From September 1939 Lindemann was the personal adviser of just one, though important, government minister. From May 1940 he was in a position to outrank scientific advisers to other ministries, especially on matters of major concern. His boss was not only Prime Minister but Minister of Defence as well. There was no great issue with the Admiralty, but neither Churchill nor Lindemann had full confidence in either the Air Ministry or the War Office/Ministry of Supply. In June 1940 Sir Henry Tizard, adviser to the Air Ministry, was to make a major mistake as an adviser, seriously downplaying in the presence of Churchill the possible significance or existence of German navigation beams. The Germans were indeed using them, as became clear the same month; they were to be crucial to the German bombing offensive later in the year.[69] He resigned from the ministry, but retained important advisory positions in the Ministry of Aircraft Production and was to return to the Air Ministry as a member of the Air Council. Churchill and Lindemann were also less than enthused with the Ministry of Supply's scientific organization and got round it in ways to be described in Chapter 8.

Because they were in government, Lindemann and Churchill did not cease to become critics of the machine; indeed, they were in a powerful position to criticize more, and so they did. They distrusted other experts – be they air marshals or scientific advisers. That made them very unpopular internally and they were both accused of being irresponsible amateurs. But more than once during the war these supposed amateurs challenged and overrode the advice of other experts to very good effect. On the other hand many of the schemes they pushed turned out to be singularly useless.

Take, for example, Churchill and Lindemann's battle against the nutritionists. In the summer of 1940 the scientific experts on nutrition, in fact taking up some elementary prescriptions familiar from the Great War, 'had set forth the governing principles of food production in a besieged island'. They proposed to increase the production of potatoes, sugar beet, cereal, vegetables and milk and a great reduction in the number of pigs and poultry (which themselves ate cereal). The lowland sheep and beef cattle populations were to be reduced, to release land for crops, including fodder for dairy cows.[70] This implied a huge slaughter programme to get rid of the excess animal population. Churchill and Lindemann fought to maintain a higher standard of food supply and quality than what they took the exaggerated demands of the military to imply, and what the policies of the nutritional experts demanded. They resisted the extreme cutting back of import of animal feeding stuffs, wanting to keep cattle alive instead of prematurely slaughtering them, and favoured chickens as 'economical converters of imported grain'.[71] 'Although a vegetarian in his own person', as far as the war effort was concerned Lindemann 'was an extreme anti-vegetarian'.[72] The Prof, as he was known, 'implied in his verbal comments that he knew just as much about the processes of human nutrition as the experts'.[73] 'The quiet, but supreme, self-assurance with which the Prof expressed his withering contempt for the views of the dietetical experts must have been a comfort to Churchill in his resolution to pursue this policy,' noted a colleague later.[74] It was in fact just as well, as the slaughter implied by the nutritionists' policies was not carried out; there was not enough cold storage capacity to preserve the meat.[75]

CHURCHILL, LINDEMANN AND THE MATERIAL WAR

Lindemann had an extraordinary facility of picturing quantities and their relationship: he had a sense of the relations, say, of shipping capacity and food supply, and how this related to troop-carrying capacity. Lindemann had 'a wonderful flair for orders of magnitude'.[76] He was supported in his examinations of war production by, from October 1939, a statistical branch (S-Branch), consisting primarily of economists. On average it had six economists, one scientific officer, one administrator and then half a dozen 'computers', that is, people to do calculations, plus 'chartists', around four at the beginning, who drew beautiful coloured charts and diagrams, as well as a few typists and clerks.[77] S-Branch used telephones, masses of documentation and slide-rules.[78] S-Branch was not the only economic or statistical unit in government, but unlike the Economic Section, for example, it was singularly concerned with examination of details of the warlike effort. The criticism of departmental estimates remained an important part of the Section's work until the end of the war.[79] Shipping was one of the key issues examined by S-Branch.[80] It was especially keen to suppress the import of timber and was very concerned when it remained higher than directed. 'Nothing raised the ire of the Prof. so much' as timber imports, recalled a colleague. He rejected the excuses:

> I think he suspected that there were some bearers of a degree in the Arts – 'classics', he always called them – who had too sentimental an eye for the preservation of the British woodlands and were working away to frustrate him. 'Timber' became almost like the name of a personal enemy; the very word stank in our department.[81]

The work of the section caused a major change in shipping. One reason for the shortage of shipping for imports was that it was needed to carry men and stores out of the UK. One of the main routes from 1940 was to Egypt via the Cape of Good Hope, the Mediterranean being blocked to such traffic. The distance was the same as that to Australia or New Zealand. Lindemann, believing that the military were getting too much and that civil imports were getting dangerously low in 1942, achieved a halving of the number of ships going to the Middle East and Indian Ocean from Britain with military stores, from 120 to sixty per month. This was, according to one of his economist assistants, deliberately using modern terminology, 'the most momentous macro-economic decision with which I have been involved'.[82] Doing this involved a detailed look at what was transported and how. One of the most bulky cargoes was military vehicles. They were sent fully assembled, thus wasting a great deal of shipping space. Lindemann discovered this and insisted on boxing the thousands of vehicles being shipped each month. By the end of 1942 it took half the space it previously had to ship a 3-ton lorry.[83] 'The immense importance of this matter,' he told Churchill, was shown by the fact that even in the first effort 'boxing one-fifth of the vehicles has in one month saved about 100,000 tons of imports. Beside this, the various civilian econo-mies are bagatelles.' In June 1942 he pointed out that the increase in the milling ratio for wheat, the rationing of clothes and soap and the abolition of the basic petrol ration had between them saved about 65,000 tons of imports in one month.[84] Roy Harrod, an economist who was a member of Lindemann's statistical branch, claimed that Churchill, Lindemann and Donald MacDougall (another economist in the statistical branch), in their joint thinking on certain economic questions, caused 'decisions of no little consequence to be taken'.[85] In this he was surely right. Lindemann's influence as an in effect economic adviser went much further than consideration of shipping quantities. He was directly involved in economic negotiations with the US and much involved in discussion of post-war economic planning.

LINDEMANN AND CHURCHILL ON NEW TECHNIQUES

Churchill's war memoirs include retrospective defences of his involve-ment with and promotion of particular techniques of war. Much of the chapter in Volume II on 'The Wizard War' is a description and defence of three interconnected anti-aircraft programmes he and Lin-demann were particularly associated with from the 1930s: anti-aircraft rockets, aerial mines and the photo-electric proximity fuze. Churchill defended his investments, noting that anti-aircraft guns only became effective as the Blitz was ending, at the same time as his favoured inventions were just starting to come in. This is another way of saying that they didn't work in time.

Lindemann had many ideas for air defence. In 1938 (perhaps earlier too) he had proposed the creation of a high-altitude balloon barrage stretching along the coast from the Isle of Wight to New-castle, up to the astonishing height of 35,000ft: aircraft would not risk flying through the steel wires hanging from the balloons; nor could they fly over it given its great height. The Air Ministry rejected the proposal as impractical and costing £100m.[86] The most notorious, to his enemies, was the aerial minefield: mines attached to wires which would be dropped in the path of bombers. In January 1939 Lindemann proposed the use of mine-parachute combinations taken up by 3in rockets rather than being dropped from aeroplanes.[87] The proposals have been the subject of damning assessments, but these typically fail to register that Lindemann thought that the problem of air defence in daytime had been solved by radar and fighters: aerial mines were intended to deal with the particular problems of night bombers.

From the beginning of the war Churchill used his position in the Admiralty to promote his favoured devices, even though, like the mole, these were more obviously the responsibility of the other war departments. His first numbered minute at the Admiralty, of 14 November 1939, was on the 'UP rocket', asking for weekly reports. 'UP' stood for 'un-rotated projectile', often used as a codeword for rocket. In the same minute he demanded plans to fit four rocket projectors each to five battleships, two each for six cruisers and one to the monitor *Erebus*. He made clear that 'I regard the whole of this matter as of the utmost urgency and importance.'[88] He was clearly concerned that warships were vulnerable to air attack, as indeed they would prove to be, yet believed that the rocket would make warships immune.[89] Churchill wanted to take over the aerial mine from the Air Ministry, and the rocket programme from the War Office.[90] Apparently, the rocket research establishment was indeed under his personal control. He also turned his attention to aerial mines to be dropped from aircraft for the defence of naval bases, increasing his order from the Air Ministry from 100,000 to one million in March 1940.[91] Thus 'such impetus and drive were put into the development that a period of less than eight months lapsed between the statement of the demand for the weapon and its operational use against the enemy' in July 1940, the first operational use of rockets by any power.[92] By April 1940 about forty projectors were ready. The rockets fired canisters containing a parachute, wire, and a bomb or mine, which all drifted down into the path of attacking dive bombers, which would catch the wires and draw the mine on to the aircraft. In practice they were not effective. For example, five projectors of twenty rockets each were fitted to HMS *Hood*; they misfired in two attacks, and in one burnt some sailors in harbour.[93]

The enthusiasm for rocket defence continued into his time as Prime

Minister, extending way beyond the defence of ships. In June 1940 he had minuted Lindemann that:

> if we had large supplies of multiple projectors and rockets directed by Radar ... and also could have the proximity fuze ... the defence against air attack would become decisive. This combination is therefore the supreme immediate aim. We are not far from it in every respect, yet it seems to baffle us.[94]

In August 1940 Churchill ratcheted up the whole rocket programme to a quite extraordinary degree. One of the junior ministers at the Ministry of Supply, the future Prime Minister Harold Macmillan, recalled that there was 'constant pressure from No. 10' on the rocket and the proximity fuze, and that Dr Crow's activities were a 'mingled source of excitement and anxiety'.[95]

The capital cost alone was £14m.[96] There was to be large-scale production of 3in anti-aircraft rockets, 5in rockets for chemical weapons, rockets to lay aerial minefields, and others. By June 1941 the requirements were running at three quarters of a million rockets per month.[97] The programme required very large quantities of special solvent-less cordite. This was produced at the huge new propellant (largely cordite) factory at Bishopton in Scotland. Around November 1940 an order was made to build an entirely new factory to make only this cordite, at Ranskill. Approval for the site followed in February 1941 and the first cordite was produced a year later, by which time of course the Blitz was itself almost a year in the past.[98] As Lord Weir, who was closely involved, put it in a report related to the weekly progress reports the Prime Minister was getting on rockets: 'we are planning well ahead of development and running big risks in our capital expenditure'.[99] Production reached 2.4 million rockets per annum in 1942, and stayed at roughly that level, well below the expectation of 1941.[100]

The rocket programme was typically cronyist in that the key figure in the deployment of rockets was Churchill's son-in-law, Major Duncan Sandys, MP. In October 1940 an experimental regiment under Sandys was formed to develop techniques for attacking dive bombers with rockets and PE fuzes. The unit also developed techniques for

attacking high-altitude bombers and Sandys was promoted to command the first battery, in South Wales, consisting of rocket projectors
and a rare gun-laying radar.[101] The so-called Z batteries appeared all
over the country and were to be manned largely by the Home Guard.

Many experts in anti-aircraft matters were opposed to the rockets.
The Director of Artillery at the Ministry of Supply, E. M. C. Clarke,
thought there was 'no justification for going to production at the stage
of development which this weapon has at present reached, as shown
by its trials' of October 1940, against a Queen Bee pilotless drone,
which he thought were fixed.[102] Clarke reported that the head of Anti-
Aircraft Command, General Pile, 'said that he would prefer not to
express his opinion on the utility or otherwise of this weapon in view
of the circumstances of its introduction'.[103] In a note added to his
papers in 1966, A. V. Hill recorded that UP was the 'dearly beloved
pet of Lindemann' and that it had been 'a most infernal waste of time,
effort, manpower and material'.[104] Hill estimated the cost of developing
and deploying over two years as between £30m and £160m, involving
87,000 tons of steel and four times more cordite than required for the
same number of 3.7in anti-aircraft artillery rounds. The rockets were
so inaccurate, he maintained, that 'very few of the proximity fuzes
would function because there would be no proximity' and in any case
there were no such fuzes: the PE fuze was 'useless'.[105] Nevertheless by
October 1941 over 100,000 PE fuzes had been made for the rocket.[106]
Sandys, in charge of the Z batteries, had to face down, in Churchill's
presence, Lindemann's continued support of the PE fuze, which was
not used.[107]

While the photoelectric proximity fuze, and the acoustic one, hardly
worked at all, a third type in development at the beginning of the war
was developed with extraordinary success – the radio proximity fuze.
Towards the end of the war, fitted into old-fashioned anti-aircraft
rounds fired from a gun, it greatly increased the effectiveness of gunfire. Used in combination with new gun-laying radars only 100 rounds
were needed, on average, to bring down a flying bomb. In 1940, as we
have seen, many thousands of rounds were needed to bring down an
aeroplane even with radar control.[108] Indeed, the great majority of
V-1 flying bombs (or 'robot aircraft' as they were then often called)
were brought down long before they reached their targets in London.

In an extraordinary operation many of the guns defending London were moved to the coast to help achieve this. One reason this is not as celebrated as much as it might have been is that the proximity fuzes were US-made, as was the SCR 584 radar (designed by the MIT Radiation Laboratory) and the No. 10 BTL Predictor (designed by Bell Telephone Laboratory). This was just one of the systems that meant a 1940-style Blitz would have been impossible in 1944.

Aerial mines, seemingly so outlandish that one might think they were never deployed at all, were in fact widely used. They were deployed on a small scale in the defence of London during the Blitz, either dropped from old bombers or hung from barrage balloons which were set free to drift eastwards. A similar device was used to attack Germany. Under Operation Outward nearly 100,000 balloons were released from Felixstowe about half with trailing wires to interfere with power lines, the other half with incendiaries, from March 1942 to September 1944, though attacks were reduced from May 1944. Launched by 150 members of the Women's Royal Naval Service, they had the distinction of being the only offensive combat weapon operated by women. The programme was opposed by the RAF, who vetoed the use of explosives, as opposed to incendiaries and trailing wires; they could only be used by day, and even then could be stopped by the RAF.[109] Aerial mines fired by rockets continued to be used not only in warships but in merchant ships too, in the form of 'Fast Aerial Mines'. Ships also carried rockets that fired PAC (Parachute and Cable), which had just cable and no mine. Their effectiveness is unknown.

The most important use which emerged for rockets was bombardment of land targets. The UK-based 3in anti-aircraft rockets were hardly needed from 1941 as there was nothing for them to shoot down. From 1942 a new demand emerged for modified versions of this rocket, for anti-tank and anti-ship rockets to be fired from aircraft; these carried 25lb or 60lb warheads. The British later supplied the Americans with this airborne rocket weapon.[110] Other uses emerged: the British and Canadian armies used the 'Landmattress', a multiple rocket launcher, from 1944. The 5in chemical rockets were adapted to fire a high-explosive barrage from landing craft, in the form of the so-called 'Sea Mattress'. On the basis of these uses Harold Macmillan deemed Churchill prescient: he was 'proved right in his steady support for the new conceptions'.[111] Yet the military value of the rockets was not as great as it seemed. For example, in April–May 1944, there was only a 0.7 per cent chance that a rocket from an eight-rocket salvo from an aircraft would hit a target the size of a tank.[112] Few tanks were destroyed in this way, but the rocket-firing fighters acquired a great reputation.

While Lindemann and Churchill were great supporters of the British rocket programme, they were, rightly as it turned out, dismissive of the claims made by the British rocket scientists about the potentialities of German rockets. During the war, intelligence suggested a new and very powerful threat from the future and from the air: the long-range rocket. Britain's rocket scientists, extrapolating from their own solid-fuel rockets, estimated that the German rockets were likely to be 80-ton monsters with a 10-ton warhead. Partly as a result, plans were made for the wholesale evacuation of London, with all the disruption this would cause to the war effort. The calls were led by the Minister of Home Security/Home Secretary, Herbert Morrison. Fortunately Lord Cherwell thought it unlikely that the rockets were so big, and thus would not cause much damage. He thought that flying bombs were more likely, and that large rockets were economically irrational.[113] Cherwell, as it turned out, was correct on both counts: flying V-1 bombs were used, and the V-2 rockets were in fact quite small (one ton of explosive each, weighing around 10 tons in total) and cost the Germans more to make than the damage they produced cost the Allies. A key figure in the behind-the-scenes analysis was R. V. Jones, who was able to piece together the intelligence to produce a picture which turned out to be remarkably accurate. In reflecting on this and other experiences he insisted on the importance of having scientific intelligence assessed by intelligence experts rather than experts on the particular machines.[114] It is notable that the impact of the V-2, despite the best efforts of some, was not overestimated in the way the bomber had been before the war. This was doubtless because of increased scepticism about the power of new weapons, and perhaps because Britain did not itself have comparable long-range rockets.

COMBINED OPERATIONS

In July 1940 Churchill created a new organization which would exemplify his offensive spirit, his love of gadgets and his use of cronies. Combined Operations headquarters was charged with devising much of the complex machinery taken to be necessary by the British for seaborne landings. Some of the most famous devices of the war would emerge from it, though all were less effective and significant than the attention they garnered suggested. In Combined Operations cronyism was endemic: the first two heads were cronies of Churchill's and both appointed their own cronies as subordinates.

Churchill appointed the retired Admiral of the Fleet Sir Roger Keyes, Member of Parliament for Portsmouth, as first head of Combined Operations. Keyes had made a devastating intervention, in full uniform, in the parliamentary debate which had led to Chamberlain's resignation. The appointment, outside the normal service channels, was not liked by the chiefs of staff. Keyes pushed his luck by insisting on direct access to Churchill and was fired in October 1941. He was replaced with another crony, the very junior, young, dashing and aristocratic naval captain Lord Louis Mountbatten, a cousin of the King's and a wireless (that is, radio) specialist. Mountbatten, waiting to take command of the aircraft carrier *Illustrious*, had previously commanded nothing bigger than a destroyer, yet he was soon to be given honorary senior rank in the other services and to become a member of the Chiefs of Staff Committee.

Mountbatten's remit was wide: he was to advise on combined operations, to coordinate activity, to study tactical and technical methods for everything from small raids to an invasion of the Continent. In addition, he was to 'direct and press forward research and development in all forms of technical equipment and special craft peculiar to combined operations'.[115] He shared with Churchill an enthusiasm for 'dash, the unexpected, the use of novel gadgets', and encouraged his scientists to 'indulge their fantasies'; he 'matched the Prime Minister in his relish for new and improbable devices'.[116] His advisers on these matters included the mad Geoffrey Pyke, the communist J. D. Bernal and the far-right Sir Malcolm Campbell.

Pyke, who had no technical training, went from being a successful metal trader in the 1920s to being a poor eccentric helping out with aid to Spain in the 1930s.[117] During the war he lived in a room in a bohemian house in Hampstead, the home of J. D. Bernal's lover. Bernal was one of his few friends and regarded him as 'the greatest inventive genius of his time'.[118] Through Leo Amery, Pyke was introduced to Mountbatten, who took him on at the huge salary of £1,500. Pyke's first idea, which he had put to Amery and was strongly taken up by Mountbatten and Churchill, was Project Plough – the idea that small groups working with fast, armoured snow-vehicles should be introduced into northern Norway; these ideas led to the M-29 Weasel, a small, tracked amphibious vehicle which proved very useful in the mud of North-West Europe, especially in the Walcheren campaign. Less successful were 'Pyke's Uphill Rivers' in which objects were transported by water through pipelines. He was spectacularly successful in getting his idea for a gigantic iceberg aircraft carrier taken very seriously indeed. The project went under the name of Habbakuk, a misspelling of the Old Testament prophet Habakkuk. In the final planned version it was to be a vessel of 2.2 million tons' displacement (say fifty times the displacement of a battleship) made of ice reinforced with wood-pulp (a material called 'pykrete'). It would also use thousands of tons of steel, and need a gigantic refrigeration plant to keep the blocks of pykrete frozen.[119] It would have been the apotheosis of British refrigeration and marine engineering. A small-scale model was built in Canada in 1943 and the Canadians were to take over the project.

J. D. Bernal, a charismatic physicist and pioneering analyst of the social, economic and political relations of science, was a key figure, it seems, in helping the project gain momentum.[120] Bernal spent a good part of 1943 in Canada directing work on it there. The proposal, backed by Mountbatten, was taken up by Churchill, despite the hostility of Cherwell and Charles Goodeve, whose 'dampening comments angered Mountbatten considerably'.[121] On Goodeve's later exaggerated but telling estimation, unsupported by evidence, Habbakuk made the most serious dislocation of Allied effort of any wartime invention.[122] Goodeve pointedly reflected that 'the voices of reason were shouted down by cries of "obstruction"'.[123] Solly Zuckerman, another of Mountbatten's scientific advisers, recalls trying to talk Bernal out of it, but to no avail – Zuckerman and Bernal, who had worked together from the beginning of the war, went their separate ways.[124] That Bernal was so sympathetic to Habbakuk is perhaps not so surprising in that he was an expert on the structure of water and gigantism was a noted feature of machines and structures in the Soviet Union he so admired.[125] That Lindemann was against it from the start makes it clear that Churchill did not always take his advice, but also that even the most fervent advocates of the application of reason to problems of war were divided among themselves.

MULBERRY AND PLUTO

Combined Operations were responsible for the early development of two of the most famous British devices of the war, the Mulberry harbours and the PLUTO petrol pipelines. Both became very well-known towards the end of the war and stood as prominent examples of British engineering genius. They were thought of as being among the greatest of British inventions of war, ranking with radar, jet engines and penicillin.

The Mulberry harbours were prefabricated ports put together on the Normandy invasion beaches. There were two Mulberrys: A for the American Omaha beach and B for British (more accurately Anglo-Canadian) Arromanches. Each complete harbour was made up of steel and concrete floating piers (this element had been suggested by Churchill himself[126]) which were called 'Whales'; reinforced concrete caissons, dropped on to the sea bed to form a breakwater ('Phoenix'); floating steel breakwaters ('Bombardons'); and sunken ships ('Goose-berries'). The cost of Mulberry A was counted as 'reciprocal aid' to the USA. Between them the two artificial harbours cost £25m, just under 5 per cent of the combined output of the two main armaments ministries in the six months it took to build them: 45,000 workers were needed at the peak of construction and they required a total of 90,000 tons of steel.[127] Some parts, notably the Phoenix, intruded very little on war production as so much concrete was available now the building of airfields and factories was slowing down. But the

Whales and Bombardons did compete with scarce resources, and the production of 'gun carriages, tanks, jerricans, steam-boilers, ammunition boxes, and, above all, Bailey bridges' was affected as a result; there was some delay in ship construction and repair, although no interference in landing craft production. But the effect was marginal, according to the official historian, because of the 'healthy condition of British supplies and stocks on the eve of D-day' and the 'efficiency and elasticity of British war production in the fifth year of war'.[128] They are a powerful indicator of the plenitude of resources available to Britain.

But was the huge investment worth it? The 'renown which soon attached itself to the name of Mulberry may have led the public to exaggerate . . . the part which the artificial harbours as a whole played in the success of D-day operations', the historian of war production noted cryptically, without adding evidence.[129] The judgement is surely right. Mulberry A was in operation for only three days when it was severely damaged by a storm on 19 June, so much so that it was not completed or repaired. Yet the Americans were able to continue to land huge quantities of materiel using landing craft and DUKW amphibious trucks on the Omaha beaches. In fact they landed more than the British did through Mulberry B, 'Port Winston'.[130] The latter was used to land a prodigious quantity of stuff: from 'four days after D-Day until the 31 October, 1944, 628,000 tons of supplies, 40,000 vehicles and 220,000 troops were put ashore'. An impressive total to be sure, but only 35 per cent of British stores, 17 per cent of British vehicles and 23 per cent of British personnel.[131] It is telling also that the most effective part of the Mulberry system was the least innovative and best-known – the old-fashioned sunken ships, the Gooseberries, were more effective than anticipated. The Bombardon floating breakwaters, which Bernal was particularly keen on, were the least effective part.[132]

The second great project originating with Combined Operations was the 'Pipe Line Under the Ocean' (PLUTO), designed to carry petrol to France before ports could be opened to tankers. PLUTO was an undersea extension of the British oil pipeline system which had itself been built in the war. PLUTO was devised in 1942 in two forms, one a hollow cable derived from undersea telegraph and telephone cables, the other a welded steel pipe of the sort used by the oil industry for land pipelines. Both types were much smaller in diameter than the new overland petrol pipelines; PLUTO pipes were 3 inches in diameter, whereas the gigantic American 'Big Inch' pipeline, which went all the way from the Gulf of Mexico to the north-east of the United States, was 24 inches; the pre-war crude pipelines from Kirkuk in northern Iraq running to Lebanon and Palestine were 12 inches; the line running from the oil fields of Burma to the Rangoon refinery was 10 inches (see maps, pp. 190–93). By 1943 there were ambitious plans for PLUTO: it was expected to supply 4,000–5,000 tons a day, around 40–50 per cent of requirements from D+12, that is, twelve days after D-Day. Yet planners did not rely on it. The Americans did not want it, seeing it as too difficult and costly an affair to get involved in.[133] The Allies were planning to provide tankers for all cross-Channel supplies.[134]

This was just as well as PLUTO was much slower to come into operation than the ports it was supposed to replace. The pipes were more difficult to lay than envisaged, and many were damaged and became unusable. Astonishingly it was not till August 1944 that any pipes were laid at all. This first line to Cherbourg (named Bambi) did not supply any petrol until mid-September. In October work started on the much shorter Dumbo, to Boulogne, which was only in partial operation by the end of the month. It supplied a mere 700 tons of petrol a day. Dumbo eventually achieved a throughput of 3,300 tons per day, but only after the war was over, but even then most of its sixteen pipes were unusable. The steel pipes were not laid till January 1945. PLUTO (and the name of the parts suggests the whole was named for the Walt Disney character and not the planet) was closed down in July 1945 to save technical manpower. Down to the end of the fighting PLUTO supplied only 8 per cent, an average of under 1,800 tons a day, of the oil products that came from Britain, and less than 7 per cent of the total supply.[135] If it was a bonus, it was a small one, though this is very rarely mentioned in the glowing tributes then and since to this supposed triumph of British science and engineering.

It is well-known that Churchill had long resisted a full-scale invasion of northern Europe, emphasizing bombing and a Mediterranean strategy instead. The Americans had initially gone along with this, but as the war progressed they got their way. One of the arguments made is that Churchill and the British chiefs of staff were right to resist an earlier attempt at a large-scale invasion. One of the reasons given is that the technical means necessary to achieve success, specifically Mulberry and PLUTO, would not have been ready. Thus it turned out to be necessary to wait till 1944.[136] The above suggests that while there may have been good reasons to wait, Mulberry and PLUTO were not, in retrospect, or even prospectively, among them. PLUTO was not believed to be necessary, and, while something like Mulberry was, it turned out not to be.

CHURCHILL AND THE ATOMIC BOMB

Churchill, Lindemann and Anderson were crucial decision makers in the British atomic bomb project. Its significance is often underplayed because of the huge scale of the later US project. In fact, as befitted the greatest military-scientific power on earth, Britain had, until 1942, the largest bomb project of any power. As with radar this was a case of not being particularly brilliant at invention, but rather of being good at development. British scientists did not invent the atomic bomb – it was in any case an idea that had been around for years. Nor was it a British experiment which started the race to the bomb; work done in Germany and France did this. Research in Britain began, under the aegis of Sir Henry Tizard's Air Ministry committee on a scientific survey of air warfare, in the summer of 1939.[137] In the spring of 1940 Professor James Chadwick, in Liverpool, and two émigré scientists in Birmingham, Rudolf Peierls and Otto Frisch, came to conclude that an atomic bomb was possible. The result was the speeding up of research. After the fall of France, British efforts were strengthened by a team of French physicists and their heavy water. Through the defeat in France, and through the Blitz, this obviously long-term project accelerated, yet another indication of confidence. In the summer of 1941 the scientists had reported that a separation plant would cost £5m, and that overall an atom bomb would be cheaper when measured by explosive power and damage done than conventional ones.[138] Imperial Chemical Industries estimated the cost at £8.5m for thirty-six bombs of 1,800 tons TNT equivalent each.

By comparison, the equivalent in conventional bombs of 1 ton each (65,000 bombs) came out at £14.15m. They concluded that a bomb could be built by 1943 and was thus worth pursuing.[139] Sir Henry Tizard, who like many others was sceptical about the timings and costings, specifically compared the proposed project with what had turned out to be the much costlier and longer process required to make synthetic ammonia using the Haber–Bosch process earlier in the century.[140] The bomb project was by no means unprecedented in its scope and scale.

The decision to go for a bomb was taken, in its entirety, by Churchill and the chiefs of staff, with Lindemann's strong support, in the middle of 1941. The Cabinet were not consulted, nor indeed the War Cabinet. The War Cabinet's Scientific Advisory Committee considered the case, without knowing that Churchill had already decided to go ahead. How the project was to be run was also decided in 10 Downing Street, the previous arrangements being summarily dismissed without consultation.[141] The project was placed in the capable hands of Sir John Anderson, where it would remain even as he changed ministries. It was Anderson who named the project Tube Alloys and appointed ICI's top scientist, Wallace Akers, to run it, with a consultative committee including Cherwell and Hankey.[142] The British Empire – for the Canadians were brought in – was on the way to becoming the world's first atomic power. The bomb was not meant just as a counter to a possible German bomb, but as a war-winning weapon in its own right. The estimate by the scientists remained that it could be available in 1943.

The United States decided on a bomb project some months later than the British, in late 1941, before it was in the war against Japan or Germany. The US had access to British research results freely handed over from the time of the Tizard Mission in October 1940. However, at this stage Anderson resisted suggestions of more atomic collaboration with the USA with the excuse that British security and secrecy were much tighter. By 1942 the boot was very definitely on the other foot. The British now wanted a joint programme, while the Americans used the same excuse and restricted information transfer very severely; for example, British scientists could not (at this stage) visit Los Alamos.[143] In December 1942 President Roosevelt expected the British to hand over all their information in return for US information, but only on topics the British were working on.[144] Churchill could not get a reply from the Americans for months as he attempted to reverse the American position in early 1943. Frustrated, he ordered studies to be made for a wholly British bomb project.[145] By the beginning of 1943 the Anglo-Canadian project was costing an annualized £430,000.[146] A full project, it now was estimated, would cost over £60m (the American project would cost fifty times this). In May 1943 Roosevelt appeared to concede full collaboration but only much diplomacy and British disavowal of post-war commercial aspirations led to the signing of the Quebec agreement in the summer of 1943, under which limited collaboration was resumed. But yet again the Americans stalled. The British were forced to replace Wallace Akers of ICI as the British technical representative in the USA before collaboration resumed. Akers remained in charge of the British project, which dwindled away to nearly nothing as the British scientists went to the USA from December 1943. Thus was a mighty empire humiliated and the atomic special relationship born.[147]

The British scientists, and exiles who had come to Britain, made a contribution to the Manhattan Project, but despite British attempts to talk it up in 1945, it was not major. It is obviously difficult to assess the contribution in the final phase, but General Groves made his own view devastatingly clear after the war: to quantify the British scientific contribution at 'one per cent of the total would be to over estimate it'. In technical and engineering terms it was 'practically nil'; overall if there had been no British contribution 'the date of our final success need not have been delayed by a single day'.[148] In theory, aspects at least of the bomb programme were overseen by a Combined Policy Committee in Washington; the British formally gave permission for the bomb's use against Japan, but this body was most definitely not where policy was either made or discussed. High policy was left to the Americans, who brooked no interference in it. Churchill's first concern in discussing the bomb was the effect any position he might take would have on the wider Anglo-American relationship.[149] As in so much else the British position at the end of the war was radically different from what it had been in the beginning.

The atom bomb was to have no effect on the European war. By September 1944 it was clear that a bomb would not be ready for a year, by which time it was assumed the war in Europe would be long over. Yet, the story of the bomb project can stand for many aspects of Britain's war effort in addition to the blinding light it shines on the supposed Anglo-American special relationship. Britain's early lead in and desire for the most powerful weapons of mass destruction reflect a more general commitment, even in the difficult period of 1940–41, to novel methods of making war. It exemplifies the importance of ministers and the chiefs of staff, rather than scientific advisers or the Cabinet, in the taking of key decisions. But those ministers were far from ignorant of science and operated in a world not just of extreme secrecy but also of mutual trust within tiny coteries. It also stands for much of wartime technical development in another way: far from advancing the date of victory, investment in the development of the bomb, as in so many other gadgets and devices, delayed it. For resources were diverted to the making of a weapon which became available too late and in this case, as in others, did not add materially to the devastating destructive power which was available anyway towards the end of the war.

While Winston Churchill's technocrat cronies don't generally figure in our analyses of the war, the parallel story for Germany is rather different. One common post-war story was that Hitler made very bad technical decisions, that the Nazis did not mobilize science properly, a story supported by the argument that the technocrat Albert Speer transformed German production towards the end of the war, displacing the armed forces and the ideologues. Hitler undermined German technical strength; Speer saved what he could. Yet those stories, just like the ones for Britain, do not convince. Hitler and Speer were cronies, both interested in machines and both committed Nazis. In 1945 Albert Speer damned, to British interrogators, the German General Staff for not knowing about or understanding modern machines, for not understanding matters technical and economic. By contrast, in his view Hitler had 'a great understanding of technical matters', and greater knowledge of weapons in use than the high commands, down to details. Speer ensured Hitler made technical decisions in conversation with ordnance officers, people from industry and officers from the Front, and not the staff officers.[150] Among the results was the V-2, a rocket that wasted German strength far more than Britain was damaged by its rocket programme or any other of Churchill's technical enthusiasms. Only the atomic bomb was a comparable waste of resources.

5
Politics and Production

The Second World War was an experts' war, but also a war between experts on the same side. Experts had enemies on their own side, and opponents among the German and the Japanese experts, to adapt the parliamentary witticism. An expert Prime Minister battled with his military and economic advisers; senior soldiers, sailors and airmen had different views as to which machines would win the war. Experts outside government challenged those inside; inventors fought scientists; scientists of the left and of the right disagreed vehemently. Many such battles were presented as something else: as battles between science and reason, between science and the classics, between the expert and the politicians, between different conceptions of science. Looking at these debates helps to illuminate the wartime role of experts but also the ways in which the war has been understood then and since: as a war of production, a total war or people's war, a war in which modern weapons, the product of the civil imagination – the aeroplane, the tank, the radio – were central. These issues had particular salience in 1942, a remarkable year in the politics of experts: there were important debates about production, the place of the scientific expert in government, about the quantity and quality of British tanks, and also the policy of bombing Germany, not seen to anything like the same extent earlier or later in the war.

1942 was a special year in British politics in many other ways. In the first half the British Empire had suffered some of the most disastrous and humiliating defeats in its history. British and imperial troops had surrendered en masse to the Japanese in the East and to the Germans at Tobruk in North Africa. The Royal Navy had recently lost two battleships to what some regarded as Japanese 'monkey-men'.

British and imperial soldiers were not fighting as an imperial race should have and had done. In contrast to the earlier optimism about Britain's capacity to build a mighty army of machines to provide the mechanical might of democracy, many now expressed the view that there was something very wrong with the country. Meanwhile the Soviet Union continued to take on the bulk of the Wehrmacht. All this gave rise to a profound national embarrassment, reflected in the campaign for a second front, and the fact that the British Communist Party, actively supporting a capitalist government, reached a record membership. As the standard bearer for a robust modernity, the party complained of the inefficiencies in British production, but in 1942 it was hardly alone in this. Indeed, the politics of production, internationalism and nationalism, of science and of machines, was on display in fascinating and unfamiliar ways.

PRODUCTION

One of the key issues debated in 1942 was production.[1] It loomed large because military defeat was interpreted in significant part as the result of failures in production (see Low cartoon). For example, the pseudonymous author Argonaut complained in early 1942 that 'Victory over the Axis forces is certain – ultimately'; meanwhile 'we go on losing battles'. But he went on: 'The common cause of all these setbacks is a lack of arms.' This was so although 'potentially the production capacity of the anti-Axis forces is immensely superior'.[2] The issue of production had many dimensions. The first was the powerful sense that the war was a war of production, and that around the world miracles of production were being achieved. Thus Sir Stafford Cripps in July 1942: 'Perhaps the most revolutionary element in our war experience is [the] revelation of an almost unlimited capacity for production.'[3] The other was the sense that Britain was not participating as fully as it might in this revolution. In *People in Production*, published in 1942, Mass-Observation noted that in the period covered by the report, October 1941–April 1942, the reader of newspapers could 'not fail to know there was some argument going on about industrial a sergeant'. This was unjustified not because there was a lack of class prejudice in the British army, but because Bevan was wrong about Rommel's class origins.[17]

Bevan concerned himself with aircraft partly because the central production and technical case made by others centred on tanks and tank guns. Wardlaw-Milne complained that British tanks in Libya were no match for the Panzer IV. British tanks were armed with the 40mm 2-pounder gun while the Panzer IV had a 75mm gun. As we have noted, and will consider in detail in Chapter 7, the comparison was highly misleading. Of particular concern to some was the new heavy tank named after Churchill himself, which also had the 40mm 2-pounder gun. Oliver Lyttelton, in his poorly received speech, developed a defence which depended on the story of a profound weakness in British tank forces: 'We started the war with no modern tanks, we lost all the armoured equipment which we had in France in June, 1940, though that equipment would by itself have had little value to-day.'[18] Lyttelton's argument was that lots of tanks were needed quickly, and that had meant deciding to produce the Churchill tank with its 2-pounder even though the tank had not yet been made or tested. High levels of production and high quality did not go together.

Churchill, in a brilliant winding-up speech, also made great play of the weakness of British arms in 1940, especially tanks, attacking one of his antagonists in 1942, the Secretary for War responsible for tanks in early 1940, Leslie Hore-Belisha. He also insisted, as he had in earlier speeches, on the massive growth in British arms production since 1940, a point on which he was undoubtedly correct. He noted the amount of materiel which went to the Middle East: 'from this country, from the Empire overseas and to a lesser extent from the United States, more than 950,000 men, 4,500 tanks, 6,000 aircraft, 5,000 pieces of artillery, 50,000 machine guns and over 100,000 mechanical vehicles'.[19] According to Churchill about 2,000 had been sent to the Soviet Union.[20] In his view the losses in the Middle East were not due to failures in production. Yet this powerful case was not made clearly, especially since the defeat in 1940 was by Churchill and Lyttelton themselves as a failure of production. Furthermore, and more importantly, Churchill's very negative private

Low's very British explanation of defeat in 1942 –
not the man, but the machine

efficiency and inefficiency', although this was not the case in the oral culture.[4] It went on to record that 'Citizens on the Home Front have been taught to believe that they can *win the war with production*. That is the basic assumption in the home war effort.' Yet it went on to note, and this is another dimension, that 'the army of war production is beginning to wonder if it is any good "winning with work" if politicians and generals don't show what to do with the finished products'. On the other hand, there was a feeling growing in the public mind that 'we are being misled over the amount of armaments we are really producing', and that 'the results of our production are being misrepresented by ballyhoo'.[5]

The focus on production is powerful testimony to the fact that the political elite and intellectuals thought of success and failure in war in precisely these terms. It is particularly telling that much of the opposition to Churchill in early 1942 was expressed in terms of production, and in critiques of British tanks and of the Churchillian policy of strategic bombing.

Sir Stafford Cripps, MP, was to be the key political figure in the politics of early 1942. Cripps was an independent MP, having been thrown out of the Labour Party in 1939 for urging unity on the left. He now stressed his independence from party. He was a 'scientific lawyer' – he had a degree in chemistry and went on to be exceptionally successful at the Bar, making huge sums in the highly specialized and technical field of patent law.[6] Like Lindemann he was personally austere and a vegetarian. Cripps was sent to the Soviet Union as ambassador in 1940 and returned when the Soviet Union was Britain's great fighting ally. He spoke out to huge acclaim for austerity and efficiency, and a more rigorous war effort. In a broadcast in the *Postscript* slot, which had a huge effect, he stated boldly that 'Had our efforts in production been greater we should not now be retreating in North Africa.'[7] Churchill was forced to reconstruct his government, appointing a Minister of Production in the War Cabinet.[8] Cripps became Leader of the House of Commons and a member of the War Cabinet.

On 1 and 2 July 1942 Churchill faced his most serious parliamentary challenge of the war, a possible vote of censure on the 'central direction of the war'. That was the title of the motion, and the debate is often thought to have been about this issue narrowly defined.[9] It was tabled by Sir John Wardlaw-Milne, a Conservative businessman who was chairman both of the Select Committee on National Expenditure (which looked into, among other things, production) and of the Conservative Foreign Affairs Committee.[10] In fact production was a critical issue in the debate, as is indicated by the fact that the reply to the motion was given not by the Prime Minister himself, or the Foreign Secretary, but by the recently appointed Minister of Production, Oliver Lyttelton, then without much parliamentary experience. This had been suggested by Cripps, the new Leader of the House. Briefing Churchill for the great debate, Cripps reported there was concern

about the suitability of British generals for 'mechanical warfare', 'the strongest line of criticism' was the quality of British equip above all tanks and anti-tank guns, and also what was percei inadequate support for research and invention.[11] These were charges, not least because Churchill had been in the forefront lar attacks on the government in the late 1930s. Yet here he w two years later, presiding not only over catastrophic military but also failures in production. The boot was now on the ot As Sir Herbert Williams, MP, a right-wing businessman and put it:

> During that period they have spent £9,000,000,000 which i convenient way to measure our war effort in man-hours spent, used, etc. With what result? One major victory against o enemy, the Battle of Britain in the air; won with what? With designed and, in the main, constructed before the present G sat on that Bench. That is their only victory.[12]

This was a stinging charge.

Another vociferous critic, and a much better-known was the Labour MP Aneurin Bevan, who acidly co 'the Prime Minister wins Debate after Debate and lo battle'.[13] The central accusation made by Bevan was being pursued was wrong, the wrong weapons w and the officers handling them had 'not studied th weapons'.[14] In particular, he claimed, 'no-one has th in terms of the last war, more than the Prime Min astonishingly silly assessment for reasons we hav was not alone on the left in this strange assessm ment the socialist J. T. Murphy claimed of Chur dawned on him that the Maginot Line was out o kept pace with the revolution in technique of fact the arguments made by Bevan were mostly that the army should have been supported w have had more transport aircraft (he was aga he wanted Polish, French and Czech general how to fight, to lead British troops, and re that 'if Rommel had been in the British Army

Low's very British explanation of defeat in 1942 –
not the man, but the machine

efficiency and inefficiency', although this was not the case in the oral culture.[4] It went on to record that 'Citizens on the Home Front have been taught to believe that they can *win the war with production.* That is the basic assumption in the home war effort.' Yet it went on to note, and this is another dimension, that 'the army of war production is beginning to wonder if it is any good "winning with work" if politicians and generals don't show what to do with the finished products'. On the other hand, there was a feeling growing in the public mind that 'we are being misled over the amount of armaments we are really producing', and that 'the results of our production are being misrepresented by ballyhoo'.[5]

The focus on production is powerful testimony to the fact that the political elite and intellectuals thought of success and failure in war in precisely these terms. It is particularly telling that much of the opposition to Churchill in early 1942 was expressed in terms of production, and in critiques of British tanks and of the Churchillian policy of strategic bombing.

Sir Stafford Cripps, MP, was to be the key political figure in the politics of early 1942. Cripps was an independent MP, having been thrown out of the Labour Party in 1939 for urging unity on the left. He now stressed his independence from party. He was a 'scientific lawyer' – he had a degree in chemistry and went on to be exceptionally successful at the Bar, making huge sums in the highly specialized and technical field of patent law.[6] Like Lindemann he was personally austere and a vegetarian. Cripps was sent to the Soviet Union as ambassador in 1940 and returned when the Soviet Union was Britain's great fighting ally. He spoke out to huge acclaim for austerity and efficiency, and a more rigorous war effort. In a broadcast in the *Postscript* slot, which had a huge effect, he stated boldly that 'Had our efforts in production been greater we should not now be retreating in North Africa.'[7] Churchill was forced to reconstruct his government, appointing a Minister of Production in the War Cabinet.[8] Cripps became Leader of the House of Commons and a member of the War Cabinet.

On 1 and 2 July 1942 Churchill faced his most serious parliamentary challenge of the war, a possible vote of censure on the 'central direction of the war'. That was the title of the motion, and the debate is often thought to have been about this issue narrowly defined.[9] It was tabled by Sir John Wardlaw-Milne, a Conservative businessman who was chairman both of the Select Committee on National Expenditure (which looked into, among other things, production) and of the Conservative Foreign Affairs Committee.[10] In fact production was a critical issue in the debate, as is indicated by the fact that the reply to the motion was given not by the Prime Minister himself, or the Foreign Secretary, but by the recently appointed Minister of Production, Oliver Lyttelton, then without much parliamentary experience. This had been suggested by Cripps, the new Leader of the House. Briefing Churchill for the great debate, Cripps reported there was concern about the suitability of British generals for 'mechanical warfare', while 'the strongest line of criticism' was the quality of British equipment, above all tanks and anti-tank guns, and also what was perceived as inadequate support for research and invention.[11] These were serious charges, not least because Churchill had been in the forefront of similar attacks on the government in the late 1930s. Yet here he was, a full two years later, presiding not only over catastrophic military defeats, but also failures in production. The boot was now on the other foot. As Sir Herbert Williams, MP, a right-wing businessman and engineer, put it:

> During that period they have spent £9,000,000,000 which is a very convenient way to measure our war effort in man-hours spent, material used, etc. With what result? One major victory against our major enemy, the Battle of Britain in the air; won with what? With machines designed and, in the main, constructed before the present Government sat on that Bench. That is their only victory.[12]

This was a stinging charge.

Another vociferous critic, and a much better-known one since then, was the Labour MP Aneurin Bevan, who acidly commented that 'the Prime Minister wins Debate after Debate and loses battle after battle'.[13] The central accusation made by Bevan was that the strategy being pursued was wrong, the wrong weapons were being made, and the officers handling them had 'not studied the use of modern weapons'.[14] In particular, he claimed, 'no-one has thought of this war in terms of the last war, more than the Prime Minister himself',[15] an astonishingly silly assessment for reasons we have explored. Bevan was not alone on the left in this strange assessment: outside Parliament the socialist J. T. Murphy claimed of Churchill that 'it had not dawned on him that the Maginot Line was out of date ... he had not kept pace with the revolution in technique of modern warfare'.[16] In fact the arguments made by Bevan were mostly very flimsy. He claimed that the army should have been supported with dive bombers, and have had more transport aircraft (he was against strategic bombing); he wanted Polish, French and Czech generals, who in his view knew how to fight, to lead British troops, and repeated the strange taunt that 'if Rommel had been in the British Army, he would still have been a sergeant'. This was unjustified not because there was a lack of class prejudice in the British army, but because Bevan was wrong about Rommel's class origins.[17]

Bevan concerned himself with aircraft partly because the central production and technical case made by others centred on tanks and tank guns. Wardlaw-Milne complained that British tanks in Libya were no match for the Panzer IV. British tanks were armed with the 40mm 2-pounder gun while the Panzer IV had a 75mm gun. As we have noted, and will consider in detail in Chapter 7, the comparison was highly misleading. Of particular concern to some was the new heavy tank named after Churchill himself, which also had the 40mm 2-pounder gun. Oliver Lyttelton, in his poorly received speech, developed a defence which depended on the story of a profound weakness in British tank forces: 'We started the war with no modern tanks, we lost all the armoured equipment which we had in France in June, 1940, although that equipment would by itself have had little value to-day.'[18] Lyttelton's argument was that lots of tanks were needed quickly, and that had meant deciding to produce the Churchill tank with its 2-pounder even though the tank had not yet been made or tested. High levels of production and high quality did not go together.

Churchill, in a brilliant winding-up speech, also made great play of the weakness of British arms in 1940, especially tanks, attacking one of his antagonists in 1942, the Secretary for War responsible for tanks to early 1940, Leslie Hore-Belisha. He also insisted, as he had in earlier speeches, on the massive growth in British arms production since 1940, a point on which he was undoubtedly correct. He noted the huge amount of materiel which went to the Middle East: 'from this country, from the Empire overseas and to a lesser extent from the United States, more than 950,000 men, 4,500 tanks, 6,000 aircraft, nearly 5,000 pieces of artillery, 50,000 machine guns and over 100,000 mechanical vehicles'.[19] According to Churchill about 2,000 tanks had been sent to the Soviet Union.[20] In his view the losses in the Middle East were not due to failures in production. Yet this powerful claim was not made clearly, especially since the defeat in 1940 was seen by Churchill and Lyttelton themselves as a failure of production. Furthermore, and more importantly, Churchill's very negative private views of the fighting qualities of the British army could not be made public.

Churchill beat the Opposition handsomely, leaving only twenty-five members voting against him, though more abstained. He survived, not because he won the argument, which continued to rage, but because there was, with the possible exception of Stafford Cripps, no alternative leader. Furthermore, the coalition, and its supporters, stood firmly behind him, including even Willie Gallagher, the single Communist Party MP. Bevan was one of eight Labour members to vote against Churchill, together with three Independent Labour Party members (a tiny left Opposition party). The rest of the anti-Churchill vote came from the right and centre: eight Conservatives, two Liberals and six independents, many of whom had recently left coalition parties, although two had recently been elected as such.

EXPERTS IN PARLIAMENT

Over a quarter of those who voted against Churchill in the July vote of confidence had a technical background, a proportion higher than their representation in the Commons. They came from all political positions. They were Thomas Horabin, elected in 1939 as a Liberal with Labour support as well as that of Churchill, the chairman of Lacrinoid Products, a company which made buttons and knobs from plastic materials;[21] the Labour member Richard Rapier Stokes, an engineer and managing director of the engineering firm Ransome and Rapier;[22] and the Conservative Sir Herbert Williams (a Liverpool-trained electrical engineer).[23] Two were doctors (Leslie Haden-Guest (Labour) and Henry Morris-Jones (Independent, formerly Liberal National)). The right-wing independent ex-Tory Captain Cunningham-Reid qualifies by having been in the Royal Engineers briefly; he was better known as a playboy and for a scandalous divorce. Lastly, there was Denis Kendall.

Kendall was the first independent to win a wartime by-election. As the 'production' candidate, he won the Grantham by-election of March 1942. Kendall was the managing director of a large arms factory in the town which made 20mm cannon for aircraft.[24] The company was called the British Manufacture and Research Co. (BMARC), but was a subsidiary of the Swiss-based Hispano-Suiza.[25] He defeated Air Chief Marshal Sir Arthur Longmore, the official Conservative candidate, who not only had stronger local connections but also the support of the Communist Party (this was no aberration – they supported official candidates even against the left).[26] Kendall was strongly backed by the lowbrow weekly newspaper *Reveille*, whose editor had previously stood in by-elections and now acted as his agent.[27] *Reveille* saw in Kendall the 'drive and indomitable personality' of 'another Beaverbrook' and 'the combined engineering ability of a Nuffield'.[28] *Picture Post* spotted his American links (he had worked in the USA as well as France), noting he named his house Brusa (from *Br*itain and *USA*): he was 'like something out the highest speed Hollywood film'.[29] Kendall also made much of the fact that he was an engineer. At a meeting on 11 March 1942 he told his listeners he was 'not a politician' but was 'concerned with production purely and simply'. Britain was losing battles, Dunkirk, Singapore, Java, he said, 'because they had no equipment and for no other reason'. He did not attack Churchill and Cripps, who were the only two good people; 'too many' in government were 'wearing old school ties', and too many retired admirals and brigadier generals were in the ministries.[30]

It is worth noting that the first serious by-election threat to government candidates came from the extraordinary Noel Pemberton Billing. He stood in four by-elections in 1941, getting nearly half the vote in Hornsey and in Dudley, though doing less well in The Wrekin and in his last effort, Hampstead. Billing was one of the founders of the Supermarine aircraft firm, which went on to make the Spitfire; he had been elected to Parliament as an independent back in 1916 on a demagogic air power ticket (he was in the Royal Naval Air Service); and in 1918 he was the successful defendant in an infamous libel trial, the result of his virulent homophobic campaign against British liberals. Out of politics from 1921, he tried and failed to recapture his old seat in 1929. He spent the interwar years as an inventor/businessman in the record and photographic industries. When war returned he revisited aeroplanes and powerboats, but with so little success that he could publish his plans in the press without compromising national security. He also returned to politics, again as a right-wing demagogue calling for the bombing of Germany.[31]

Following Kendall's sensational victory, three more independents were elected in 1942, and five more in the rest of the war, including three for the leftist Common Wealth Party.[32] In early 1942 seven sitting MPs left government-supporting parties to become independents.[33] In the same year the Liberal MP Richard Acland and the Independent MP Vernon Bartlett formed the Common Wealth Party.

The election of independents in 1942 has been taken as a key bit of evidence for a move to the left in the electorate.[34] Yet in Parliament Kendall teamed up with another newly elected independent, William Brown, who although once a Labour MP was now definitely on the right.[35] Together with two other independents, who had defected from the Conservatives and the Liberal Nationals, Capt. Alec Cunningham-Reid and Eric Glanville, they formed 'The People's Movement', which aimed for 'total efficiency in this total war'.[36] According to the historian Angus Calder, some 'gloomy spirits on the left saw Brown and Kendall as portents of a new British Fascism', but he rejected the idea.[37] Yet the suspicions of the left were justified: MI5 surveillance of BMARC and Kendall shows that Kendall was in touch with former Mosleyites, associations which continued after the war.[38] Kendall was re-elected to the Grantham seat in 1945 and had a colourful career as a promoter of a new car and tractor.[39]

The number of technical professionals in the Commons was larger than one might expect. The biggest group were the doctors, of which there were eighteen in the war years, mostly Conservatives.[40] Leaving aside men like Anderson, Reith and Grigg, who were parachuted into seats when they became ministers, there was a substantial cadre of men with a scientific or engineering background, usually businessmen. The most senior politically was Neville Chamberlain, the only British prime minister until Margaret Thatcher to have had a university education in science and the only university-educated twentieth-century prime minister to have studied entirely outside Oxbridge. Not showing much aptitude (though he had a lifetime passion for natural history), he trained as an accountant and went into business, starting with the growing of sisal in the Bahamas.[41] Chamberlain's industrial interests included Elliott's Metal Company, which made brass, Birmingham Small Arms and the family-owned Hoskins & Co., which made berths for ships' cabins.[42] Among the others was Cripps, who maintained: 'I do not personally believe in government by specialists ... we do not want to see our Parliament composed of experts and technicians' – Parliament needed to form a general judgement, though often on the basis of 'scientific evidence or technical knowledge'.[43] The National Liberal Sir Murdoch MacDonald, a past president of the Institution of Civil Engineers, became an independent during the war and abstained in the vote of censure. Others were the Conservative Alan Chorlton, who was a past president of the Institution of Mechanical Engineers, and the Conservative Captain Leonard Plugge, a pioneer of commercial radio, broadcasting to the south of England through Radio Normandie.[44] Plugge chaired the Parliamentary and Scientific Committee during the war.[45] The inventor, engineer and broadcaster Group Captain William Helmore was elected as a Conservative for Watford in 1943. Helmore, with a Cambridge Ph.D. in engineering, was an official RAF broadcaster up to 1943, specializing in the romance of flight and heroic deeds of British airmen: in one broadcast he claimed that 'Excluding the aeroplane, there are three things alone, according to the proverb, which reach out beyond the edges of the world – the light of the sun, the darkness of the night, and the long arm of God.'[46] The second Common Wealth Party candidate to be elected, Hugh Lawson, in January 1944, was an engineer who was serving in the Royal Engineers.[47] The Liberal MP and businessman Geoffrey Mander of Mander Brothers, paint, ink and varnish makers of Wolverhampton, was a Trinity College, Cambridge, science graduate. The most senior scientist was Professor A. V. Hill, one of the MPs for Cambridge University, a powerful critic of the government, who also abstained in the vote of censure. In 1942 Hill and Richard Stokes distinguished themselves by their attacks on production, strategic bombing and in Hill's case the role of science in government as well.

TWO EXPERT CRITICS

Perhaps the least remarkable feature of Richard Stokes was that he was an engineer trained at Trinity College, Cambridge. There were hardly any other Labour MPs who were either businessmen or received the accolade of being a 'convivial West End clubman'.[48] Since 1927 he had been chairman and managing director of the Ipswich firm of Ransome and Rapier, makers of railway equipment, cranes, earthmoving machinery, sluice gates and much else. His maternal grandfather, a Rapier, had been one of the founders.[49] His uncle, and predecessor as chairman and managing director, was Sir Wilfrid Stokes, the inventor of the simple but highly effective Stokes mortar, which became the standard British mortar of the First World War. An active Catholic, Richard Stokes stood unsuccessfully in Glasgow Central in 1935, but was elected in a by-election in Ipswich in 1938, supported no doubt by his employees, a singular phenomenon for a Labour MP.

He was to become a persistent rebel who championed seemingly contradictory causes. Together with the Labour Parliamentary Peace Aims group he had called for a peace conference in October 1939, on the basis of disarmament, a new Polish state, a general settlement of minority issues, universal free trade and a single world currency 'possibly based on Sterling'.[50] This classical liberal internationalism fed into a demand after May 1940 for the publication of British war aims, in order to woo the German people away from Nazism. This in turn was connected to a powerful critique of strategic bombing. Speaking in June 1942 he called for aircraft to be switched to the navy and 'not to the lunatics who think Germany can be bombed into subjection'. The bombing of Cologne in the famous 1,000-bomber raid of May 1942 was 'morally wrong' because it was not carefully directed against armaments 'but it was also strategic lunacy' because it took aircraft away from the protection of the Fleet and of shipping. For Stokes it was by 'spectacular work on the high seas that we shall bring the war to a successful conclusion and not by the sensational flattening of continental cities'.[51] In March 1943 he spoke out in Parliament: 'I protest once again against the steady destruction of Europe by bombing', although not against 'the wiping out of Krupps with which

I agree'. 'I protest in the name of humanity,' he continued, against the 'cold-bloodedness' and the 'perfectly merciless destruction and slaughter of women and children'.[52] Stokes has an honourable place in the history of opposition to strategic bombing.

Stokes was a relentless critic of the quality of British tanks. Oliver Lyttelton recalled that the chief parliamentary troubles of the Ministry of Production concerned tanks, and that Dick Stokes was his principal opponent.[53] As Stokes put it in a debate on production in July 1942:

> I should not feel comfortable if I were sitting in a British tank with a 2-pounder gun with an effective range of 600 yards, or an even bigger tank with a 6-pounder gun which has an effective range of 1,200 yards, taking on a German tank with a 13 or 14-pounder gun with an effective range of 2,000 yards. A Minister who has no better to tell us than that, stands condemned, and the Government stand condemned, and the sooner those responsible get out the better. I want to know who was responsible for nothing being done from the period September, 1939, to July, 1940, and who has been responsible for apparently doing nothing from June, 1940, to the present time to bring out something equivalent to the German Mark IV tank.[54]

Stokes was a supporter of 80-ton tanks designed from 1939 by a committee headed by the Great War tank design veterans. The actual design was by Sir William Tritton and his firm William Foster of Lincoln, the man and firm who had designed the first British tank of the previous war. The group were labelled 'The Old Gang' and the prototypes therefore labelled TOG 1 and TOG 2. In response Lyttelton accused Stokes of supporting a tank that the army did not want, which was too slow and unreliable.[55] Stokes continued to attack on the issue of tanks right through the war. Such was the continuing pressure that the government conceded a Secret Session debate on the subject in March 1944, and after the war a parliamentary report was produced.[56] These attacks were very influential in creating the general perception, which endures to this day, that there was a serious problem with British tanks.

The Cambridge University by-election of 1940 was the fourth contested by-election of the war. Unusually, the electoral truce effectively broke down, for the Conservatives and Labour parties both fielded proxy candidates. The Conservatives proposed A. V. Hill but he stood as an Independent Conservative.[57] In response Labour supporters put up an 'independent progressive', the Regius Professor of Physic at Cambridge, John Ryle, who despite his archaic title was a pioneer of social medicine.[58] Ryle stood on a platform close to that of Stokes and his peace aims campaign.[59] Hill won two thirds of the vote, on a turnout of 42 per cent of the electorate of around 36,000.[60] The voters in university elections were the graduates of the relevant universities, and they thus give us an indication of the politics of this group.[61] Graduates voted overwhelmingly for Conservatives and independents.[62]

Hill, Professor of Biophysics at University College London, was one of the most important academic politicians of his day. Elected young to the Royal Society, he won a 1923 Nobel Prize in Medicine/Physiology. He was the only scientific laureate ever to sit in the Commons, and indeed the only laureate to be elected after winning the prize.[63] Hill was very different from the caricature of the absent-minded, long-haired professor of the 1930s. His son would recall that 'He was tall, slim, upright in bearing, his clipped moustache, short hair were those of a soldier and reflected his often-stated liking for military men.' Not only did he have 'presence' but he 'had influential friends in high places who greatly admired him for his good looks, his vigour – he was very good company'.[64] In the Great War he had worked on anti-aircraft gunnery and in the 1930s had been an adviser to the Baldwin and Chamberlain governments, specifically the Air Ministry and the Committee of Imperial Defence on air defence. He was Secretary of the Royal Society and a member of the Scientific Advisory Council of the War Cabinet and that of the Ministry of Supply.[65] He was very much an establishment figure, an insider in the pre-war and wartime military-scientific complex indeed. Yet he was to become perhaps the most important public scientific critic of the Churchill government, including its commitment to strategic bombing. Lord Strabogli, a Labour peer interested in science, claimed in a Lords debate on science and the war in July 1942 that 'whenever the Government are asking for a vote of confidence in another place, Professor Hill always votes against it'.[66] While not accurate, there was an element of truth in this charge, as Hill did not vote in either of the two great confidence motions of the war – 8 May 1940 and 2 July 1942.[67] In 1943 he became a founding member of the Tory Reform Committee.

In 1942 he claimed in a private paper that:

The Prime Minister and Lord Cherwell between them have been responsible for vast expenditure, tens of millions in chasing wild geese. This must be stopped, and Lord Cherwell must be prevented from interfering in technical matters: he is nothing but a liability. The Prime Minister's fertile brain is quite unsuited to well considered technical decisions on weapons, and Lord Cherwell's judgement on them has been shown by events to be disastrous, as his scientific colleagues foresaw.[68]

He alluded to this when he complained in Parliament in February 1942 about too many new weapons going into production. As he put it:

If production, however, is to be rapid and as efficient as possible, new types must be few and fancy weapons must not be allowed to clutter up development and supply against the best advice of collective expert opinion.[69]

And the problem was that:

There have been far too many ill-considered inventions, devices, and ideas put across, by persons with influence in high places, against the best technical advice. One could tell a sorry story of them. They have cost the country vast sums of money and a corresponding effort in development and production, to the detriment of profitable expenditure of labour and materials elsewhere.[70]

It is not clear exactly what Hill was referring to, but we know from his papers that he thought the greatest waste of money was the rockets.[71]

Although he did not participate in the censure motion debate, he added his weight to the attack with a letter to *The Times* during it. 'The defeat in Libya,' he claimed, 'is due largely to a single cause, the inferiority of our tanks', a case of failure to anticipate future requirements, a failure to collect and analyse evidence, and 'unjustified optimism'. He called for a central coordinating body of 'research, design, development and the quantitative planning of technical resources' to parallel the chiefs of staff machinery and the Ministry of Production. He complained that technical issues were left to departments, where they were 'subordinate to administration'.[72] Warning of the need for 'more critical and far-sighted technical policy', he claimed that otherwise there would be 'continuing failure and disaster'.[73]

Hill was typical of many critics of Churchill in 1942 in his attacks on strategic bombing. In the Commons in February 1942 he attacked 'the exaggeration of the importance of bombing an enemy country'. He looked back to the Blitz, commenting rather brutally that the number killed, over many months, was less than the number of prisoners of war taken by the Japanese in Singapore. The loss of production in the worst month of the Blitz was 'about equal to that due to the Easter holidays'. The greatest cost of the bombing was in the defensive measures provoked, not the damage done. Germany was further away than Britain was from German air bases, better aircraft and navigation were needed to attack it than the Germans had needed, and German defences would be more effective. 'The idea of bombing a well-defended enemy into submission or seriously affecting his morale, or even of doing substantial damage to him, is an illusion. It may be persisted in by those who use big and beautiful adjectives; its futility is recognized by those who prefer arithmetic.' We know, he said, 'that most of the bombs we drop hit nothing of importance'. The policy of bombing, he argued, was futile and wasteful, leading to lack of aircraft for army support and especially for Coastal Command. He recognized some merit in the bombing offensive against Germany, which he wanted to keep 'within reasonable limits'. Bombing had the limited function of making the enemy 'waste his substance in defending himself'.[74]

EXPERT VERSUS EXPERT

Hill's criticisms of Churchill and Lindemann were subtle and important. He was attacking over-enthusiasm for new gadgets, not lack of enthusiasm for them. Like an eighteenth-century natural philosopher wishing to interpose himself between scheming projectors and gullible investors, he stood for the rigorous, conservative assessment of novelties. The expert was contrasted with the inventor and the enthusiast, as much as the supposedly conservative administrator. Hill was speaking up for something called operational research, the scientific investigation of military operations and the effectiveness of weapons in the field. He called for the 'scientific analysis of the results of past operations and the critical study of current operational plans and methods', since it seemed to Hill that the authority of facts and arithmetic was not recognized.[75]

In thinking through these issues, Hill came up with a novel constitutional analysis. In draft documents from 1940, he claimed that there existed 'an admitted principle in British forms of government, in which the work of permanent officials is always guided, helped and criticized by elected representatives, or by other independent persons'. His primary concern was with the direction of government research, which without external control tended 'to become "set" and absorbed in routine: even the liveliest minds in [government research establishments] are liable to lose their freshness and initiative'. Political control was deficient, given the lack of interest in science in the political class, with the result that 'permanent scientific and technical staffs' either had their own way, or were the servants of the administration. What was needed was 'sound, critical independent advice' at the high level of the minister and his colleagues, which they would be 'expected to consider'. He claimed that the new Ministry of Supply, on whose advisory committee he was, worked well, but that the Ministry of Aircraft Production and the Admiralty did not.[76]

These were interesting and important arguments, but in public reflections by scientific intellectuals the impression often given was that there was political control by men ignorant of science, that there were no independent advisers, nor large numbers of government scientists and experts. The government was anti-scientific, run by amateurs, administrators and politicians trained in the classics. For example, in 1942 Bernal attacked the 'old-fashioned humanistic bureaucracy as an enemy of science'; in the time before a new scientifically minded bureaucracy could be created, science should be in the hands of scientists, representative of 'junior scientists and technicians'.[77] Thus what was in effect the most technically oriented British government ever was portrayed as being beyond the pale of science. Not only that:

the critics ignored the extent to which Lindemann was in many respects an expert on experts and expertise, and a critic of the over-inflated claims of enthusiasts, and of civil service science. He was indeed in his use of statistics, his close study of what was going on, not so far removed from being an operational researcher. What critics said could never happen was often in fact a matter of routine.

1942 was a significant year in the administrative politics of science. As Hill attacked the Churchill government as an MP, the Royal Society, of which Hill was a key officer, was agitating to force the government to make better use of elite scientists, and to give them a voice in production matters in particular. The government response was to appoint a group of three full-time scientists to the Ministry of Production, thus potentially overseeing the supply ministries. Their influence on war research programmes was also meant to be, and was, minimal.[78] Internally the government noted:

> We have been concerned for some time as to the best way to deal with the agitation in Parliament and in the press about our scientific and technical organization in the production sphere. We think that this agitation has been to some extent ill-conceived and largely based on ignorance of our organization and its achievements. At the same time we have not felt able to ignore the general feeling of uneasiness and unrest in scientific circles, nor have we felt that our existing organization enabled us to answer adequately all the criticisms that have been expressed.[79]

Yet the government also made very clear that the scientific panel was there to advise and should and would not interfere with existing lines of responsibility. As had been the case following a similar agitation in 1940, the external scientists were given a sop.[80] Oliver Lyttelton's memoirs give us a frank insight into the politics of science at ministerial level in 1942. He recalled that he had found it difficult to get 'the leading scientists and academic figures ... to sing in harmony'. He 'hoped to make them sing in unison'. To this end he proposed his triumvirate of scientific advisers. Yet the proposal 'irritated the Prime Minister, who wrongly thought that I was trying to expand my "empire" and invade some of his'. Lyttelton recorded Churchill as saying after dinner to some 'pretty and charming ladies' they were both talking to: 'Here's Oliver, always avid of power: now wanting to run the scientific side of the war; he's going to take it over from me: he first has a spearhead of three graces, and so we may expect to see everything in the scientific field better run.' The next morning Churchill received his Minister of Production from his bed.

'Good morning, my dear. I hope you are not vexed with me?'

'Well I was vexed, but I've got over it,' say I.

'I know I shouldn't have chaffed you so much, but why do you want to run everything?'

'I don't, really I don't, but it does require people of the *métier* to compose some of the quarrels amongst the scientists. A won't talk to B, because B has invented something A thinks he ought to have invented, and so on. That's all there is to it.'

'Oh well, go ahead. I don't like it, but if you want it I suppose you had better do it.'[81]

And indeed that was probably all there was to it. However, there were less visible changes which were important. 1942 saw some major state laboratories coming under civilian leadership, and indeed a significant infusion of academic talent into some of them, notably the Woolwich research and design departments, as we shall see in Chapter 7. This is not to be interpreted as a victory for the external scientists – it came at least as much from dissatisfaction with the status quo from Churchill and Lindemann.

PARADOXES OF FREEDOM AND PLANNING

Scientists and scientific intellectuals were divided politically. Their portrayals of what was happening were political statements, imbued with very particular assumptions about what constituted science, the place of science in government and much else besides. It is a mistake to think that when 'science' spoke, it spoke its own distinct language, for scientific intellectuals generally used the standard languages of British public intellectuals, whether Conservative, Liberal or Marxist. It would be a mistake to think they spoke more empirically than other

intellectuals. Indeed, scientific intellectuals produced remarkably partial and ill-informed pictures of the place of science in government,
industry and elite culture.

Scientific intellectuals of the left set out a manifesto outlining the
contributions they might make to war. *Science in War* (published in
August 1940 as a Penguin Special) was remarkable for the extent to
which it saw the war as a war of civilians and a war of production –
the military and weapons hardly figured in it.[82] It is an example of a
serious lack of frankness by scientific intellectuals about the relations
of science and war. In 1942 a Penguin Special on *Science and World
Order* set out the claims of the scientific left for a greater role in government for science, the need to plan science and for the power of
operational research.[83] Throughout the war the scientific left argued
not only for more research, but for more planning and for the greater
power of scientists in running human affairs, nationally and interna-
tionally. There was a move to the left in the scientific community, as
is clear from the rise in membership and the radicalization of the
Association of Scientific Workers (which became a trade union in
1941) – its membership in 1939 had been 1,000 concentrated in London and Cambridge; by 1944 it was up to 15,000, mostly in industry.

The socialist scientists were not the only ones in the game. There
was a counterblast from a new scientific right, influenced by the anti-
planning economists, notably, but not only, Friedrich Hayek, a professor
of economics at the LSE, which was evacuated to Cambridge during
the war, who wrote *Road to Serfdom* (1944). Their line of thinking
was to be very well represented in a new journal launched in January
1942. Imperial Chemical Industries funded a new glossy, *Endeavour:
A Quarterly Designed to Record the Progress of the Sciences in the
Service of Mankind*, edited by the historian of chemistry and former
head of science at Clifton College, E. J. Holmyard.[84] 'Our purpose,'
declared the chairman of ICI, Lord McGowan, was 'to enable men of
science and particularly British men of science, to speak to the world
in an hour when not only nations but the internationalism of the sciences are threatened by a recrudescence of barbarism in its grossest
and most destructive manifestation.' *Endeavour* was published in
English, and four foreign languages, and celebrating British science
was a key part of the brief.[85] But the magazine had a clear and strong

line against the 'influential scientists' who wanted science determined by social conditions, attacking the idea of planning of science – cast as the 'mechanization of the whole army of workers', which would result in 'dreary mediocrity'. It claimed that under planning 'scientific genius would fail to flower', invoking an image of Newton, Faraday and Einstein being forced to undertake mundane tasks. It was untouched by the irony that it was a company magazine, and that of a company that told its army of scientists what to work on.[86] ICI was not willing to present itself as a planner of research. On the other hand, the scientific left were not interested in discussing industrial research and how it was planned, only in talking about the importance of research *for* industry. It concerned itself with academic and with government civil scientific research and increasingly with the use of scientists as operational researchers, rather than laboratory workers.

SCIENTIFIC INTELLECTUALS AND OPERATIONAL RESEARCH

The assessment of the effectiveness of weapons and tactics, and advice on changes to improve performance, became the concern of groups of scientific advisers who undertook what they called 'operational research' during the war. It was the work of a rather special minority of scientists. Many of them came from specialisms not obviously suited to wartime research needs.[87] Many of the leading figures were or became scientific intellectuals, and many had been advisers to the Chamberlain government, as were Sir Henry Tizard, A. V. Hill, P. M. S. Blackett and J. D. Bernal. One of the ways in which they were special was that they criticized the powerful tendency to over-invest in new weapons, as did, as we have seen, A. V. Hill.

Already in 1938 Sir Henry Tizard, who was to become a key figure in operational research, as he was in scientific advice more generally, stressed that in the war 'the main things to do will be to get on even faster with the job of applying the results of research'.[88] Tizard thought that a large proportion of scientific men at universities would be used on anti-aircraft gunnery and similar tasks, and that, for example, the

RAF would need a lot of meteorologists. He explicitly rejected an overemphasis on research: 'I think,' he told his long-standing associate A. V. Hill, 'that most people . . . are liable to exaggerate the demand for scientific people for research. I am inclined to think that we have got to get most of the research over before a war starts. When it starts it will be too late, but there will be much to be done in hastening application . . .'[89] This was a theme that Tizard would echo again and again, not only during the war, but after it.

Sir Henry Tizard and Patrick Blackett both opposed the building of a British atomic bomb, on the grounds that it was likely to take longer and cost more than promised.[90] In this they were to be proved correct, as we have seen. Blackett engaged in a general critique of the pursuit of novelty. Writing in December 1941, in a paper setting out the nature of operational research, Blackett held that 'the scientist can encourage numerical thinking on operational matters, and so can help to avoid running the war by gusts of emotion', an accusation which went much further than the serving officers who might seem to have been the target of this attack.[91] But Blackett was specifically concerned to challenge calls for 'New weapons for old', as a form of 'escapism'. He argued that for existing weapons it was known that they didn't work well, and that there was a lack of training and spares. Yet in thinking about new weapons it was assumed that they were in full production, with spares available, and operated by fully trained crews. This tendency could be countered by 'continual investigation of the actual performance of existing weapons' and 'objective analysis' of the likely performance of new ones. This was necessary because in his view too much effort was going into new weapons and 'too little on the proper use of what we have got'.[92] Changing tactics could be more effective than changing weapons.[93] More strongly still, at the first meeting of the 'independent scientific advisers' – as they revealingly called themselves – in June 1942 their informal leader, Sir Henry Tizard, told them that 'by far the greatest contribution that scientists could make at this stage was to do everything possible to improve the operational efficiency of equipment and methods now in use'. They suggested the transfer of researchers away from development to the study of operations: researchers 'were better employed' on operational research 'in the interests of the war than on experimental work however important it might appear'. Blackett clearly agreed, supporting 'taking the best people away for operational research as soon as possible'.[94]

Of course, the armed forces had themselves studied the effectiveness of operations. The military intellectual Basil Liddell Hart suggested before the war, when he advised the War Office, that the army start research on military operations. The minister, Leslie Hore-Belisha, accepted there was a lack of 'pure military research' (that is, research on military matters) and proposed to establish a small unit 'to study the practice and lessons of actual warfare'. A unit was indeed established, but separate from the scientists; Liddell Hart was nearly appointed 'Director of Military Research' to oversee the branch and to link it to the scientists. In 1942 there was a War Office plan to bring Liddell Hart in as the head of an operational research section of the army to parallel a scientific section, but in the event he was again left out in the cold.[95] Liddell Hart tried, without success, to make himself into the father of operational research, claiming that 'I constantly tried to get Operational Research started in the army, during the twenty years before the war, and on several occasions did pieces of such research myself.'[96] Liddell Hart spent the war on the margins of influence, and away from London, increasingly hostile to the mass war developing around him and to strategic bombing.[97] The other important military intellectual of the interwar years, Major-General J. F. C. Fuller, was lucky not to be interned for his association with British fascism and was silent.

Operational research was done in close collaboration with the military.[98] It was started by state researchers. Fighter Command was the first command to get operational research, under the scientific civil servant Harold Lardner. The practice started with new techniques of war, above all radar, but soon extended far and wide across all the services. It was often brought into play when the machines were obviously not working well. Danger, defeat and obviously poor performance were the main stimulants.[99]

The Army Anti-Aircraft Command, which had a very poor hit rate, even with radar-controlled guns, was an important early case of the application of OR. Academic scientists, led by the physicist (and gunnery expert) Patrick Blackett, were brought in to examine the problem and went around the AA emplacements attempting to work out how to increase the hit rate. 'Blackett's circus' had some success – the rounds fired per aircraft shot down fell from 20,000 to 4,000 by the spring of 1941, by which time the Blitz was of course over.[100] The team moved on to the RAF's Coastal Command, whose major task was hunting and destroying U-boats at a time of danger at sea. The best-known improvements came (the extent of their success is not known) from painting the undersides of aircraft white, setting depth charges to explode less deeply than had been the practice, the discovery of a systematic aiming error and improvement in maintenance scheduling.[101] Blackett transferred to the navy in January 1942, where the 'tall, dark and very serious'[102] physicist worked on convoy size, arguing that large convoys were preferable. This argument met resistance, but since the number of cargo ships was increasing more rapidly than the number of escorts, convoy sizes were forced to increase. Blackett's evidence and argument were much refined over time and gave confidence in a policy decided on essentially for other reasons.[103] This was a nice example of the complexities of learning, and of the relationship between advice and action.

As we have seen in Chapter 2, the greatest machine failure was in offence. British bombing was spectacularly ineffective in 1940–41: while the British night-bombers could get through they could not find targets. One of Lindemann's assistants was put on to the task of assessing bombing photographs, and reported in August 1941 that one in four claimed successful attacks on German targets were within five miles of the target, discounting all the aircraft that had to turn back, got lost or could not find the target.[104] On one estimate 5 per cent of sorties led to bombing within five miles of the target. A huge operational research initiative, along with new navigational aids, was to help transform this situation. Basil Dickins, a scientific civil servant trained at Imperial College, became the head of the new OR section of Bomber Command in September 1941. There were no fewer than fifty-five scientists working there at the 1943 peak.[105] The section was at the Bomber Command HQ at High Wycombe, and worked very closely with the leadership of the command in building up its effectiveness. Operational researchers asked many important questions which led to improvements in practice. Which was the best bomber?

The Lancaster. Was it worth adding the weight associated with a nitrogen supply to keep petrol vapour out of petrol tanks? Yes. Just how precise was the bombing? How many bombers would be needed to achieve a particular mission?

The bombing effort was supported by many organizations. The Research and Experiments Branch of the Ministry of Home Security shifted its work from defence to offence.[106] One section, formed in 1942, to be called RE8, worked for the Air Staff on bombing, and it was transferred to the Air Ministry in 1945. It used very extensive data collected about German attacks on Britain to model attacks on Germany. And by comparing actual damage in Britain with aerial photographs of that damage it developed techniques for estimating damage from aerial photographs of Germany, and developed standardized methods of photo-interpretation. Americans who had been in RE8 formed the nucleus of the Joint Target Group in Washington, which was to advise and analyse targets in Japan, and the Group used 'many of the methods and procedures first established' by RE8.[107] One of its key members was the mathematician-poet Jacob Bronowski, though it was hard to guess his role in the strategic bombing of Germany and Japan from his post-war moralizing about science.[108]

As we have seen, during early 1942 there was a serious debate as to whether strategic bombing should continue to be expanded. As Lord Hankey, ejected from the government in 1942, put it: 'Many authorities hold that the long-range bombing policy has been carried too far at the expense of the Navy and Army.' There was force in the argument, he suggested, that the effects of bombing were outweighed by enemy success in Libya, Malaya, Burma and in the war on shipping.[109] We have already noted how strategic bombing was criticized in Parliament by A. V. Hill and Richard Stokes and others on similar lines. There was also an internal debate involving estimates of effectiveness among scientific advisers.[110] The supposed decisive encounter between Sir Henry Tizard on the side of the angels and Lindemann on that of the bloodthirsty ignorant militarists recounted by C. P. Snow never happened.[111] Tizard had no objections to strategic bombing in principle. How could he, as a long-standing adviser to the air force? He believed that to be decisive bombing would have to be 'carried out on a much bigger scale than is envisaged', pointing to an arithmetical error in Lindemann's calculations on future bombing; once this was resolved Lindemann was able to respond: 'I am glad to see that we do not differ in arithmetic, or even in our general conclusion.'[112] Patrick Blackett, now at the Admiralty, was, however, clearly concerned to get aircraft switched from Bomber Command to anti-submarine and anti-shipping duties in the Atlantic and Mediterranean. But this was not a matter of gifted scientists against obscurantist administrators. On the other side of the argument were not only the Air Ministry but also Lindemann's economists and statisticians. One of them, Roy Harrod, recorded that 'in the judgement of our office, which was as highly skilled as any group in the country in the interpretation of statistics, Blackett's case was completely demolished'.[113]

Scientific advisers, and operational researchers, were as much advocates and defenders of particular positions as independent analysts. They supported their own service and, indeed, often had intimate connections with it. Blackett had been a career naval officer during the Great War; Charles Ellis, the scientific adviser at the War Office, was a Harrovian who went to the Royal Military Academy, Woolwich, in 1913. At the War Office he 'found the high-ranking military very congenial', recalled his technical assistant.[114] George Thomson, son of the physicist J. J. Thomson and scientific adviser to the Air Ministry from 1943, had spent most of the Great War at Farnborough, the research and experiment centre of the air force.[115] His (in effect) predecessor, Sir Henry Tizard, had also been associated with the air force from the Great War. Wartime scientific advice came with service badges. Scientific advice was linked to service interests.

BOMBING AND LIBERAL INTERNATIONALISM

Yet the case of bombing cannot be understood just in terms of either party politics or service politics. There was an important ideological dimension which cut across the former, and even the latter too. It was certainly not a matter of an anti-bombing left against a pro-bombing right, or a matter of scientists being against it and old-fashioned militarists for it. Although Stokes and Bevan were of the left, Hill was not. There were plenty of military men and naval men who were against it. The air force was of course in favour, sometimes expressing that support in extreme terms. In a newsreel in 1942 Sir Arthur Harris told the British public that 'there are a lot of people who say that bombing can never win a war. Well my answer to that is that it has never been tried yet and we shall see.'[116] Many progressives agreed. In the interwar years many suggested that an international body, the League of Nations, say, should have at its disposal an international air police. This world force would have a monopoly over military aircraft, and civil aircraft too would come under international control. Yet while the League of Nations was discredited by the late 1930s, the idea of an international bomber force to extirpate militarism, nationalism and war was not.[117] These themes were taken up in semi-official propaganda. J. M. Spaight, one of the leading British air propagandists, a retired Air Ministry civil servant, wrote a book called *Bombing Vindicated* in 1944 which claimed: 'The Bomber Saves Civilisation'. For Spaight, the bomber was 'a murderous weapon. Its only merit is that it can murder war.'[118]

In 1941, and thus before the US entered the war, Stephen King-Hall, a National Labour MP (from February 1942 an independent), wanted an Anglo-American naval fleet and air force, a 'Peace Force' as he called it. The fleet would be three times larger than any other, and twice as strong as three other fleets; the air force would be four times larger than any other, and twice as large as two others combined.[119] The idea was that 'unilateral aggression be crushed instantly by unilateral counter-aggression'.[120] Others dilated on the theme of an Anglo-American air force which would 'lead the way ultimately to a World Federation of the Air . . . a guarantor of a co-operative international system'.[121] David Davies, a Liberal Welsh peer and coal-owner who had made the international air police his great cause in the 1930s, got the Liberal Party conference in September 1942 to vote for an international police force, the internationalization of civil aviation and a federation of English-speaking peoples.[122] It is interesting as well as fitting that Churchill himself should have been a supporter of this notion, preventing British diplomats from turning down a Soviet proposal for such a force at the 1944 Dumbarton Oaks conference, reminding them he had supported such an idea in the 1920s and noting that the 'proposal will gain very great acceptance in Great Britain and it certainly seems by weaving the forces of different countries together to give assurances of permanent peace'.[123] Churchill continued to believe in it, and it would form a crucial but neglected part of his famous 'iron curtain' speech in Fulton, Missouri, in 1946, where he proposed the temporary assimilation of national air squadrons into a United Nations force. Otherwise, he claimed, in his inimitable style, but in a clichéd sentiment, 'the Stone Age may return on the gleaming wings of science, and what might now shower immeasurable material blessings upon mankind, may even bring about its total destruction'.[124]

Especially from 1942 the RAF was itself seen, with some justice, as itself an international force, made up as it was of pilots and aircrew from many nations, mostly, but far from exclusively, the white Commonwealth. An anonymous letter to the *Manchester Guardian* in May 1942 claimed that 'Under the political direction of the United Nations, and with the R.A.F. as nucleus, an international air force has actually come into being . . . There is no doubt that it can be – and will have to be – maintained and consolidated as an integral element in the peace settlement.'[125]

Wartime bombing got support from some of the most important left intellectuals of the 1930s, at least as they moved to the centre. When in the War Cabinet Cripps strongly supported strategic bombing, even redrafting a Cabinet paper in its support by Sir Arthur Harris.[126] The RAF's wartime PR man, Squadron Leader John Strachey (who in the 1930s was Britain's leading Marxist intellectual), discussed bombing in a thoroughly economic analysis of war which was typical of official thought:

the German army, and the German economic system which supports it, are inseparable; they stand or fall together; an attack on one is an attack on both. No one who has thought about the economics of modern warfare can doubt that. The simplest of all economic principles is that of supply and demand. Bombing cuts down the enemy's supply of all the materials of war. Land attack vastly increases the enemy's demand for all the materials of war ... The object of war is to bankrupt your enemy. He will go bankrupt when his supplies fail to meet the imperative demands of his war machine.[127]

This was modern, total war, thought of in material terms drawn from civilian life.

The characterization of modern war as civilian, industrial and determined by the most advanced machines was shared even by the most passionate and outspoken of anti-bombing campaigners. Vera Brittain reflected in 1940–41 that the new war was very different from the Great War, which she had experienced at first hand as a nurse. It was a 'struggle of production versus production rather than of men versus men'. It was 'an enormous, illimitable struggle, fought with economic and political weapons reinforced by the perverted ingenuity of countless human minds'.[128] Great inventions had the capacity to unite the world in peace but were instead used as powerful and vicious weapons. For Vera Brittain, as for many others, modern war was defined by chemicals, aeroplanes, tanks and the radio. For her the aeroplane had become the 'cruellest vehicle of destruction ever conceived by perverted genius', while the tank 'turned the roads of Europe into pathways of blood'.[129] That wars still happened was explained using one of the great clichés of the age, the cultural lag argument. A tedious commonplace, it sees modern war as a tragic clash of old and new. It was expressed well by Vera Brittain:

Many of us are sadly aware that the fiercest problems of the present age, whether international, national, or merely personal, have been caused by the degree to which material development and scientific invention have outstripped man's knowledge of his own nature and his power to control it. This outpacing of moral by intellectual capacity is a phenomenon of the past two centuries and especially of the last fifty years.[130]

INTERNATIONALISM

One of the least discussed features of the politics (and ideas) of 1942 is the rise of internationalism. From its beginning the war was routinely presented in liberal internationalist terms. The war had never

been anything like a war between nations; rather it was a war for freedom and civilization against barbarism. It was in some crucial respects
a war against nationalism. The internationalism of the war was highlighted in the coming over to Britain of governments in exile, in Lend-
Lease and the Anglo-American Atlantic Charter of August 1941 and
the Anglo-American supply of armaments to the Soviet Union in late
1941 and beyond. But from the beginning of 1942 a whole new
emphasis on internationalism is evident. As of January 1942 the title
of 'United Nations' was given to those who pledged to fight the Axis
and subscribed to the Atlantic Charter: the original signatories were
the big four, the USA, the USSR, China and Britain; the four dominions, Australia, Canada, New Zealand and South Africa, and India;
eight European governments-in-exile; and nine Caribbean/Central
American states, most of which were US-backed dictatorships. Henceforth the war was, in much propaganda, a war of the United Nations.
This is the context in which a British newsreel described the Lancaster
bomber as the 'Finest heavy bomber operated by the United Nations'.[131]

Internationalism was an important and an underrated feature not
just of propaganda, but of wartime political and intellectual activity.
Scientists came together to proclaim what was a general internationalism as peculiarly their own. In September 1941 the new Division
for Social and International Relations of Science, of the venerable
British Association for the Advancement of Science, organized a meeting of scientists of twenty-two nations – the USA, the USSR, Britain,
the dominions, allies and neutrals – to discuss 'science and world
order': 'on the programme of the proceedings, there were no national
labels. Science knows no frontiers. And the statesmen acknowledged
that commonwealth,' enthused the Penguin Special based on the
conference.[132]

The production and raw material aspects of this United Nations
effort were lauded. Thus Sir Stafford Cripps could say in July 1942 that:

> We have today reached a degree of co-operation between the united
> nations in the use of raw material resources and in the employment of
> finished products greater than has ever been reached before in the his-
> tory of the world. Selfish and national interests have been over-ridden
> by the insistent demand for efficiency by the fighting peoples.[133]

The plans of the 'United Nations for their war effort', claimed Cripps, involved 'nothing less than the subordination of private and national interests to the public and international good'.[134] This was not a straightforward matter of subservience to a greater power, but an attempt through a division of labour to maximize the exploitation of common resources.[135] Indeed, that internationalism was often associated with bodies like the Combined Chiefs of Staff, the Combined Munitions Assignment Board, the Combined Production and Resources Board, the Combined Food Board, the Combined Raw Material Board and others, which were set up from 1942. They were in fact Anglo-American bodies, with a Canadian presence, but this crucial detail was often ignored.

Internationalism was everywhere in 1942. In a broadcast to accompany the publication of his December 1942 report on *Social Insurance and Allied Services* the liberal intellectual Sir William Beveridge started by invoking the internationalist Atlantic Charter, which, he said, 'speaks of securing for all improved labour standards, economic advancement and social security', and going on to say that his scheme gave effect to those last two words, 'social security'.[136] The term 'People's War' had a crucial internationalist dimension. The extent to which it was used is hard to judge; indeed, Angus Calder gave very little evidence for its use, and neither have historians who use the term more than he did.[137] Yet it seems it was not associated with Britain in particular – it carries little hint of 'British People's War'. On the contrary it was used to describe the war in general, as a synonym for total war, for intensity of warfare, for a breakdown of the differences between the civilian and the soldier. It was used in an internationalist context and about and in many countries (the USA, China, India, the USSR and Germany). Goebbels was cited using the term, though it usually carried a hint of democratic commitment to war too.[138] Thus, when in its National Campaign in August 1942 the new radical political party Common Wealth proclaimed, 'This War must be made a People's War', it is difficult to know what was meant, other than the war was *not* a people's war.[139] E. H. Carr rightly noted in 1945 that what was extraordinary about the Second World War was not only the planned economy with a 'nationalist as well as a socialist face', but a new internationalism.[140]

A NEW NATIONALISM

Yet for all of the very great importance of internationalism in 1942 and for the rest of the war, nationalistic and conservative currents ran strong. The nationalism of the left became marked, and was powerful even in the case of the Communist Party, so intimately associated with its own kind of internationalism. One important theme in communist thinking before and after 1941 was that the British ruling class was a *Traitor Class*, to give the title of a 1940 book by the well-known communist film maker and class traitor Ivor Montagu.[141] From 1941 there was a new emphasis on criticizing British business for not being efficient enough (and shop floor Trotskyites as fascists and wreckers).[142] Here the party drew on a critique of British capitalism dating back to the work of J. A. Hobson in the Edwardian years. British capitalism, they argued, was oriented to owning assets abroad, which only paid their way if they resulted in imports into Britain.[143] It was hardly surprising if production was insufficient and inefficient if government controls were in the hands of monopolistic importers and restrictionist producers.[144] The national and nationalist focus of the critique was to inform socialist critiques in the rest of the war and beyond. For example, in 1944 Aneurin Bevan attacked the idea that exports needed to be increased after the war and foreign investments built up. He saw this as a Tory plot to deprive British workers of goods and to allow capitalists to benefit from higher rates of profit abroad. His alternative was to invest at home and stimulate home demand.[145] J. T. Murphy, an ex-communist, called in 1942 for a 'full-blooded patriotism' to 'transcend the vested interests of all classes and parties and subordinate all property and all service to the all-in war against Nazism'.[146] Murphy wanted a full-scale takeover of private property as well as a general militarization and bringing together of the military and civilian.[147]

Of course, a reinvigorated nationalism need not only have a socialist face, and in 1942 it did not. A particular kind of right-wing nationalist politics was very evident, as we have seen, in the by-elections of 1941 and 1942. Humphrey Jennings's documentary *Listen to Britain* of 1941 ends with the familiar strains of the nationalist hymn 'Rule, Britannia'. It is also present in some of the most popular biopics of

…, *One of Our Aircraft is Missing*,
…ressburger, features one Sir George
…with a military past. It is a conservative
…the character of upper-middle-class leader-
…ctively because unofficial – Sir George is not in
…, and his Dutch opposite number is the female and
…er of a Dutch resistance group. The film is notable for
…ing connections between Britain and Europe before as well
as …ing the war. But the most interesting thing about it is that the
central character was modelled on Sir Arnold Wilson MP, who died
as an RAF rear-gunner in action over Europe in May 1940. Wilson
was a notably independent and intellectual Tory MP, an English
nationalist, former soldier and senior oil man (with Anglo-Iranian)
with clear pro-fascist leanings, and an enthusiasm for science which
led him to chair the Parliamentary Science Committee between 1933
and 1938.[148] Of course, it would have been impossible to make Wilson's
politics overt in a film of this sort, yet it is remarkable that such a film
was made at all if one considers the years 1940–42 as involving a
marked shift to the left. It is telling that it was not the only significant
film of 1942 which featured a positive character from the extreme
right without the politics being made clear. Leslie Howard's *First of
the Few*, telling the history of R. J. Mitchell, designer of the Spitfire,
sympathetically portrays Lady Houston, the benefactress of the 1931
Schneider Trophy-winning British entry, designed by Mitchell. Her
attacks on the British left and her pre-war support for Mussolini
and Hitler are entirely eliminated, but her call of 'England Awake'
is very present. Otherwise, the film is a vindication of the armament
firm Vickers and the need for strong armaments in peacetime.[149] It
was the top British film of 1942, and Leslie Howard, who directed,
produced and starred, was the second-most popular player of the
time.[150] The other great success of 1942 told the story of a warship
and its commander. *In Which We Serve*, directed by Noël Coward and
David Lean, focused on a leader (very obviously Lord Mountbatten)
and his destroyer HMS *Kelly*, called HMS *Torrin* in the film. It was a
celebration of a professional Royal Navy embedded in an existing
and old-fashioned England.

THE END OF THE AFFAIR

The critiques of early and mid-1942 were to cease with the reversal of British fortunes in the Middle East in the autumn. The 8th Army under Montgomery, supplied with huge quantities of equipment, including Sherman tanks and lorries from the USA, overwhelmed the weak German and Italian forces at the Second Battle of El Alamein. The imperial army marched west to meet up with US forces which had landed in Morocco, and the mostly British 1st Army, in Tunisia. In a great and conclusive victory in early 1943 North Africa was cleared with vast numbers of Germans captured, causing losses to Germany comparable to those of Stalingrad at a tiny fraction of the cost in lives. Churchill survived and prospered, never to be under such pressure again. Complaints about the quality of British tanks died away, though Stokes would take up the fight again in 1944. Attacks on strategic bombing fell away too, just at the moment when raids were beginning to get heavy, though they too would revive in 1944. Complaints about the lack of respect for science died away also, as did those about the production record. The moralizing independent Stafford Cripps had by the end of 1942 been demoted to Minister of Aircraft Production outside the War Cabinet. There, with god as his co-pilot, he managed the huge continuing expansion of bombers which were beginning to destroy German cities.

And yet, some of the critiques of early 1942 were hugely important and powerfully affected how the war was to be understood. Some, but certainly not all, of the debates of 1942 were to shape important post-war traditions of interpretation of Britain at war, particularly with regard to production and expertise. These perspectives have proved to be much more influential than Churchill's own positions, despite his having written the history of so much of the war. The criticism of tank production and quality, in particular, and of Churchill and his government's supposed hostility to experts are key instances. So too was the assumption that military defeat was due to a failure in production. As far as the politics of science is concerned, scientists' opposition to strategic bombing, and to Lindemann, and an association of all this with operational research, was put at the centre of the story. The critiques also fed into a continuing tendency to downplay British strength not only in 1940–42 but also in 1940; here the productionist critique and the government's defence that there were hardly any tanks sent to France in 1940, and none left afterwards, combined to give a wholly erroneous picture of the British position. This picture might have had a certain authority at a time when debates took place under a regime of strict state secrecy.[151] At the beginning of the war journalists were, for example, refused copies of the leaflets which were being dropped on Germany, a classic early instance of many cases where governments used secrecy legislation to keep from its own people information which was available to the enemy.[152] The routine statistics of production, trade, economic activity and much else besides were pulled back into the confines of the state machine. During the war the *Economic Digest* was a secret document.[153] Of course, such information was even more difficult to obtain for Germany. Yet after the war it was known that Britain was the world's greatest tank producer in the years 1941 and 1942, and indeed, as Richard Overy has noted, the British economy of this period 'outproduced Germany and its new European Empire in almost all major classes of weapon'.[154]

The debates of 1942 cannot be read through the forms of political division of either the 1930s or the later 1940s, or indeed the frameworks through which historians have usually discussed the war – the supposed move to left between 1940 and 1942 and the debate on reconstruction.[155] Take the example of the character of the conservative and backward Colonel Blimp, central to the wartime work of the cartoonist David Low. His character was the protagonist of a famous film made in 1942 and released in 1943, Powell and Pressburger's *The Life and Death of Colonel Blimp*. This connection, and the fact that in the special context of 1942 Churchill was exceedingly hostile to the film as it was being made, has given it a wholly undeserved reputation as a radical and critical film, or one expressing a progressive vision.[156] Yet it is not by any stretch of the imagination an attack on the wartime British army or government. It has a powerful and clear message: to fight the Germans, especially the Nazis, one must and should play dirty. It is a critique of what it takes to be an old moralistic England, and a celebration of a fast, jukebox-listening and ruthless new England.[157] It was about the need to be beastly to the Hun. As in Noël Coward's song of the same year, it made its argument by criticizing the views of opponents taken to be excessively humanitarian.

6

Sons of the Sea

In his history of the war Winston Churchill made the remarkable claim that 'The only thing that really frightened me during the war was the U-Boat peril ... I was even more anxious about this battle than I had been about the glorious air fight called the Battle of Britain.' Even more notable was that he was most frightened of U-boats in late 1940. For Churchill the 'British Battle of the Atlantic', the one that really counted, was won by July 1941. This may seem surprising given the great losses at sea of 1942, and the standard story that the Battle of the Atlantic was won in 1943, but, as Churchill pointed out, the 1942 losses were along the American coast and temporary.[1] Although historians have insisted, with some justification, that the Battle of the Atlantic came to a *crisis* in early 1943, Churchill's view that the critical victory came much earlier has been strongly restated and defended.[2]

Although there were great Allied losses at sea, particularly in 1942, the British Isles were never seriously under siege by U-boats; it is quite inappropriate to think of Britain as having been blockaded, even though this is the impression that could be got from some of the literature. Even in 1942 the economy 'managed quite well' even with imports lower than the previously stipulated minimum and there was no evidence that the shipping shortage affected major decisions.[3] The shortage of shipping was real and a powerful constraint, but one that could be and was met.[4] Yet this shortage should not be equated with losses from or caused by U-boats. The loss of near supplies, the closure of the Mediterranean and the demands of the military were in themselves major causes of reduced shipping capacity.

The key point, however, is that huge quantities of civil and military supplies kept moving throughout the war. In the last two years of the war vast Allied fleets moved at will along nearly all the seas of the world, with the Axis fleets nearly completely destroyed. Command of the seas of the world was achieved by the United Nations to a degree unprecedented in world history. As far as Britain was concerned, continued availability of sea supply had profound implications for how it fought the war. Imported food and oil were available in large quantities in Britain, certainly enough to maintain an exceptionally motorized set of armed services, and also to keep the population very well fed. Britain did not need to rely on its own resources, but could draw what it needed to fight most efficiently and effectively from the rest of the world. It did increase national production in many areas, but it continued to rely on imports, which in some commodities and categories were greater than before the war. Much changed in the nature and origin of imports into Britain, and indeed the flow of materials around the globe, but they remained crucial to Britain's ability to live and fight.

IMPORTS INTO BRITAIN

Britain started and ended the war as the world's largest importer. The bulk of imports dropped, but not by the half suggested by the usually quoted figure for dry-cargo tonnage (essentially solid raw materials and most food). Raw material imports dropped radically, food imports less so, but oil imports doubled, becoming considerably larger than food imports. Including petroleum reduces the wartime tonnage fall to around 30 per cent. In value terms the fall, excluding munitions, was similar at 29 per cent at 1935 prices (food down 25 per cent, raw materials down 39 per cent).[5] In real terms imports (excluding munitions) were down compared with 1938 by 30 per cent in 1942, 23 per cent in 1943 and 20 per cent in 1944.[6] Putting in munitions imports into Britain, excluding those for US forces (as well as imports direct to British forces outside the UK), changes the picture radically.[7] Munitions imports were running at around one third of the total value of imports in 1943.[8] More than half of 1944 imports were munitions.[9] This suggests that in value terms British imports towards the end of

Table 6.1 Imports into the United Kingdom at 1938 values.
(Index: 1938 = 100)

	1938	1940	1942	1944
Total	100	94	70*	80*
Food, drink and tobacco	100	78	73	74
Raw materials	100	105	63	61
Manufactures	100	112	72*	102*

* Arms excluded

Source: R. S. Sayers, *Financial Policy 1939–45* (London, 1956), Table 6, p. 495.

the war at least, were running at about 100 per cent of 1938 levels. In other words, it is likely that wartime imports overall were at the same level in terms of inflation-corrected prices as before the war, though less in quantity.

The story of wartime imports is thus not a story of a fall so much as an increase in the value per ton of imports. What Britain did was to substitute cheap bulky imports with expensive less bulky ones. Raw materials were replaced by semi-finished and finished products, notably munitions. In the case of food, too, materials of low nutritional density were replaced by elaborated and expensive industrial foods. Sometimes saving shipping implied increasing the proportion and sometimes the quantity of overseas supply, as in the cases of meat and weapons, which were less bulky than the materials needed to make them. In some respects wartime Britain was more national in supply, in others more international, shaped by the nature of the product, by German and Japanese conquest of suppliers, the nature of its alliances and the tyranny of distance. It is thus mistaken to characterize the wartime situation as a fall in imports partially made up for by domestic supply.

Britain was not cut off from the world; rather it was integrated with it in new ways. There was a dramatic east to west shift in origins of imports. In 1944 the proportion of imports from foreign – that is, non-Empire – countries was still the 60 per cent it had been before the war. But there was now practically no trade with Europe, while imports from the USA had ballooned. The USA quintupled its non-munitions exports to a remarkable 40 per cent of British imports. North America as a whole now provided 53 per cent of British non-munitions imports, whereas it had represented only 23 per cent in 1938. Imports from Empire hardly increased at all, with the key exception of Canada, which like the US was close to Britain.

There was also a dramatic change in the nationality of ships entering Britain. More than 44 per cent (see Chapter 2) of pre-war imports by bulk came in foreign ships. At the beginning of the war this proportion was to fall markedly, though from May 1940 many foreign ships would come under British control. Later in the war the most important change was the hugely increased presence of US ships in British ports, reflecting its rise to become the largest maritime power in the world. In 1938 only about 3 million net 'tons' (a measure of the cargo-carrying volume of the ship, always less than the gross tonnage) of US shipping entered British ports, a negligible fraction of the 94 million total, yet by 1944 it was up to 22 million net tons, over a third of the much reduced total of 57 million. The British-registered net tonnage entering fell from 50 million to 27 million.[10] British-controlled tonnage entering was higher: in December 1944, 418 large non-tankers arrived in British ports, of which 336 were under British control, while seventy-two were US-controlled, and of these latter ships sixteen carried British cargo, around 5 per cent of the total.[11] US vessels carried material for their own forces, as well as some of the Lend-Lease supplies for Britain.

The Liberty ship was central to the expansion and transformation of the Allied fleets. They were about the same size as existing cargo ships and tankers and can therefore stand for wartime shipping as a whole. Roughly speaking they could carry around 10,000 tons of cargo each. By the standards of a late-twentieth-century bulk carrier that is a small load, but it corresponds to some 300 heavy tanks, perhaps 3 million rifles, or 1kg of meat each for 10 million people, easily enough to keep London supplied for a week at more than the ration. In the whole of 1944 some 1 million tons of bombs were dropped on

Germany, a load which could have been carried by 100 Liberties – a task which required around 100 Liberties of aviation spirit to be shipped from the USA. In 1944 Britain imported 20 million tons of petroleum products, the equivalent of 2,000 Liberties-full – that is, an average of over five tankers arriving per day.

SHIPS

While the British-flagged and British-owned fleet fell during the war, the tonnage controlled by the British actually increased. In September 1939 the British-controlled dry-cargo fleet was 18.7 million dead-weight tons, overwhelmingly British-flagged and British-owned. By 1943 it had increased marginally, to 20 million DWT (that is, about 2,000 Liberties), but only about half was British-flagged and British-owned. The balance came from foreign-owned ships with British flags and crews, which peaked at 3 million DWT (ships did not generally go down with all their crew so crews were more available than ships), and chartered foreign-flagged and crewed ships (which peaked at 4 million DWT).[12] The great bulk of this additional capacity was European, from shipping organizations in exile. The biggest was the Norwegian Nortraship, the Norwegian Shipping and Trade organization, most of whose ships were chartered to the British. Indeed, Nortraship had British and US ships transferred to it, because it had spare crews saved from sunken ships. In the case of tankers British-controlled DWT increased from 5 million in late 1939 to no less than nearly 6.8 million by September 1941, largely through chartering of foreign ships, many Norwegian. At the end of the war the tonnage was comfortably above the pre-war level, though British-flagged tonnage was somewhat lower.[13]

At the beginning of the war a national focus caused problems. The Ministry of Shipping in the early days was made up of British shipping people who were experts in British ships and what they carried; timber came on foreign ships and was thus a 'pariah commodity', one whose significance was not understood. Furthermore while the Ministry of Shipping thought in tons, timber people thought in 'standards, fathoms, cubic feet and square feet'; the Ministry of Supply's Timber

Control had to succumb to an unfamiliar measure as 'the whole hierarchy thought in tons and compelled the Control to think in tons'.[14] But as the war progressed the issue became 'British-controlled' shipping, and later 'Allied' shipping. Indeed, many of the great battles over shipping to Britain were later necessarily to take place in Washington rather than London.

The stories of shipping capacity available to Britain and the British fleet were also rather different, and should not be conflated. The British 'merchant navy' as it came to be called during the war, replacing 'Merchant Service' or 'Mercantile Marine',[15] suffered very major losses, hence its share of world shipping and indeed shipping entering Britain fell. The merchant navy lost about the same number of British seamen as the RAF members of Bomber Command. The merchant navy was much smaller than the Royal Navy or the Royal Air Force, or needless to say the army, but it lost a higher proportion of personnel than any of the armed services. Its rate of sacrifice was beaten only by exceptional front-line parts of the armed services like Bomber Command aircrew. The losses of ships were significant too, with huge tonnages of expensive ships (and cargo) lost. Some 1,500 Liberty-ship equivalents, worth hundreds of millions of pounds, were lost by British owners. Replacements were on hand for many. For example, the Prince Line, of twenty pre-war ships, lost fourteen ships during the war, a net loss of four since it had three small and five large ships built for it in Britain during the war, and two in Canada. In addition the line managed at least eight new ships: one Empire ship built in Britain, two Oceans and five Sams.[16] The Blue Star Line lost twenty-nine ships, and had thirteen pre-war ones left at the end of the war. It lost all five of the South American passenger-cargo liners built in the 1920s. Three were lost returning from the River Plate: *Avalena Star* carrying meat and oranges; *Andalucia Star* with meat and eggs; *Avila Star* with meat. The *Almeda Star* was lost going to the River Plate. The *Arandora Star*, which had been refitted as a luxury cruise ship before the war, was torpedoed taking interned Italians and Germans to Canada: 805 passengers and crew were lost. But Blue Star also managed at least eight mostly refrigerated Empire ships, the *Empire Wisdom*, *Empire Might*, *Empire Camp*, *Empire Galahad*, *Empire Glade*, *Empire Highway*, *Empire Strength* and *Empire Talisman* (all survived the war, though two were damaged).[17] The Shaw Savill Line operated a total of thirty-one ships of which it lost fourteen; it started the war with twenty-four ships and ended with seventeen. It got five newly built ships: two Sams, the *Empire Grace* and the *Empire Hope*, and one built on its own account to the same design. It also got back three ships it had owned before the war as Empires.[18] The British Tanker Company (Anglo-Iranian) entered the war with ninety-three tankers (1 million DWT), and lost forty-four plus six they managed; they added around twenty British-built ships to the fleet during the war.[19]

The danger faced by merchant shipping led to many different sorts of defence measures. The ships were demagnetized, as a counter to the magnetic mines the Germans laid at the beginning of the war. A magnetic mine was recovered from the mud of the Thames Estuary in November 1939 and defused. Magnetic mines were well-known to the British – indeed they had been invented by the Admiralty, but Admiralty scientists needed to know in detail how the German mine worked to devise countermeasures. In what was soon trumpeted as the first great move of the scientific war the Admiralty scientists did just this – devising means of exploding the mines at a distance with magnetic sweeps from minesweepers, and then aircraft, as well as protecting all ships. This was done by a procedure known as degaussing (the Gauss was the unit of magnetic field strength) by which entire ships were fitted with electrical coils to counteract the existing magnetic field.

Many, though not all, ships sailed in convoys protected by naval ships. Ships were themselves given protection, first sand-bags and then concrete were added, to no great effect, followed from the end of 1940 by the extraordinary plastic armour (see Chapter 8). They were given machine guns, rocket-powered anti-aircraft devices, and in some cases heavy guns. They were painted grey and major efforts went into the tedious but important task of smoke reduction. Smoke could be seen from many times further away than the ship producing it. From the height of a submarine conning tower a ship was visible twelve miles away; its smoke would be visible at eighty miles.[20] One measure taken from 1942 was, at great cost, to supply the right types of coal in bunkering stations; types which depended on the kind of furnace in use. Another was to instruct the stokers and engineers in traditional naval practice of little-and-often stoking; a third to increase the forced draught, the flow of air into the boiler, at appropriate moments, with a magical device called a smoke eliminator. Later a similar device for ships without forced draught was developed and was used with a system of mirrors that allowed stokers to see what was issuing from their funnels. A vast programme of installation, conversion and training was undertaken, including the translation of material into Urdu, Gujarati and other languages of lascar seamen.[21]

Defensive measures were immensely costly. Convoying slowed ships down, and it concentrated port activity in time, increasing turn-around times for individual ships. The closure of the east coast ports for fear of bombers attacking ships in the North Sea and then their operation at much reduced capacity was of very great significance. London had been the world's largest port. Imports, and dockers, were shifted to the west coast, to ports like Glasgow, Liverpool and Bristol, which became busier than ever. More than that, specialized equipment had to be built in the west to replace capacity in the east. This applied, for example, to facilities for the import of petrol. The shift to west coast ports added hugely to congestion not only in these ports but also in the inland transport system, in which railways were central.

Many different factors caused shipping capacity to fall in the beginning of the war. Losses at sea, while significant, need to be kept in proportion. In 1941, a bad year, 5 per cent of British food imports were lost at sea.[22] The loss rate in convoys between November 1940 and June 1941, a bad period, was 3.4 per cent, while that in a later very temporary bad patch – March 1943 – was 2.7 per cent.[23] These rates were comparable with loss rates of aeroplanes on bombing raids, though they applied to voyages of weeks, rather than flights of a few hours. 677 shiploads got to Russia by the North Cape; 90 per cent of the cargo got through.[24] In late 1942 and early 1943 around half of all North Atlantic convoys were attacked, but the number of convoys which lost one or more ships was usually significantly less than one in three. There were only eight months in the war with losses to U-boats (loss and damage) of over 100 ships: seven were in 1942, the other was the blip in March 1943.[25] The March 1943 losses were concentrated in mid-Atlantic but the great 1942 losses were much further south and west, off the US coast, in the Caribbean and off the north of South America. These arose because US ships were now targets and because ships were not adequately convoyed and protected in these seas. To cap it all there was a 'Great Blackout' of signals intelligence – Bletchley Park could no longer read U-boat traffic – between February and December 1942.[26]

It is noteworthy that on the main North Atlantic convoy route losses fell continuously through the war. One in 181 ships was lost in 1941; 1 in 233 in 1942; 1 in 344 in 1943, falling to less than 1 in 1,000 in the latter half of the year.[27] The deadliest bit of ocean was the U-boat operational area AM, to the west of Britain – 294 ships were lost to submarines here, mostly early in the war. Fewer than twenty were sunk in 1943 or later, and few in 1942. The next most deadly area was AL, to the west of AM, where 229 ships were lost, though only twenty-four in 1943 and none later.[28] In these waters the decisive victories against the U-boat came early, as Churchill suggested.

For some of Britain's most important routes losses hardly came into the picture. The Mediterranean, deemed too dangerous for British merchant traffic between 1940 and 1943, was largely closed. Instead ships took the much longer route to Egypt, indeed to India and Australasia, via the Cape. The main supply route for the expanding forces in Egypt from Britain via the Cape was the same distance as the journey to Australia and New Zealand – indeed Egypt was now closer to Australia than to Britain. On these routes losses were minimal. That Britain could supply a whole army in a place so far away as Egypt is a testament to the extraordinary strength of British sea power. Yet it needs to be remembered that these long journeys meant a serious reduction in the quantities which could be carried with a given number of ships. The closing of the Mediterranean and the loss of European supplies in themselves had a huge effect on British cargo-carrying capacity.

TROOPERS, REEFERS AND TRAMPS

British shipping needs to be thought of globally, not merely in terms of supplies into Britain. British merchant ships had multiple uses and operated all over the world. British passenger liners, together with those which came over from conquered territories, gave the British the ability to move troops across Australasia and Asia and from Britain to the Middle East. The 'monsters' like the *Queen Mary* and the *Queen Elizabeth*, the new *Mauretania, Aquitania, Ile de France* and the *Nieuw Amsterdam* were in a class of their own. These were the six over 35,000 'tons', and the two largest, the *Queens*, were gigantic, costing as much as a battleship. These and other ships would become vital in transporting USA and Canadian troops to Europe. The two *Queens* could make three transatlantic trips a month, carrying around 15,000 troops. Between them they carried some 425,000 US troops to the United Kingdom, nearly one quarter of the total. Many other British and British-controlled passenger liners were also involved.

A second class of ship used for trooping was the distinctive refrigerated cargo liners. Reducing the meat ration rather suddenly in January 1941, the Minister of Food, Lord Woolton, asked the public: 'Would you rather have a little less meat . . . or would you rather have Bardia?', referring to a town recently captured by British forces as they advanced into the Italian colony of Libya. As he explained to the House of Lords:

> Ships were required of a certain size and of a certain speed in order to take troops to the Mediterranean. Those ships to a considerable extent were ships that had been accustomed to bring meat to this country, and we had to face the question whether in war we would pursue the military effort and have a little less meat. The Government decided that they would take the ships, and it was because they took those ships that we have had those grand victories. When I told an audience that at one time we had the choice of beef or Bardia, they thought I was joking, but I was telling the quite simple truth, and I think the choice that was made was the correct choice. So long as that Mediterranean campaign goes on, I see no possibility of our getting those ships back, and I see no possibility of our increasing the meat ration from its present position of 1s. 2d.[29]

Indeed, refrigerated liners did play a key role in transporting both meat and troops. The Nelson Line's five Highland ships all became troopships, one of which was lost in 1940. The Shaw Savill Line had seven of its twenty-six ships taken to be troopships. The young J. G. Ballard sailing to England in 1945, with a thousand other British internees of the Japanese, remembered the decks and holds of the *Arawa* lined with piping.[30] The gigantic *Dominion Monarch* was requisitioned as a troopship in August 1940 and altogether carried more than 90,000 military personnel and over 70,000 tons of cargo (including 51,500 tons of butter, cheese and meat) between Australasia and the UK, covering 350,000 miles.[31] Indeed, 'As far as possible the troopships retained their insulation and refrigerating machinery in good order and, Service requirements permitting, it was often possible to use these vessels on the homeward journey to pick up cargoes of foodstuffs.'[32] Thus troops were transported out to Egypt and on the return journey the ship could stop off and collect meat in Buenos Aires. However, pressure of trooping requirements in 1942 led to a decision to reduce imports from Australia and take more from the USA. Precious refrigeration ships were diverted but unfortunately wasted as additional meat was not available in the USA. The plan broke down in the early months of 1943.[33]

The shortage of British shipping had global and often terrible effects. For example, the British masters of the sea decided not to supply the shipping that might have averted the Bengal famine which took millions of lives in 1943. Parts of the world which exported food were sometimes no longer able to on the same scale – Britain no longer imported bananas, and less sugar, with serious consequences for Caribbean producers. These places, like many others exporting cash crops, suffered particularly in that they relied on food imports which no longer arrived so easily. All over the Empire there was a push for the diversification of agriculture to ensure local food supply. Little wonder that a British junior minister in charge of the economic department for the colonies stated that 'The farmer is the best ship-builder.'[34] But all over the world shipping restrictions imposed rationing of goods and food. Petrol was rationed in many parts of the Empire, from Nigeria to Australia. Food was rationed in the Middle East. Meat was rationed even in meat-exporting areas, to maintain exports to Britain and elsewhere, as was the case, for example, in Australia from 1944.[35]

FOOD

Long-established accounts of British food in wartime stress that it was rationed, that it was in short supply and that the health of the people improved.[36] However, most food was not rationed, most was in plentiful supply and health improved because more people could afford more food. In terms of calories only one third of food supply was rationed.[37] Rationed food accounted for only a quarter of food expenditure by 1940, and a third from 1942, with a further quarter or so controlled by points and in other ways.[38] Basic foods like bread, potatoes and vegetables were plentiful and, as the older histories acknowledge, 'There was plenty of food; at least there was plenty of bulk.'[39]

Elsewhere in the world the position was different. On the Continent, staples like bread and potatoes were rationed: in France at around 2kg of bread per week, and 4–6kg of potatoes per month.[40] In the East the Nazis planned for the systematic starvation of millions and millions of people. In the wake of the invasion of the USSR, Winston Churchill's propagandistic enthusiasm turned out to be prescient: 'So now this bloodthirsty guttersnipe must launch his mechanized armies upon new fields of slaughter, pillage, and devastation. Poor as are the Russian peasants, workmen and soldiers, he must steal from them their daily bread. He must devour their harvests, he must rob them of the oil which drives their ploughs, and thus produce a famine without example in human history.'[41] Famine was indeed the result. But it was not the only place where the increased consumption of armies meant famine. Although where the British army operated abroad there was often shortage as a result, famine was rare, with the great and important exception of the Bengal famine of 1943, when millions perished.

In Britain rationing was restricted to a few foods. They were industrial foods and largely or completely imported: meat (but not fish), cheese, butter, margarine, cooking fat, sugar, sweets and tea. These were not all luxuries in Britain, where, for example, sugar was consumed in very large quantities in many forms. But one important aim of rationing was to prevent the large rises in imported food consumption which would have followed from rising wartime employment and wages. The rationed items were exactly the kinds of food a more affluent population would turn to.[42] Rationing was not merely a cap on supply, but was meant to provide a guaranteed minimum supply, and that minimum was usually generous. For rationing did not imply drastically cut supplies: the worst cuts, of 30 per cent, applied to sugar. The suggestion that 'in mid-1941 the weekly ration of certain basic foods' amounted to no more than a 'single helping' in a comfortable pre-war household, while perhaps not wrong, is highly misleading in that rationed foods were not 'basic', and because the ration was more than a single daily helping for much of the population.[43] Rationing, price control and full employment ensured that the middle-class diet lost calories and proteins, but that for the working class (the largest group by far) calorie intake fell only a little, and the protein intake was stable or up. The result was the equalization of the diet at this chemical level of analysis so characteristic of the era.

Some groups – children, nursing mothers, the infirm – had special supplements and concessions. There was, for example, a huge increase in school meals during the war, and in the supply of milk to schoolchildren. Pregnant women were also getting milk preferentially.[44] Yet large numbers of particular classes of healthy adults had access to additional supplies. This was true to such a great extent that the idea of rationing as imposing an equal distribution of rationed foods looks distinctly strange. Many factory workers could get additional quantities of otherwise rationed foods in canteens, just as the well-off could go to restaurants.[45] A Mass-Observation reporter noted of the EKCO radar factory in Malmesbury that the women workers there ate a great deal while at work – tea and cheese rolls at ten, 'a good helping of roast meat and vegetables, followed by pudding, for under a shilling' at one, and a high tea of something like cheese pie or fish cakes at six. This was usually preceded by a hot breakfast at home, and followed by a hot supper at home, and supplemented in most cases by cakes and sandwiches brought from home. It seems meat was usually available; if not it was replaced by proteinaceous beans.[46]

Secondly the millions of servicemen and servicewomen, even if billeted with civilians, were on a quite different regime from civilians.[47] Their food allocation, known as the Home Service Ration, was very generous and included significantly larger quantities of foods covered

"nous vaincrons parceque nous sommes les plus forts"

SOUSCRIVEZ AUX BONS d'ARMEMENT

France, March-June 1940) in a poster of early 1940. His optimism was well-founded

2. Empire and Modernity: the Glasgow Empire Exhibition, 1938, which highlighted the modern industries of Britain and the Empire

3. A snapshot of Britain's global shipping distribution

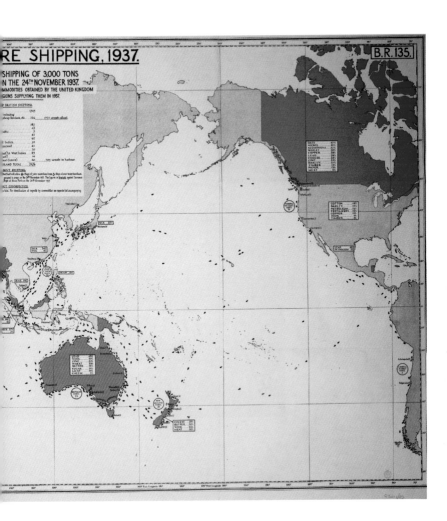

4. The MS *Dominion Monarch*, the largest motorship in the world just before its maiden voyage to New Zealand. As well as passengers she carried refrigerated food

5. A Liberty ship. Ships of this type were made, to a British design, in the USA, Canada and Britain in such large numbers that they transformed the world merchant fleet

6. Imperial production: Filling shells in India

7. In 1940 Britain wanted to show that it was not alone

8. Churchill at the first test of his great trench-cutting machine, NELLIE, November 1941

9. A Z-battery of anti-aircraft rockets, a project strongly supported by Churchill

10. British tanks in 1941, a year in which Britain produced more tanks than Germany

MIGHTIER YET!

In ships, tonnage and gun-power, the Navies of the British Empire are the most powerful sea force in the world — and another million tons of British warships are building

11. An accurate propaganda poster of late 1940 alluding to the words of the Edwardian jingoistic hymn 'Land of Hope and Glory'

12. Berlin or Bust: 'Tanks for Russia' week, September 1941. A Valentine tank on its way from a British factory to the Soviet Union

13. A 1942 poster on production, the great theme of the politics of that year, together with calls for a Second Front

by the civilian ration. At the beginning of the war male servicemen at home and not on active service had a daily ration of 12 ounces of meat per day, cut in December 1940 to 10 ounces, and in March 1941 to 6 ounces, where it stayed more or less to the end of the war. In addition there was a ration of bacon and offal or sausages.[48] For civilians meat was rationed by price not quantity, but one estimate of the average consumed in weight is 19 ounces per week, that is to say 2.7 ounces per day, less than half the much reduced Home Service Ration.[49]

Army catering was transformed during the war. For example, in 1941 an Army Catering Corps was formed under the Royal Army Service Corps and brought in a lot of civilian catering expertise, led by Sir Isadore Salmon, chairman of the Lyons caterers and teashop operators and a Conservative MP. The British armed forces were themselves part of the global British cold store chain. In the Middle East there was large-scale building of cold stores for meat, which was then taken frozen to advanced bases by refrigerated rail car, ship or lorry.[50] The army had a standard design of refrigerator unit which could be hooked up to twenty insulated containers each carrying 1.75 tons of meat. They were roughly cubical with sides about 7ft square. Five could be carried on a railway flat car, and one on a 3-ton truck. The same refrigerating unit could be attached to a 50-ton static cold store, or a 50-ton refrigerated hold on a ship. Smaller insulated containers were available for mules and men to carry.[51] In North Africa, for example, the aim was to provide, 'fresh food, particularly meat and bread', as far forward as possible, and this was very far forward: 'there was never any real lack of food in the Middle East [for British troops]' and 'at no time did stocks of frozen meat run out'.[52]

Apart from the steady supply of food to be cooked in field kitchens, the army started providing 'composite rations' for the field, that is to say a complete ration for a day for a small number of men, packed in one or a few cases, to be prepared by the men themselves. This was first tried in Norway in 1940, and by the end of the war some 40 million of these ration packs were made. The greatest achievement was apparently the fourteen-man pack, which replaced the earlier twelve-man pack. Some 10 million fourteen-man packs were produced between 1942 and 1945, in seven varieties, which included puddings.[53] Canned meat of one sort or another was standard. Smaller packs were made for tank crews. There were also twenty-four-hour rations for individual soldiers.

Of course, it is not surprising that a rich nation at war with reasonable control of the sea and access to finance should be able to feed its people and forces well or indeed discriminate strongly in favour of the armed services and key industrial workers, between them around 20 per cent of the population. What is surprising is the image of privation, equality and the near universality of rationing that prevails.

FEEDING THE NATION

But how exactly was this achieved? The answer is by adapting to reduced shipping space by reducing the bulk of imports.[54] The method was well-known, indeed obvious.[55] It was to substitute bulky imports with home production where possible, and with less bulky imports when not. Indeed, while bulky imports fell, the import of nutritionally dense foods increased. British consumers were exhorted to eat potatoes, instead of bread, some of which was made from imported wheat or flour.[56] Ship-saving did not mean that domestic production became more significant across the board.

It has been routinely stated, from wartime propaganda films directed at the USA to historical works, that the British 'cut down their food imports by a half'.[57] This is strictly true, but highly misleading. Of the wartime fall of 11 million tons of food imports, leaving another 11 million tons or so of imports, no less than 7.4 million tons was animal feedstuffs, including that element of wheat (wheat offal) that went to animals; most of the rest was maize. As a result the British animal population, with the exception of dairy cows, fell. The story of human food imports was very different. It fell overall by only around one quarter. But not all types of human food imports fell. Imports of grains and vegetables fell, as they were easily replaced by increased home production, 'ship-saving food' as it was called. Fruit (mainly apples, bananas and oranges) and vegetable (mainly potatoes, onions and tomatoes) imports, which were very substantial before the war, fell to practically nothing.[58] However, to a considerable extent, decreases in imports of food for animals were replaced by increased imports of

Table 6.2 Selected food imports, millions of tons

	1938	1940	1942	1944
Wheat	5.100	5.800	3.500	2.800
Barley	0.990	0.460	—	—
Maize	2.900	2.100	0.100	0.100
Beef	0.613	0.566	0.489	0.352
Lamb	0.274	0.291	0.326	0.282
Tinned and canned meat	0.073	0.123	0.282	0.213
Total Meat	1.500	1.400	1.600	1.800
Cheese	0.146	0.156	0.315	0.251

Source: CSO, *Statistical Digest of the War* (London, 1951), Table 147.

animal products, from meat to cheese, despite butter imports (Denmark and Holland had been important suppliers) falling, while lard imports increased. It was clear that 'considerably more first quality protein is imported than before the war'. The calorific value per ton of imported human food increased 25 per cent over the pre-war figure, a measure of the effect of a constraint on bulk and no constraint on cost.[59]

The quantities and the proportion of meat and cheese from abroad increased during the war. As domestic production fell, cheese imports increased markedly, meat imports less so. In 1944 meat imports amounted to 1.755 million tons (excluding poultry and game), compared with 1.507 million tons in 1938.[60] Imports of lamb and mutton, from New Zealand mostly, increased, as domestic sheep flocks fell. Imports of bacon and ham remained roughly constant, but those of pork and other pig imports increased hugely, again to compensate for falling domestic production. Canned meat imports increased very significantly too. Beef and veal imports, by contrast, fell, while domestic production increased slightly as a by-product of increasing milk production.

The quality – in the sense of taste and texture – of some imported

food fell notably. The need to ensure long life and more compact storage and to maximize production compromised quality. Frozen beef
kept indefinitely, but it was much less appetizing than perishable and
space-consuming chilled beef. The elimination of the chilled trade
increased the volume which could be carried by 30 per cent. The
reason was simple – frozen meat was simply piled up in holds, while
chilled meat was hung.[61] On the first day of the war 'wireless orders
were sent to all ships at sea that immediate steps were to be taken to
freeze down any meat which was then being carried in a chilled condition.'[62] Cheese 'was the only cargo which the authorities still
permitted to be carried chilled'.[63] The import of chilled beef did not
resume until the 1950s. More-compressed methods of packing frozen
meat themselves could cause quality problems. Mutton and lamb,
previously packed whole, were cut in half and telescoped, such that
the legs were put into the body cavity. Beef was boned, which caused
particular difficulties as warm air pockets were left that affected
the thoroughness of the freezing.[64] Quality was degraded in other
ways. The proportion of inferior canned meat increased. Corned beef
imports were stepped up (largely for troops), as was the import of
tinned steak and kidney pie, meat and vegetables, etc., much of it sent
direct to concentrations of troops. Corned beef, which once came in
various grades, now came in one low-quality grade, which allowed
the maximum possible output. War stimulated experiments and development of other forms of preserving meat. Dried American eggs
replaced fresh Dutch eggs, just as Spam substituted for Danish bacon.
Margarine, to a very great extent, replaced butter.

One type of high-value food which came in by sea was badly
affected by the war: fish. Supplies halved.[65] Many of the most modern
Hull and Grimsby trawlers were taken over by the Admiralty for use
as minesweepers, patrol vessels, etc., usually with fishermen in the
reserves as crew. Older vessels, crewed by those often too old or too
young for naval service, continued to fish, though many were transferred from east coast to west coast ports. Furthermore some
continental fishermen made their way over with their boats – for
example, Belgians to Brixham in Devon, and Danish fishermen to
Grimsby and Whitehaven. A moving film was made of Dutch *Fishermen in Exile* (1944), pointing out that they fed the British people,

while their families in the Netherlands went without.[66] Fish were not rationed because the supply was very volatile – fish could not yet be effectively frozen.

DOMESTIC AGRICULTURE

There was a remarkable expansion and change in British agriculture. This had been called for from both the left and elements of the right in the 1930s: 'science can make us largely, and perhaps entirely, independent of imported food, and can also raise the health of the people ... The Government must mobilize science for this great triumph,' claimed scientists of the left.[67] The new atmosphere is conveyed well by an official pamphlet from 1945 on wartime agriculture, which oozed contempt for pre-war farming, when farms were 'allowed to slip back from cultivation to ranching', despite the lushness of the land and the immemorial skills of the British farmer; farmers had forgotten how to use the plough, and turned to 'livestock farming on the ranching system' where in some cases fields were merely exercise yards for cattle who turned 'raw material from abroad into milk and meat'.[68] The war and the future would be different.

Domestic wheat production doubled in two years and by 1943 there was more home-grown than imported wheat (which fell 40 per cent). Potato and barley production both nearly doubled. British gross calorie output increased 55 per cent (with a much lower increase in inputs); net of imported inputs, calorie output increased 91 per cent; that of proteins, 106 per cent.[69] This was no scientific miracle, but the result of a simple expansion of acreage and other inputs, and a change in crops. Land yields increased only a little, and labour productivity changed to an unknown degree. The point was to get food, and that meant maximizing yields by choosing the most economic crops.[70]

In wartime, the inputs into agriculture were generally increased (with the important exception of the near elimination of imported animal food). The amount of land used increased with pasture and unused land ploughed up for planting. The number of farm workers jumped 20 per cent from 806,000 to 981,000 between 1940 and 1945.[71] Though the agricultural horse population, still comparable with the human labour force, fell from 642,000 to 545,000 in the same period,[72] the number of tractors jumped rather suddenly, rising from around 50,000 to 150,000.[73] The Ford works at Dagenham supplied the great majority of pre-war and wartime tractors. Production had been around 10,000 a year, mostly for export; during the war exports ceased and production nearly trebled.[74] Only the Fordson tractors, the Model T of the tractor world, and of similar vintage, were made, machines no longer produced in the USA or the USSR. This old-fashioned machinery allowed some 2–3 million acres to be switched from producing fodder for horses to producing food for humans.[75] On the down side, tractorization implied greater oil imports, and indeed the use of other imported materials like iron and steel, which could have been used elsewhere. On balance, however, it was deemed a worthwhile change, another indication of control of the seas.

The one kind of animal husbandry which grew during the war was dairy farming. The production of liquid milk was seen as a more efficient way of producing food than making meat, though meat was also got from dairy cows. Problems came with using the low-quality home-grown fodder which replaced imported feed. Leading vets made cow productivity central by proposing themselves (a profession hitherto concerned primarily with horses and a few other valuable sick animals) as providing a health service for cows. They wanted to train vets in Danish techniques for dealing with what they called 'temporary infertility' in order to increase milk output. This bovine health service reached only 10 per cent of the dairy herd.[76]

Standardization of high-output processes, in larger plants, became central to food processing. Domestic slaughterhouses and cheese production facilities began to look like those of the River Plate and Australasia. Around 16,000 pre-war slaughterhouses were reduced to 600–800, of which over 100 were municipal, following the introduction of meat rationing and bulk buying of livestock by the Ministry of Food, in January 1940.[77] One great advantage of concentration was the greater recovery and use of offals (which were not rationed).[78] After the war the number of slaughterhouses remained low.[79] In cheesemaking there was a dramatic shift to factory production as a result of a ban on farm-based production. Industrial 'creameries', as the cheese-plants were known, made standardized types of cheese, with cheddar increasing its dominance. It took decades for artisanal cheese-making to re-establish itself.[80] Flour, which once came in various grades, was from the beginning of the war nearly all straight-run, as it is still called; that is, made in one run through the mill without regrinding. The making of different grades was outlawed.[81]

It wasn't just British agriculture that was affected by the war. Britain had a million troops in the Middle East. At first they were largely supplied from overseas, but the exploitation and development of local food supply for the troops was later organized on a very large scale. It is claimed it saved 1 million tons (more than 100 Liberties) of shipping in the year to March 1943, that is to say, more than was saved by the prohibition on white flour imposed in Britain in March 1942. Perhaps the most important source of saving, according to the history of the Royal Army Service Corps, was the growing of potatoes in Egypt and Syria, potatoes 'without which the British soldier does not consider himself properly fed'. Local production of cooking oil, oil cake and soap, was undertaken, dairy farming and fishing were promoted, as well as jam- and marmalade-making, and dehydration and canning of vegetables. Cattle were imported on the hoof from the Sudan, saving refrigerated space.[82] The economy of the whole area was profoundly affected by the shortage of shipping and the closing of the Mediterranean to British and Allied traffic, and the lack of exports from Britain. Shortages of all sorts appeared, made much worse by the purchases of the British forces. Shortage of imported fertilizer became a key issue.

To try and improve matters, an extraordinary organization, the Middle East Supply Centre, first British then Anglo-American, was created. This coordinated civil imports and promoted local supply. For example, it shifted Egyptian railways from British coal to local oil burning; it advised on schemes for food rationing, widely imposed in the area. The forms of rationing imposed were quite different from the egalitarian sort found in Britain – here it was based on what households already consumed, not what they needed.[83] British planners, advanced farmers, agricultural economists and biologists were taken out. Here in a new economic empire forged in war, planning and modernization were imposed by British technocrats, indeed overseen by a British government minister resident in Cairo. The Middle East

Supply Centre was a pioneering transnational body concerned at least in part with the economic development of the poor world, a harbinger not only of the British technocratic imperialism of the post-war years focused on Africa, but of development initiatives more generally. It also seemed to pioneer new forms of transnational organization.[84] The very term 'Middle East' in its modern sense was created: it came to mean roughly the area covered by the new Middle East Command based in Cairo, and the associated Middle East Supply Centre: North Africa, Greece, the Levant, and territories all the way to Persia.[85] Similarly, South-East Asia probably did not exist as a geographical term before the creation of South-East Asia Command in 1943.[86]

EXPERTS VERSUS SHIPS

The issue of shipping was, as is clear, central to food supply policy. It determined much that happened, and was overall a more significant factor than the advice of nutritional experts. The object of food policy was 'to feed the nation efficiently with the minimum amount of shipping'.[87] What that meant was the subject of much wartime debate between experts, but in the final analysis the shipping allocations were the critical element.

For some of the nutritionists the vitamins and minerals were the key to analysis of diet, together with calories. Foods, and food policies, were thought of in these particular terms. Perhaps the most famous and influential dietetical expert was Jack Drummond, Professor of Biochemistry at University College London and from February 1940 the scientific adviser to the Ministry of Food.[88] In Drummond's analysis of the war food situation in May 1940 there was no sense of crisis or of a major shipping shortage. The theme was the need to import calories, 'energy foods' as he called them, notably by increasing wheat imports, and to increase home production of 'protective foods' (foods which, in his understanding, provided vitamins and calcium), like vegetables and milk. He stressed the importance of vitamin and calcium supplies, especially for the poor. His May 1940 report envisaged major reductions in meat imports and significant increases in the supply of wheat, oats and milk.[89] In fact this plan was, as far as imports were concerned, soon null and void.[90] The shortage of shipping led to rapid cuts in wheat imports and as we have seen meat imports increased. The nutritionists did not write the script of wartime nutrition.

The strange and convoluted affair of the extraction rate of flour for bread shows the complexities of the relationships between shipping, the programmes of experts and the politics of morale, and indeed the politics of food more generally. It provides a fascinating case of the tensions between narrow nutritional and broader conceptions of the healthiness of food. Flour is made by mechanically milling wheat to different degrees, the higher the extraction rate, the browner the flour. In the case of wholemeal flour all the grist was milled to flour, giving an extraction rate of 100 per cent. Wheatmeal flour was milled to retain some 85 per cent of the grist. White flour represented an extraction rate of 70 per cent. The balance between the grist and the flour was known as 'wheat offal' and was used for animal feed. For some nutritionists there were powerful arguments in favour of moving from the typical white flour/bread consumed in Britain to browner varieties, both wheatmeal and wholemeal. Raising the extraction rate produced flour and bread with more nutrients, and also reduced the amount of wheat to be imported. Indeed, at the beginning of the war, the minimum extraction rate was raised from 70 to 73 per cent (it was raised to 75 per cent in 1941).[91] But it was the shipping shortage of early 1942 which was critical: the decision to increase the extraction rate in 1942 was a shipping not a nutritional one.[92] Britain's bread was henceforth brown. Towards the end of the war, the rate of extraction was reduced, only to be increased again after the war, at one point to 90 per cent.[93]

For some nutritionists the question of 85 per cent extraction was a *cause célèbre*. 'We still have white bread . . . instead of highly extracted bread enriched with yeast and added vitamin concentrates,' a group of radical scientists complained in mid-1940.[94] Later some lamented that, while called for by the scientists in 1940, it took nearly two years to institute a policy which was sensible from both a nutritional and a shipping point of view, saving hundreds of thousands of tons of shipping. They saw scandal (the influence of the millers) and ignorance holding back the tide of reason.[95] The saving in imports (even compensating for reduced 'wheat offal' with increased maize, bacon and egg imports) were considerable: annual savings were 400,000 tons a year (or forty Liberties) in imported grain.[96] Other estimates suggest 700,000 tons per annum.[97] The story was more complicated than these experts' morality tale. For example, as Lindemann claimed (see Chapter 3), there were bigger savings in shipping to be made in other areas. There were also contrary scientific views. One scientist complained that the bread was less digestible, others objected to the use of bread as an agent to supplement diets. Lord Woolton, the Minister of Food, was given contradictory advice.[98] John Maud, an official in the ministry, told a scientist that 'Many complaints and attacks' on the ministry were 'due to rivalry and spite between scientists'.[99]

Jack Drummond proposed increasing the extraction rate of flour, to increase the amount of vitamin B1 and of iron in the British diet. Against this was the argument that the public liked and was used to white bread, and that white flour kept better.[100] The result was a compromise in July 1940 in which it was decided to add vitamin B1 and calcium (also thought by the nutritionists to be underconsumed) to white bread, an idea supported by the millers. In addition it was decided that brown bread made with high-extraction flour (80 per cent raised to 85 per cent in 1941) would be available at the same price as the white loaf, as an alternative to branded but more expensive brown bread like Hovis. The addition of calcium to flour became a controversial issue, not least because of the discovery of an unexpected feature of brown bread. Research in 1940 by nutritionists at Cambridge who tested the calcium flows through volunteers eating different kinds of bread showed wheatmeal bread reduced calcium intake.[101] The Minister of Food and his scientists disliked the idea of adding more calcium to brown bread, partly because it implied that it was less healthy than white. The result was another compromise in which the same quantity of calcium (in the form of chalk) was to be added to both white flour and wheatmeal, even though the latter needed more. To complicate the story further, vitaminized white bread was widely available before white bread was banned. From March 1942 flour would have to have a minimum of 85 per cent extraction, and soon after the amount of calcium to be added was increased. Extraordinarily, specialist brown breads were not required to be fortified at all, even though some wholemeal loaves (of 90 per cent) needed even more calcium than wheatmeal loaves. The reason was reluctance to adulterate what was taken to be a purer bread by the community that bought it.

Nutritionists might hail fresh fruit and vegetables for their particular vitamin and mineral content and rail against industrialized food and specious claims made by manufacturers on its behalf. But they also favoured the fortification of industrialized foods with minerals and synthetic vitamins. The effects of war were seemingly contradictory in a similar way: there was greater consumption of rustic staples like potatoes and vegetables, while at the same time there was a marked industrialization of other foods. Drummond helped push the supply of milk and vitamin supplements to expectant and nursing mothers and children, and later the supply of concentrated orange juice. He was also keen on dehydration, and the importation of dried eggs, milk, and vegetables and soups.[102]

OIL

Pre-war Britain was a net exporter of energy, but the import of petroleum products was such as to make Britain the largest oil importer in the world. This was no surprise as Britain was the most motorized large country in the world apart from the US (a net exporter of oil). The scale of pre-war oil imports was comparable to the other great bulk imports – timber, iron ore, grain. As in most of these cases, Empire sources were of small significance. Imperial oil, mainly from Trinidad, but also Burma, accounted for only 8 per cent by value of imports.[103] Oil tended to come, however, from places which while not formally part of the Empire were under varying degrees of British influence. By the late 1930s Venezuela/Dutch West Indies was by far the largest supplier and together with the smaller neighbouring Trinidad accounted for over half of all imports.[104] Persia and Iraq were under direct British influence and were significant suppliers, as were the Dutch East Indies. Britain imported not crude oil (except in small quantities) but refined products. Its refineries were close to the oil fields. Venezuelan crude was refined in the Dutch colonies of Curaçao and Aruba; other important refineries were in Trinidad, Haifa (connected by pipeline to northern Iraqi fields), Abadan, Rangoon, and Pelambang and Balikpapan.

By the end of the war the oil position was to change very dramatically. Imports were to double such that at the end of the war Britain was importing 20 million tons of petroleum-derived fuels, much more than food imports. Most now came from the United States. The forces dominated consumption. Although the level of motor spirit consumption was about the same as before the war, private motoring was severely cut back, and the balance was taken up by transport and the services. However, for other oil products demand increased very greatly. The Royal Navy, which for decades had been largely steam-driven and oil-powered, was now consuming around 4 million tons per annum of fuel-oil (about the same as the total UK motor spirit consumption), some ten times its peacetime requirement. The Anglo-American air forces based in Britain were by 1944 consuming similar quantities of 100-octane aviation spirit, which from 1943 was produced to the same specification for both forces.[105] In May 1944 consumption was running at an annual level of 5.2 million tons divided about equally between the RAF and the USAAF in British bases alone. This may be compared to a total German supply of aviation spirit in all theatres which peaked at below 2 million tons in 1943, at a time when total oil supply was at its maximum at 5.6 million tons. The figures for 1944 were 1.1 million tons of aviation spirit and total oil supply of 3.8 million tons.[106]

Table 6.3 Imports of oil products to the UK (all users, including US forces), millions of tons

	1938	1944
Motor spirit }	4.699	4.773
Aviation spirit }		4.751
Admiralty oil	0.403	3.912
Gas oil/diesel	n/a	2.211
Total	11.618	20.344

Source: Payton-Smith, *Oil*, Table 52.

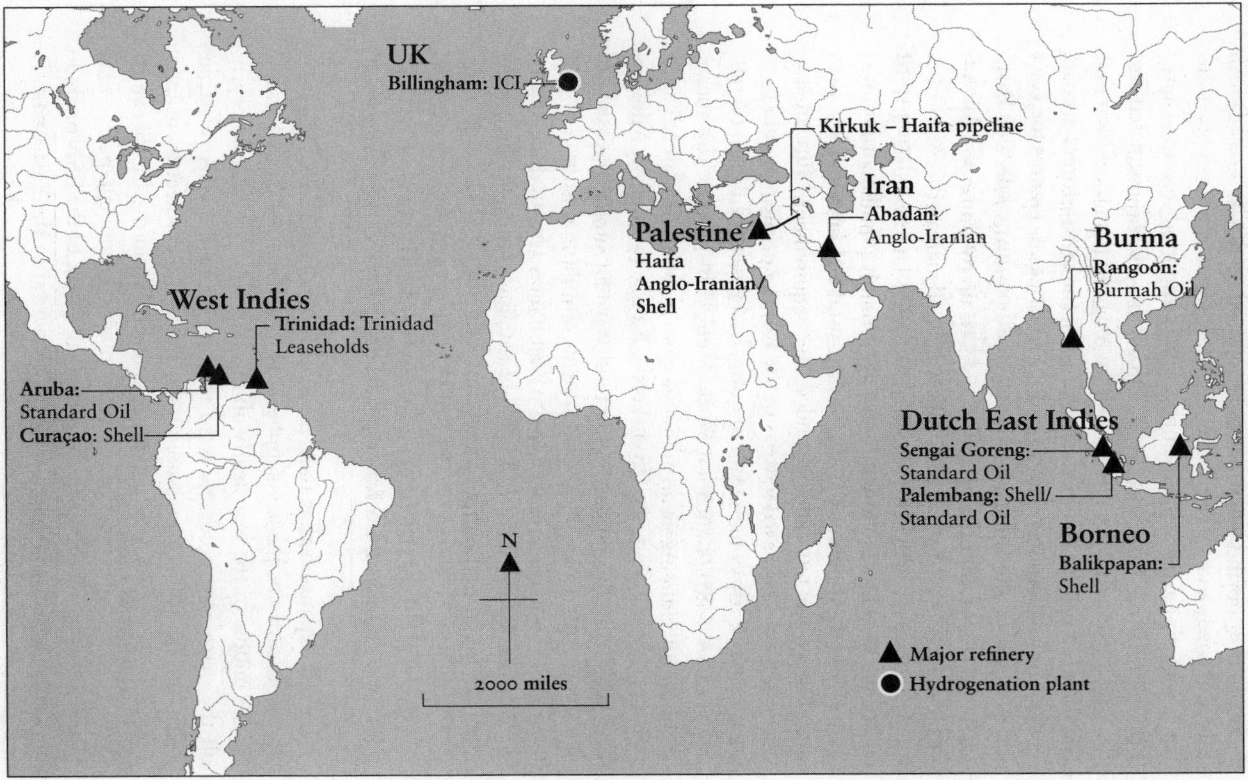

UK
Billingham: ICI

West Indies
Trinidad: Trinidad
Leaseholds

Aruba:
Standard Oil
Curaçao: Shell

Palestine
Haifa
Anglo-Iranian/
Shell

Iran
Abadan:
Anglo-Iranian

Kirkuk – Haifa pipeline

Burma
Rangoon:
Burmah Oil

Dutch East Indies
Sengai Goreng:
Standard Oil
Palembang: Shell/
Standard Oil

Borneo
Balikpapan:
Shell

▲ Major refinery
● Hydrogenation plant

N

2000 miles

British oil sources in 1940

Imports into the United Kingdom tell only half the story. British forces abroad, and in British imperial territories, were supplied directly, with about the same quantity as was imported into the United Kingdom. Thus for the last nine months of 1942 it was estimated that the United Kingdom needed 9.5 million tons and that 10 million tons were needed for the rest of the Empire and for British forces abroad, most to be supplied from Persia and the Americas.[107] Figures for 1943 are similar: UK consumption was 12.6 million tons while 'The East', including the Middle East to South and East Africa, India and Australasia, consumed a total of 14.5 million tons.[108]

In 1939–40 Britain did not intend to produce all its aviation spirit in Britain, or even in the Empire. But supply controlled by Britain was understood as crucial. In 1939, 'to have accepted anything less than absolute certainty, to have depended on the goodwill of foreign suppliers to meet the essential needs of the Royal Air Force, would have been a radical break with traditions that had governed British oil policy since long before the First World War'.[109] In practice this meant that aviation spirit would be obtained from plants on British or Dutch territory, but not from the US or Persia. From 1937 the Air Ministry was agreeing to buy large quantities of iso-octane (a key ingredient of aviation spirit) from Trinidad, Curaçao and Aruba; later it would include the Dutch East as well as the West Indies. The producing firms were not necessarily British: Trinidad Leaseholds was, but Shell, operating in Curaçao, was Anglo-Dutch and Standard Oil (Esso), operating in Aruba, was American.[110] The British state-owned Anglo-Iranian in Abadan was not part of the plan, for this was on foreign land. In 1940 additional supplies from Britain, to be discussed below, were seen as crucial.[111]

For all expansion that took place in British-controlled, including domestic, capacity, the key supplier was to become the United States. By the middle of the war nearly all the aviation spirit, indeed nearly all the petroleum products, were to come from the USA. First, in 1941, the US relieved pressure on the British tanker fleet by using a shuttle service of its own tankers to move oil from the Caribbean to New York.[112] New pipelines were to play a prominent part – first pipelines that took the spirit from the Gulf Coast of the USA (including spirit landed from the Caribbean refineries) to the East Coast. It was then shipped to west coast British ports, where it went by pipeline (like the American ones, newly built) to the places it was used, including aerodromes, British and American, around the country. From late 1944 some would continue by PLUTO pipelines across the Channel to France. Britain imported on Lend-Lease nearly all the oil consumed from the British Isles, yet a good quantity was notionally then given to US forces, accounted for as reciprocal aid. This came to £214m in the UK through the war, and £83m in overseas theatres (£60m in India).[113] On a different basis, it seems that petroleum on the UK government account (including that outside the UK) received $1.8bn and gave back in reciprocal aid $1.1bn.[114]

In aviation spirit the US achieved an extraordinary dominance, producing around 20 million tons of 100-octane per annum at the end of the war; while British operations in Britain and abroad managed about 2 million tons.[115] The investments in alkylation and related projects by the USA for the manufacture of 100-octane amounted to over $1bn, half the total investment and operating cost of the Manhattan Project. It was a gigantic effort, mostly in the US, but $260m was spent abroad, much of that in Abadan.[116] In 1939 lawyers for all the major patentees had signed cross-licensing agreements which made the processes workable, and a wartime agreement divided the alkylation royalties as follows: 24.6 per cent each to Shell and Texaco; 32.7 per cent between them to Anglo-Iranian and Standard Oil (New Jersey); and 18.7 per cent to Universal Oil Products, a Chicago-based research and development operation jointly owned by many oil companies.[117]

The most spectacular British developments came at the British state-owned Anglo-Iranian refinery in Abadan in Iran, which, like the US supply, had been shunned at the beginning of the war. This refinery became, unexpectedly, by far the largest British producer of aviation spirit, with a huge plant for iso-octane production using the alkylation process. The new plant used scarce US equipment and personnel and received the highest priority from the US as well as the UK; output would reach nearly 1 million tons of 100-octane per annum, for use in the area and in the Soviet Union.[118] It was shipped by sea, by pipeline and by train north to the Soviet Union. Britain's greatest refinery became a crucial part of a US-financed global war.

Table 6.4 Allied 100-octane production (excluding USSR), millions of tons

Place	Firm(s)	1942	1944
US	various	3.075	15.125
Aruba	Esso	n/a	0.505
UK	ICI/Trimpell/Shell	0.184	0.549
Curaçao	Shell	0.057	0.230
Trinidad	Trinidad Leaseholds	0.088	0.165
Abadan	Anglo-Iranian	0.258	0.858

Source: Payton-Smith, *Oil*, p. 384.

That the US would become Britain's supplier of oil and aviation spirit was unthinkable even in 1940. Great efforts were made to avoid US purchases. In fact Britain was investing a great deal in new capacity for aviation spirit, especially in Britain. In 1938 Sir Harold Hartley, chairman of the Fuel Research Board, director of research of the LMS railway, and a former academic chemist and chemical warfare expert, together with Shell, Trinidad Leaseholds and ICI, proposed an ambitious imperial scheme. It involved a consortium of the three companies, which was to be called Trimpell Ltd, to build and operate three plants, two in Trinidad and one in Britain. The idea was to use ICI's hydrogenation experience to hydrogenate base-oil made from petroleum (and *not* coal) to make 'base petrol' and butane gas. Using a process developed by Shell and Universal Oil Products, butane could be de-hydrogenated, polymerized, and then hydrogenated to make iso-octane. The iso-octane was to be blended with base-petrol and tetraethyl lead to make high-octane fuels.[119] Trinidad Leaseholds would supply the base-oil to all three plants, which would make a total of 720,000 tons per annum of 100-octane aviation spirit.[120] In 1940 it was estimated Britain would need, in the future, 850,000 tons of 100-octane fuel per annum. British-controlled resources, from plants in Britain, the Caribbean, Persia and the Dutch East Indies, would produce up to 500,000 tons, leaving a deficit of 350,000 tons.

The new plants, it was hoped, would avoid the need to buy from the USA.[121]

Work on the first hydrogenation plant, at Heysham in Lancashire, started in the spring of 1939.[122] It was to be the most expensive single plant built for the government by ICI – it cost about as much as a battleship. In July 1939 the scheme was altered to two plants, but both in the UK, with Shell building the second plant at Thornton, adjacent to its existing facilities at Stanlow on Merseyside. It was to use a different process for making iso-octane, the much more efficient alkylation process.[123] The two hydrogenation plants in Trinidad did not go ahead because of difficulties in construction.[124] However, there was already a small iso-octane plant in Trinidad and a large 100-octane-producing complex was working by 1941.[125] Among the paradoxes of this national-imperial policy was the fact that it depended in part on intellectual property from abroad. Before the war, there were 'intricate links' between the ICI experts in Billingham and those of IG Farben, the vast German chemical combine building many coal hydrogenation plants in Germany. The collaboration lasted, it appears, up to the outbreak of war.[126] In December 1938 and early 1939 the British government was contemplating paying £3m–£4m in licences to International Hydrogenation Patents, which included IG Farben as a key member.[127]

By April 1941 the position had changed radically. Lend-Lease had now abolished the dollar problem, and assurances about future supply were made by the USA. Furthermore increasing shipping constraints made importing aviation spirit rather than the greater quantity of raw material to make it much more attractive. The Shell Thornton plant then under construction would require an additional annual 185,000 tons to be shipped to the UK, additional to, that is, the aviation spirit it would produce.[128] In June 1941, once assurances had been received from the US about supplies, the order to stop work went out. The Thornton alkylation plant was shipped to the Shell plant at Curaçao, but it was not until the beginning of 1945 that it was fully working – the result of low US priority for refinery capacity in the area.[129] The same shipping logic threatened the further advanced Heysham plant as well. In shipping terms, it was argued, it would be much better to turn over the plant to hydrogenating nitrogen rather than base-oil, in order to make fertilizer, which would increase domestic food production and save even more shipping.[130] Heysham did in fact produce both. Production of hydropetrol started in July 1941 and of iso-octane in October 1941. In August 1941 it was decided to run Heysham as two thirds aviation spirit and one third ammonia.[131] In 1944 Heysham produced 344,000 tonnes of hydropetrol, 55,000 tonnes of iso-octane and 22,000 tons of ammonia, reflecting increasing demands for aviation spirit. To do this it needed 466,000 tons of imported gas-oil, 100,000 tons of coal, 204,000 tons of coke, and a large supply of electricity and water.[132]

Heysham was not to be alone in making aviation spirit in Britain. ICI Billingham started making 87-octane spirit from 1939 (from creosote) and built an iso-octane plant. From March 1942 its only output was 100-octane aviation spirit.[133] Shell Stanlow was the third domestic producer. Total British domestic production of 100-octane was over 500,000 tons at the end of the war. It was a higher total than envisaged in 1940, but a smaller fraction of consumption.

That Britain was producing aviation spirit domestically owed a great deal to the nationalism of the 1930s, and yet also revealed its limits. Imperial Chemical Industries built an oil-from-coal hydrogenation plant alongside its Teesside synthetic ammonia plant at Billingham. It started operating in 1935, supported by a large indirect subsidy.[134] There were pressures to increase production in a world becoming increasingly autarkic, the technical term for economic self-sufficiency, not least in petrol production: Germany had a huge synthetic petrol programme under way. In 1937 a committee under the 8th Viscount Falmouth argued it would be hugely expensive (needing a £5 per week per employee subsidy) and that the plants were more vulnerable to air attack than tankers, ports and dispersed storage. A plant of the size of Billingham, which produced 150,000 tons of petrol per annum, would cost the same to build as twelve tankers capable of bringing in twelve times that amount of petrol into Britain annually.[135] Through a neat coincidence, Billingham's annual production would fit into roughly twelve tankers. Despite its rejection of oil-from-coal in general, the Falmouth committee suggested using oil-from-coal processes as a supplementary source for aviation spirit.[136]

Getting all the motor spirit (half of net imports of all oil) from coal would mean building thirty plants the size of Billingham, at a cost of £160m, and, given the higher cost of production, the loss of another £44m in tax, when the total cost of imports of all oil was only the same £44m per annum. Making oil from coal would be, given a growing navy, a 'double insurance policy' which made no sense.[137] In energy terms, too, the economics were very poor – it took 6 tons of coal to make 1 ton of petrol, including the coal needed for the process itself.[138] Yet the pressure for oil-from-coal remained, especially from the right. Indeed, some hard-core naval nationalists argued for reversing the pre-Great War policy of oil-firing the navy, and for returning to coal.[139] The British Union of Fascists attacked the Falmouth committee, claiming that 'Again we find the profits of financial investment abroad put before the welfare of British miners and the safety of the British people.'[140] At the beginning of the war the Mines Department set up a committee under the President of the Royal Society, Sir William Bragg, to inquire into domestic production and this spawned six committees looking at various aspects of this complex subject. The war did see increasing use of creosote and benzole (both products of coal-tar distillation) as fuel oils and motor spirit, but nothing particularly significant.[141]

What if Britain had, like Germany, to rely very significantly on domestic oil from coal? Although coal was by far Britain's main supply of energy, ten times the bulk of peak oil imports, the cost of getting the oil from coal would have been prohibitive. It would have required a 50 per cent increase in coal production, which would have meant an additional 250,000 miners and a huge investment in plant. That in itself would have greatly reduced Britain's capacity to fight.

We might also imagine a Britain that fed itself during the war. This would have needed a doubling of the agricultural workforce, to perhaps 2 million, though it is not clear there was enough land for them to work. Some of this would be compensated for by the reduction of the merchant navy and the Royal Navy, which could certainly have replaced the miners by halving its size. But one million more agricultural workers would have had to come from a reduced army and air force. Assuming they came out of the army, that would mean an army of under 2 million men (assuming the additional agricultural workers were all men). That was not a force which could have had a great impact, even alongside an air force of one million.

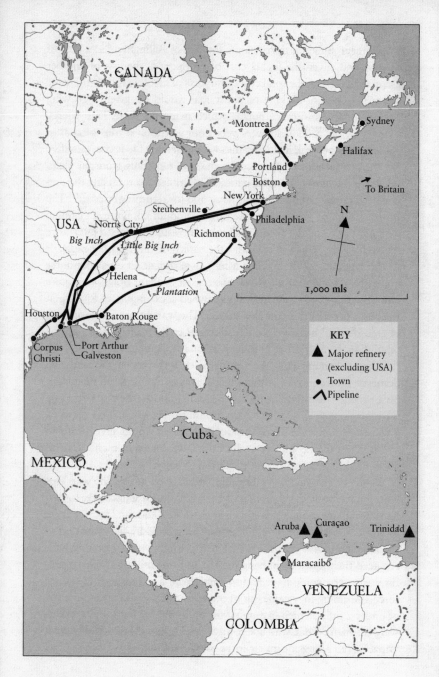

Oil pipelines supplying Britain by the end of the war

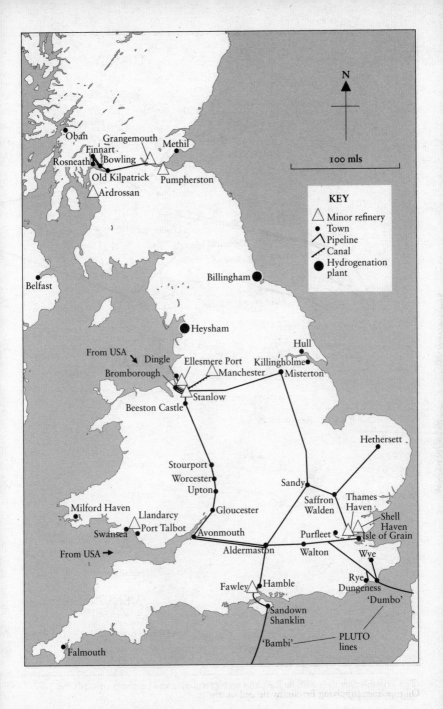

N

100 mls

KEY

△ Minor refinery
● Town
⋀ Pipeline
╱ Canal
⬤ Hydrogenation plant

Oban
Grangemouth
Methil
Finnart
Rosneath
Bowling
Old Kilpatrick
Pumpherston
Ardrossan

Belfast

Billingham

Heysham

From USA
Dingle
Ellesmere Port
Killingholme
Hull
Bromborough
Manchester
Misterton
Stanlow
Beeston Castle

Hethersett

Stourport
Worcester
Upton
Sandy
Milford Haven
Llandarcy
Gloucester
Saffron Walden
Thames Haven
Port Talbot
Swansea
Avonmouth
Purfleet
Shell Haven
Isle of Grain
From USA
Aldermaston
Walton
Wye
Fawley
Hamble
Rye
Dungeness
Sandown
Shanklin
'Dumbo'
'Bambi'
PLUTO lines
Falmouth

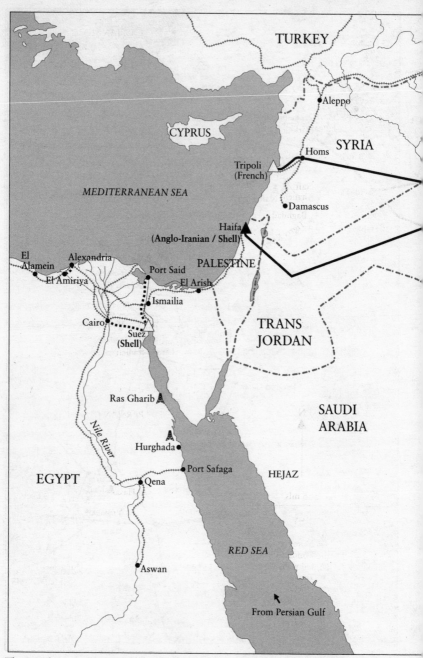

The British-controlled Middle East and its oil refineries and pipelines towards the end of the war

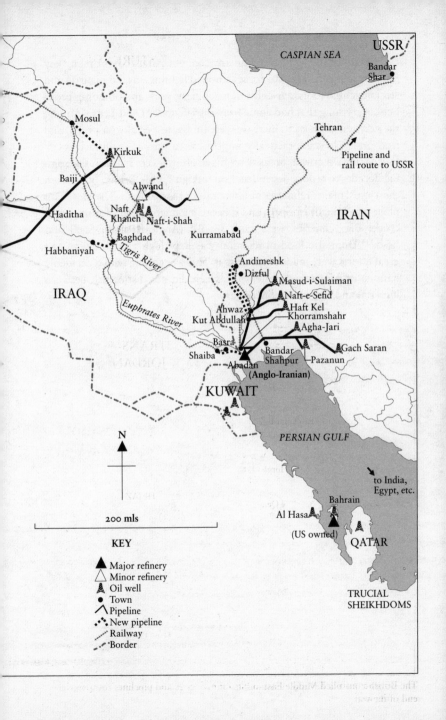

USSR

CASPIAN SEA

Bandar Shar

Mosul

Kirkuk

Baiji

Tehran

Pipeline and
rail route to USSR

Alwand

Haditha

Naft
Khaneh   Naft-i-Shah

Kurramabad

IRAN

Baghdad

Habbaniyah   *Tigris River*

IRAQ

Euphrates River

Andimeshk
Dizful

Masud-i-Sulaiman

Naft-e-Sefid

Haft Kel

Ahwaz
Kut Abdullah   Khorramshahr

Agha-Jari

Basra   Gach Saran

Shaiba   Bandar
Shahpur   Pazanun

Abadan
(Anglo-Iranian)

KUWAIT

PERSIAN GULF

N

to India,
Egypt, etc.

Bahrain

200 mls

Al Hasa

(US owned)   QATAR

KEY

▲ Major refinery
△ Minor refinery
⚒ Oil well
● Town
⌐ Pipeline
⋯ New pipeline
⁄ Railway
·–· Border

TRUCIAL
SHEIKHDOMS

The end of the war brought many changes. After the war many key imports fell: petrol (as the forces stopped fighting), as did pig iron and steel, machine tools, agricultural machinery, pork and other pig products, everything that had come from the US under Lend-Lease. Among the very few imports to increase were timber (especially pit props) and iron ore.[142] It was practically impossible to import manufactures.[143] Indeed, in two crucial areas determined efforts were made in the coming decades to reduce dependence on foreign supply. Britain built, from the 1940s, major refineries at home rather than abroad. The Heysham plant was bought by Shell, and it became a huge refinery for imported crude. Big refineries were built by BP (Anglo-Iranian), Shell and Esso.[144] Domestic food production was kept high after the war, and great efforts were made to increase it. This went along with a decidedly nationalist propaganda offensive, which did not distinguish between imperial and foreign suppliers of food.[145]

7
Worlds of War

The British war machine fought and produced all over the world; it had many fronts and many home fronts too. There were training grounds for Empire aircrew in Canada, arms factories in the USA, Canada, India, and Australia, and fighting fronts from the North Cape to North Africa to Burma. The machine was supplied with material from around the world, extending its reach ever further. An extreme case was the scheme to develop plantations of Quipo trees, to replace balsa wood for the Mosquito bomber, in Darien in Panama, an 'area unhealthy and held in superstitious dread', whose people were 'unused even to the handling of a spade'.[1]

The British Isles were, within this global system of supply and combat, both a War Front and a Home Front. Huge numbers of front-line forces served from Britain. Much of the air force and navy operated from British bases, and most of the British army was based in Britain for most of the war. A large chunk of the imperial supply of arms came from Britain, and most of the imperial procurement and military bureaucracy was in Britain. On top of this Britain had many imperial forces and Allied forces stationed here. On the eve of the Normandy landings, the US forces alone peaked at 1.5 million in the UK.

The 'home front' was also where 'war work' was done in 'war factories'. Such imagery spread well beyond the arms industry. Thus the plough, rather than being a symbol of peace, became according to an official account of 1945 a weapon of war.[2] In this world, tractors were 'leading the mechanized regiment of agriculture' and the humble rabbit was 'the object of a sustained and well-organized punitive expedition, embracing methods of extermination unknown to his ancestors . . . he was trapped and gassed in his millions', reported a Ministry of Information pamphlet written by Laurie Lee.[3]

As obvious as it seems, it is necessary to point out that the most significant feature of wartime industry was the expansion of the military-industrial complex. For, as we have seen, contemporaries and indeed historians tended to discuss the war as one of production, indeed of the production of particular weapons created by civilians, scientists, engineers and industry. Yet war production was as much a matter of new specialist industrial capacity being built as of civilian industries being turned over to armaments. The military, and its associated institutions like the supply ministries and the specialized arms industry, were creators of the radio and the aeroplane, as much as of artillery pieces or poison gas.

The expanded British national armed services and arms industry especially accounted for some 45 per cent of the British working population during the war, some 10 million people. The two groups were about the same size, just under 5 million people each.[4] Both groups were made up of relatively young people, with young men going mainly to the forces and young women to the factories. The increase in male employment in metal manufacturing (aircraft, ships, motors, engineering) plus chemicals, explosives, etc., from June 1939 to June 1943 was 700,000, but the increase in women was 1.42 million. The armed services peaked at around 4 million men and about 460,000 women.[5] Young women, like young men, did not generally replace men; rather they were recruited into new military and civil jobs.

Both worlds involved new things, structures and processes. Nearly all the machinery in use in the forces, the buildings and everything else was new. Much of the equipment in the front line in 1940 was no longer there in 1944, whether in terms of tanks, aircraft or artillery. Among the rare exceptions were some of the battleships. The new arms industry which made this equipment was the product of huge state investment from the mid-1930s. Its new equipment and buildings were in stark contrast to the merchant shipyards and the cotton mills, many of which had not seen investment since the early 1920s.

The newness of most elements of the warfare state, and the generosity with which it was supplied, were in sharp contrast to the rest of the economy. Such distinctions applied in other ways. War workers got better wages than those stuck in the mines or the cotton mills. Service personnel and war workers got better food than the average civilian. In some cases, however, sections without doubt directly engaged in dangerous warlike operations were hardly different from before the war – the merchant navy stands out here.[6]

The radical difference between the new worlds of war and the older civil world was expressed in many ways. There were important regional variations too: in 'key aspects of economic administration, there was no "United" Kingdom during the war'.[7] In Northern Ireland there was no conscription, and relatively few arms orders, with the result that the workforce remained more skilled, more male, and much more likely to be continuing to work single shifts than in the rest of the United Kingdom.[8]

There were, however, all sorts of other emergent distinctions of great importance. There were importance differences indeed between those in the forces and those in the arms industry. As the socialist J. T. Murphy reported in 1942:

> It is impossible to move among the soldiers and sailors and airmen of all ranks without hearing scathing comments on the civilian population: on the munitions workers who take home £10–£15 a week, on the joy-riding and wining and dining in the West Ends of our big cities, on the huge profits of the manufacturers and the gambling on the Stock Exchange.[9]

But experiences in the forces were radically different. The majority serving in the RAF were hardly ever even near danger, while for the minority in aircrew, especially bombers, the story was, for a few months of operational flying, very different. The danger from bombs for civilians was far from universal, concentrated in time and space – to late 1940 and early 1941, and the poorer districts of London (above all), and a few other industrial and port cities. War was a radically discriminating rather than unifying business and was sometimes perceived as such.

For every seemingly general development there were important exceptions. The blackout, which lasted for most of the war, meant no streetlights, no advertising in lights, no light spilling out of windows. Night brought a gloom to public spaces not known for many generations of city dwellers. At the same time the war ushered in twenty-four-hour working on an unprecedented scale. More warships were at sea for longer, day and night. British bombers, unexpectedly, largely operated at night. Continuous night-time warfare was a novelty.[10] In the arms industry, too, night-working, very rare in engineering before the war, became standard, with two-shift working, and sometimes three-shift working. In other sectors, too, pressure of work led to, for example, ploughing by night. But it was certainly not the case that industry as a whole went over to shift working.

One of the key images we have of the Second World War is of increased speed. This was the essence of 'Blitzkrieg' or 'lightning war', the image of which was of aircraft flying at hundreds of miles an hour, and tanks racing across fields. In the newsreels everything was in motion: rapid movement of aircraft, tanks and ships, together with quick-firing artillery, chimed in with rapid-fire delivery of the commentary. Some aspects of the war were like this, with prodigious quantities of fuel oil, petrol and aviation spirit consumed. But there was also a notable slowing down of much transportation. The great passenger train expresses were a thing of the past as the system got clogged up with more passengers and slower journeys.[11] At sea the movement of goods became slower, as ships waited longer in harbour, waited to be convoyed and were then convoyed at the speed of the slowest ships. The *Empire Liberty*, which could sail at 11 knots, managed, in a period of just over a year, to arrive in Britain only four times, with two cargoes of grain from Canada, one of phosphate fertilizer (and landing craft) from Florida and one unknown cargo from the USA, picked up on a trip that took her to West Africa and the Caribbean.[12] At full speed a London–New York round trip (excluding time in port) amounted to three weeks; the one-way trip now took about the same time. The Royal Navy itself, on average, slowed down, as it deployed more and more escorts powered by the same sort of reciprocating engines as tramp steamers rather than the turbines of fast warships and liners.

NEW FACTORIES AND INFRASTRUCTURES

The building of a warfare state involved a huge construction effort. In the first years of the war over 300,000 building workers were engaged in work for the service and supply ministries, the number falling from early 1943.[13] The military and the arms industry were taking around half of all construction expenditure, with military installations costing about twice as much as industrial ones.[14] New military installations included barracks, training facilities and airfields. In 1944 the cost of a bomber airfield was £1m.[15] In 1939 Bomber Command had twenty-seven grass-covered airfields; in 1944 it had 128 airfields, of which only two were grass, about half for operations, the other half for training.[16] Concrete runways had a dramatic effect on the ease of take-off, allowing much greater loads to be borne aloft. The scale of the building of new technical infrastructure was also prodigious. From 1938 a new communications system was under construction, first as a telegraph and then as the Defence Teleprinter Network, which grew to be two and a half times larger than the slowly expanding civil system, also used by civil departments.[17] A new telephone system around the same size as the existing one was created: the number of defence telephone circuits over twenty-five miles in length exceeded the number in the public system by 1944; it had taken twenty-five years to build the main underground system, but this was almost doubled in size in five years of war.[18] Another example is the thousand miles of petrol pipeline from west coast ports to the east, linking to the great aerodromes, as well as the PLUTO pipelines (see map, p. 191). All this building was at the expense of new roads, utilities and houses, though considerable expenditure had to go into repairing bomb damage.[19]

Preparation for the future was a difficult and necessarily uncertain business. The very uncertainties meant that investment had to be carried out by the government. It was obvious that the government should build and own airfields, and military communication infrastructures, but not so obvious that it should own the arms industry. Indeed, much of it was private. If in the 1930s it had been certain that a long and victorious war was coming the arms industry would have invested in the specialist capacity to supply arms. As it was, there was a strong possibility of no war, and thus of warlike spending falling drastically after a few years of fast growth. In such circumstances the government had to build huge new munitions works, airframe and aero-engine factories, and more besides. As it was, these factories would be fully used. One half of all workers making armaments, some 1.6 million in number, were doing so using new, government-supplied machines and factories. Building factories and acquiring plant for this vast new industry cost £1bn between 1936 and 1945, mostly in aircraft and armament factories.[20] The cost was very roughly the equivalent of 100 battleships, or 100 extremely large factories. This new building was needed, it needs to be stressed, not because there was a shortage of industrial capacity in general; indeed, in the case of naval shipbuilding and armaments relatively little investment was needed. What was needed was new specialized capacity to make weapons. Aircraft could not be made on machines for making motorbikes; aero-engines could not be made even on machines used for car engines. Guns, propellants and explosives more obviously needed special equipment.

The most visible part of the new arms industry was the new Royal Ordnance Factories, extensions on a vast scale of interwar government armament factories: Enfield, Woolwich and Waltham Abbey. The Ordnance Factories employed hundreds of thousands of 'industrial civil servants' in the war, for they were part and parcel of the government. By contrast, the so-called shadow factories making aircraft and aero-engines set up in the late 1930s, though also owned by the government, were run by car makers. The idea was that they would bring experience in mass production to bear on arms production, and they would learn the peculiarities of arms production. And so they did, but they often found it difficult to cope with the exceptional complexity of aero-engines and airframes, and as new capacity was demanded during the early part of the war it was more likely to be given in the form of extensions to arms makers rather than new shadow factories. Most airframe firms saw about a tenfold increase in employment over 1936, and a six- to tenfold increase in floor space. At the peak of war production the pre-war airframe firms employed no fewer than 225,000 people,[21] and the engine firms at least 120,000.[22] Similarly, extensions and agency factories run by private firms were important in the case of the Ministry of Supply.

The story of the pre-war and wartime factories is immensely complex. We can get a reasonable picture by focusing on an example in the north-west of England, a particularly important centre for such factories. We have already discussed the Heysham aviation spirit plant run by ICI and others, but owned by the Ministry of Aircraft Production, but this was just one of many huge new plants in the region. One of the very largest of them was the Royal Ordnance's Chorley filling factory. The decision to build at Chorley in Lancashire was made in the summer of 1935, a clear instance of the seriousness of rearmament. Preparatory work followed and building started in January 1937. Limited production began at the end of 1938 (the first shell to be filled was a 3.7in AA round in December) and the official opening by the King was in spring 1939. The builder threw a gigantic party for the 12,000 building workers, to which the Queen of Lancashire, Gracie Fields, came along. It was ready for full production at the beginning of the war. In terms of those details beloved of the era, the quantities of material involved in making Chorley were prodigious: 30 million bricks; 1 million cubic yards of concrete, 15,000 steel window frames, fifty miles of road, twenty-five miles of railway, 1,500 buildings. The work involved building the world's largest concrete mixer, which turned out 5,000 tons of concrete a day, moved by thirty trains and 300 lorries on site.[23] Chorley was very different from an engineering factory in that it was a vast complex of many small plants, each surrounded by large earthworks to contain blast in the case of accident. From the air it looked like a cratered lunar landscape. Thirty-five thousand people were to work in this strange world, taking in – in one week in 1942 – 500 tons of TNT, 250 tons of ammonium nitrate and 150 tons of cordite, and sending out, among other things, 210 truckloads of 25-pounder ammunition, and seventy trucks of 5.5in howitzer shells.[24]

The filling factories were very expensive to build. By 1940 the building costs of Chorley came to £11m.[25] This was far from unusual. Two other filling factories cost over £13m.[26] Some of the explosives and propellants factories were equally or more expensive.[27]

The seriousness and depth of building for war can be further illustrated by the case of poison gas. In the 1920s the government plant at Sutton Oak in Lancashire could manufacture 20 tons of mustard a week. At that time a new plant for HS mustard was built. In rearmament, as one would expect, the fact that a small British army was planned meant that army gas capacity was not much needed. However, there was to be great expansion in the ability to use gas from the air, in bombs and as a spray. Indeed, a new kind of mustard, 'HT', was developed for the purpose. Production facilities were built on a very large scale: a plant at Randle, near Runcorn, was started in March 1937. A second factory at Valley (near Wrexham) was finished in 1940 (and as Rhydymwyn Valley Site is now open to the public); first built as a safe storage site for mustard gas, it developed into a factory for putting mustard into weapons. From early 1941 it was itself producing mustard in large quantities. The factory cost £3.2m, and in 1943 there were about 2,200 people working in it.[28] A third mustard plant was completed in Springfields in early 1942. Most were operated by ICI and there were three new factories for making 'intermediates', chemicals used for making the poison gas, all in the north-west. The chemical warfare plants as a whole cost nearly £20m.[29] The result was that at the beginning of the war hundreds of tons per week could be made, though the offensive capacity was limited by the bombs and spray tanks available.[30] Britain was soon ready for immediate use of gas on a very large scale, not only from the air but from 25lb shells, Livens projectors and 5in rockets.[31]

During the war there were some twenty major aero-engine factories, nearly all new. They each cost around £1.5m to build, but the machine tools typically cost around £4m initially, but with additions and replacement by the end of the war several factories had spent around £10m.[32] The total cost of building and equipping them came to over £100m, the same as airframe plants. In the latter the total was about 40 per cent plant; in the case of engines, 70 per cent.[33] The new aero-engine and airframe plants were typically made up of a very few vast single-storeyed steel-framed buildings, sometimes just one. They are hardly known in what literature there is on industrial architecture.

Some of the largest plants were in the north-west. One of the largest of the engine plants was the Ford Motor Company-run shadow factory in Manchester. It was built on unused land next to the Ship Canal, starting in May 1940, and Ford estimated the cost at £7m, including machine tools. The factory was finished in May 1941, reaching its target of 400 Rolls-Royce Merlin engines per month in 1942. This level of anticipated production was based on two-shift working. With an extra shift, higher line speed and small extension, the factory reached the highest production level of any aero-engine plant at 900 engines per month, in the last quarter of 1944. By the end of the war, it was employing 17,316, of whom 7,200 were women. Here the 'mountain-high fence between skilled and unskilled labour had been surmounted' and the factory employed 'men, women, girls and boys – clerks, typists, retail store salesmen and women, hairdressers, manicurists, domestic servants and many others who had never seen the inside of a factory'.[34] The factory produced 34,000 Merlins of the 165,000 produced worldwide,[35] 1,400 inspectors were employed and 30,000 precision instruments used. This wasn't quite standard mass production. Although many multiple spindle borers and drills and many other specialized machines were used, the making of a Merlin was a skilled affair in many respects. For example, the carefully assembled engines were, after a test run, disassembled, all the key parts were tested individually and then they were reassembled.[36]

The factory was enormous, with 100 production bays and 6,700 machine tools. It consumed as much gas as a town like Canterbury and electricity as Reading: 5 tons of coal equivalent of energy for each aero-engine. The workers came in 450 daily bus journeys to the factory, 400 motor cars and 2,000 bicycles. Sixteen acres of land surrounding the factory grew potatoes and other root crops, cabbages and cauliflowers; the land was ploughed by a Fordson tractor.[37]

Rolls-Royce also made Merlin engines on a huge scale in the north-west, in a new government-owned factory in Crewe, built and completed before the Ford plant.[38] The other great engine manufacturer, Bristol, ran another new factory in Accrington, where it made the sleeve-valved Hercules.[39] Napier had a large factory for making the Sabre in Speke. All these factories cost in the region of £5m.[40] The first jet engines were made by Rover and then Rolls-Royce in Barnoldswick, Lancashire.

Airframe production in the north-west was radically extended during the war. It was very largely a centre of bomber production. The rather insignificant A. V. Roe company, part of the huge Hawker-Siddeley group, ended up producing the most important bomber of the war, the Lancaster. A. V. Roe employed 35,000 during the war, mostly in new factories: the Greengate/Chadderton factory employed around 20,000 in January 1944, the Yeadon factory (near Leeds) 10,000.[41]

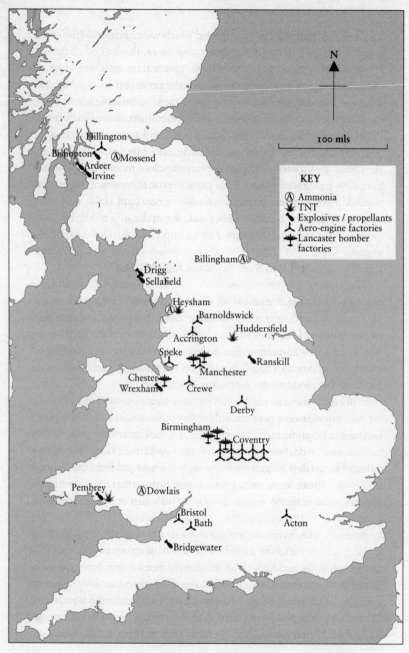

Locations of British aero-engine, Lancaster bomber, explosives and
propellants, TNT and ammonia factories

Other plants making aircraft in the north-west included Vickers at Broughton near Chester (Wellingtons) and at Blackpool (Wellingtons); Metropolitan-Vickers at Trafford Park (Lancasters), English Electric at Preston (Hampdens and Halifaxes), and the car maker Rootes at Speke (Blenheims and Halifaxes) and the Fairey Aviation company at Errwood Park, Stockport (Halifaxes and naval aircraft). With the exception of Chester and Trafford Park, each of these plants employed more than 10,000 workers in early 1944.[42] To take one specific example – Metropolitan-Vickers' new aircraft factory, close to its original works and to its new radar plants, employed over 8,000 people, more than 3,000 of them women, and produced over 1,000 Lancasters, in a single factory, with a shop length of more than a quarter of a mile.[43]

FILMS AND FACTORIES

During the war the British people were given some idea of the material constitution of society, for the material mattered. There was much propaganda on food and salvage, what would now be called recycling, and other campaigns against waste, for example of paper and food. War artists painted factories – a whole book was devoted to paintings of 'production', with an introduction by the photographer Cecil Beaton, where he notes that 'mines, foundries, workshops' were, 'for technical reasons, beyond the power of the camera . . . to reproduce in [their] full significance' – the eye but not the camera was attuned 'to the nuances of darkness amid a strange world that is spasmodically suffused by flashes of green, magenta, puce and golden light'.[44] Certainly the artists were much more successful than the camera in capturing the scale of war factories. People also learnt of weapons and production through newsreels and to a lesser extent through documentaries. The main images conveyed were of the brilliance of British machines and the armed forces that used them, and the precision, scientificity and up-to-dateness of the arms factories. Newsreels, particularly it seems in 1942, focused on the productive processes in arms factories, particularly aircraft factories. There was a powerful invocation of the power of the new, the heavy machine tool, the press, the precision production which went into making an aero-engine.

Some films show production processes in detail, for example *Speed Up on Stirlings* of 1942, produced by Edgar Anstey of the Shell Film Unit. In this film, men – managers, designers, skilled fitters – dominate.[45]

During the war, one of the great issues of interest and propagandistic focus was women workers, and this in itself directed the camera and the investigator towards productive processes. Paul Rotha's film *Essential Jobs* (1942), for the Ministry of Information, written by V. S. Pritchett, sets out to show workers in purely civil industries that they are contributing to the war effort by illustrating the links between industries: a one-inch nail, used to box cocoa, the making of soap, used to make leather, which makes special leather gloves, which are used on an anti-aircraft battery, whose personnel drink cocoa. Rotha's *Night Shift*, shown in 1942, gives a particularly vivid picture of productive processes. It was filmed in November 1941 at the No. 11 Royal Ordnance Factory in Newport in South Wales, a new engineering factory – built in 1940–41 – which made the 40mm 2-pounder anti-tank gun (whose manufacture was shown in the film) and the 40mm anti-aircraft Bofors gun. The modernity of the factory impressed:

> The first, and most lasting impression you get is of brand newness, and youth: you see girls and young women in navy blue boiler suits working in long factory bays extending away almost to vanishing point. Dark blue figures against the creamy-grey concrete floors and brand new wooden benches. They are wearing smart blue 'Sister Dora' caps to match their boiler suits, and their curls showing under their caps sway and clasp as they work energetically.[46]

Night Shift tells us 2,000 women work in the factory, working two ten-hour shifts. They worked two weeks on the night shift, then two weeks on days. Earnings were shared between four workers, two on each shift working one machine each.

The factory became famous again in 1943. It was home to Britain's answer to Rosie the Riveter, the symbol of the new woman war worker in the USA. Ruby Loftus was painted at her lathe by Dame Laura Knight 'screwing a breech ring', as the title of the painting has it, in this case of the 40mm Bofors anti-aircraft gun, the factory's other product. The painting was celebrated at the time, as was Ruby

Loftus herself, and has become the best-known image of the British factory front.

A second detailed investigation in relation to women workers came with the Mass-Observation study called *War Factory* (1943). This could not reveal that the factory was a highly secret installation making airborne radar, and indeed worked closely with the Ministry of Aircraft Production boffins to design radar equipment, notably it seems the 10cm AI Mark VIII set; the firm housed the Malmesbury Development Unit. The factory was run by EKCO, an innovative radio company of the 1930s, who had brought in avant-garde architects to design radio sets. One of their key staff, a Leeds University-educated engineer, M. J. Lipman, was an unusual engineer and businessman in that he was a socialist. Unfortunately, neither *War Factory* nor Lipman's fascinating autobiography has any detail on production.[47]

COMPARATIVE PRODUCTIVITY

The Second World War was, as far as Britain was concerned, a war not only of production but of productivity, a measure of efficiency. As labour became short, so making more efficient use of what was available became important. Comparisons of 'labour productivity' started to be made, using such measures as 'production per man-hour', overwhelmingly by comparison with the USA, the land of efficiency and mass production. In general US industry did produce much more efficiently in terms of labour than did British industry; for example, its textile industry was perhaps twice as efficient in terms of physical output per worker as the British.[48] Yet while it was the case that overall labour productivity was much higher in the USA than in Britain, in key areas, shipbuilding and airframe manufacture, the USA was not ahead.

The aim of war production was not merely to produce efficiently but to produce weapons that were effective in operations and available in good time. There was a very well-known trade-off between scale of efficiency and flexibility: large-scale production was more efficient, but flexibility in production was essential to accommodate new weapons and modifications. The Minister of Production, Oliver Lyttelton, told Parliament on 14 July 1942: 'mass production and flexibility are opposite terms. One is the antithesis of the other.' The military wanted flexibility; the right weapon available at once, the lessons of war 'translated into steel in a few minutes'. But production, especially with diluted labour (that is, a workforce diluted with a high proportion of semi-skilled and unskilled workers) and specialized equipment, meant stability in the product: 'elaborate tooling is the enemy of flexibility'.[49] In fact few armaments in production in 1941 were still in production in 1944: examples would be such things as the Lee-Enfield No. 4, the Sten gun and some artillery pieces, like the 25-pounder. Even when there was apparent continuity, as in the case of Spitfires and Merlin engines, one finds major changes were incorporated in both, reducing the rate of production, but increasing the quality of the output.

Britain and the US developed at least in part different means of combining flexibility with mass production in aircraft. Broadly speaking, in Britain production was (at first accounting) less efficient and more flexible, turning out a better final product straight off. In the US production was more efficient and less flexible. A comparative measurement of British and American productivity was made during the war by a senior Ministry of Aircraft Production engineer (Cambridge-educated, like many aeronautical engineers). He found US production to be 75 per cent more efficient than British in terms of weight of airframe per worker, but this difference was accounted for by the facts that American production runs were longer than British ones and that British factories tended to be smaller.[50] The implication of the study is that the underlying productivity was the same, and that the British chose to forgo some to achieve flexibility, and perhaps also to make production more resilient to bombing by limiting the size of factories.

The length of the production run was important for two distinct reasons. The first is that long runs made it economical to install specialized equipment to make that one type of aircraft only. The second is that it is one of the peculiarities of aircraft manufacture that productivity increased rapidly with volume and rate of production, independently of the machinery used and of the initial skill of workers and management. This was the origin of the since famous term, the 'learning curve'.

Table 7.1 Output of military aircraft in the United Kingdom and German Europe, 1940–44

	UK production		Production for German forces	
	Numbers	Structure weight, millions of lb	Numbers	Structure weight, millions of lb
1940	15,000	59	10,000	59
1941	20,100	87	11,000	64
1942	23,600	133	14,200	92
1943	26,200	185	25,200	138
1944	26,500	208	39,600	174

Source: Sir Charles Kingsley Webster and Noble Frankland, *The Strategic Air Offensive against Germany 1939–1945*, Vol. IV: *Annexes and Appendices* (London, 1961), Appendix 49, Table iii. The differences in numbers are explained by the increasing average weight of British aircraft, and the decreasing average weight of German, as the former turned towards bombers and the latter towards fighters, from a similar starting point. It is important to note that the German production came from a population at the very least 30 per cent larger than that of the UK. Note that the data ignore the aircraft production of the Empire and of aircraft production for the Empire in the USA. At the end of the war Canada alone added 4,000 aircraft per annum to the total; Australia about 1,000. Supplies from the USA were even greater See also Table 9.1.

The British paid the known price of lower initial productivity to produce the latest aircraft and latest versions. In fact the British official history reckoned that 'most British operational aircraft were never allowed to be produced undisturbed in quantities large enough to reap the full advantage of their jigs and tools' such that production might have been more economical if it had been optimized for low levels of production. It noted that 'very few unmodified batches of Spitfires were greater than 500, so that many components must have been produced under conditions which were better suited to bench methods than they were to the jigs and tools actually used'.[51] By contrast, the

American firms 'were allowed to produce large quantities, varying from 500 to as many as 15,000 aircraft, without any modifications in the production line'.[52] For modifications which might become inevitable in the intervening period special 'modification centres' were set up that 'were soon choked up with aircraft awaiting modification'. The result was that 'the flow of aircraft to squadrons was much more meagre than the impressive figures of production suggested'. The quantitative advantage was not what it promised, and 'the sacrifices in quality were probably greater than they would have been under the more flexible and looser British system'.[53] Up to one half of all the labour in US military aircraft production was in modification centres, thus making overall labour productivity up to half what it appeared to be in the main plants.[54] This seems to suggest greater overall British productivity. Furthermore Britain introduced and/or developed a whole range of new aircraft and engines, including thousands of 30-ton bombers as well as many other different kinds of aircraft. The British design, production and use of aircraft were more 'flexible, technologically sophisticated and effective' than those of Germany or the USA, according to one historian.[55]

The story in merchant shipbuilding is rather different in that the underlying productivity – a matter of labour skills – was much higher in Britain than in the USA. In shipbuilding British yards before and after the war beat US yards in labour productivity as well as in costs of production. In 1939 the USA was not a major force in shipbuilding. But under wartime conditions the US built ships because they were needed and because it had the resources to do so quickly. The US effort was stupendous. While it built only 0.5 million DWT of merchant ships in 1940, in 1943 it built the remarkable total of 19 million DWT, overwhelmingly of Liberty ships.[56] But as the *Economist* noted in 1942, while Liberty ship production was exceptionally fast, British output per man-hour was 'appreciably higher'.[57] In 1942 British yards needed an average of 336,000 person-hours to build a Liberty-type ship, while US-built Liberties at first needed 1.1 million person-hours, falling later to a minimum of 486,000. Only the most successful US yards 'approached and occasionally overtook the British average after producing a hundred or more similar ships'.[58] That is, they moved down the learning curve, for here too it was discovered that the effect obtained.

It made perfect sense for Britain to use its skilled workers and highly constrained shipyard capacity to build the most complex ships and to shunt the relatively unskilled task of building Liberty ships to the USA. Britain built 277 standard tramps of the Liberty type,[59] and acquired about the same number from the US and Canada; US production was ten times greater. In naval building Britain also had a considerable lead in tonnage produced per shipbuilding worker; it was to build practically all the ships it needed.[60]

That the issue of comparative productivity is not always what it seems is also shown in the crewing of Liberty ships. British Liberty ships, the Sams, identical to the US ones, had slightly smaller crews than US-operated ships. The US-operated *Arthur Middleton* had 42 sailors (officers and crew) and 27 gunners; the *George G. Meade* 41 and 25, the *Robert Gray* 39 and 23, the *George Cleeve* 41 and 28. The standard was probably 8 officers, 33 crew (total 41) and 25 gunners. By contrast the British-operated *Samouri* sank with 39 total crew and 10 gunners; the *Samsuva* 37 and 20 gunners. All the figures may mean is that the US ships had better service for crews, as well as greater protection. The Oceans, which were coal-fired, seemed to have somewhat larger crews, around 48 crew plus around 10 gunners.[61] British ships manned with lascars had crews twice the size of the British crews.[62]

BATTLESHIPS AND BOMBERS

The Anglo-American dominance of economic resources and distance from the exigencies of the front were reflected in their dominance in both battleships and bombers. Here Britain led and the USA followed as in so much else concerning modern war. Into the Second World War the battleship remained a great emblem of national productive and naval power; from the launching to trials to action, the battleships featured as newsreel staples, their names known to all. With them went a recitation of quantities, speeds, indices of power. They were factories, nations, worlds in miniature: they were living things, not lumps of steel, 'possessing the soul and the voice of Britain'. These were obviously masculine worlds, but the feminine was not absent, for 'deep in the glistening engine room the engineers tend her with the care of mothers' and cooking and cleaning are part of the routines of life. They were never merely machines, never wholly masculine, and not even wholly warlike either.[63] They were, for example, compared to large hotels in the extent of their catering operations.

The productive effort that went into battleship production was enormous. In 1941 two battleships were completed in Britain. Their total 70,000 tons' displacement compares with the 226,000 tons' displacement of all major combat vessels completed and 364,000 tons' displacement of all warships, including sloops, corvettes and mine-sweepers, that is, 30 per cent of major warships and 20 per cent of all warships.[64] Twenty of these monsters served in the Royal Navy during the war: there were fifteen at the start, five were added, and five were lost (two to U-boats, one to the *Bismarck* and two to Japanese aircraft, in 1941). The Germans had only two battleships, *Bismarck* and *Tirpitz*, and two smaller battlecruisers, *Scharnhorst* and *Gneisenau*. All were destroyed or put out of action by the British between 1941 and 1944. The Italians had more than the Germans, and the Japanese many more still, but between them the Anglo-American forces utterly dominated.

At the cost of £7.5m, *King George V* was, turning around the comparison already made, as expensive to build as some of the contemporary gigantic new ordnance or aero-engine factories, and took longer to build. *King George V* had a crew of around 1,500 men, less than a large wartime factory, but on a different scale from a Liberty ship. She could carry over 3,000 tons of fuel, the same as a small tanker. Her steam turbines generated 125,000hp (93MW), roughly the output of a contemporary power station.[65] Only the very largest passenger ships (the only ships which compared with them in size) had more engine power. A Liberty ship, with around 2,500hp (around 2MW), was a mere boat by comparison, equipped as it was with an engine which had the output of the very largest of the Second World War's aero-engines, like the Bristol Centaurus, or a modern wind turbine.

A battleship's guns were immensely larger than standard field artillery. *King George V* had ten 14in (356mm) guns, which each weighed around 80 tons and could fire a shell weighing more than half a ton (721kg) for a distance of thirty-five kilometres. Compare that with the standard 25-pounder artillery piece, so called because it fired a 25lb (11kg) shell from a barrel 87.6mm in diameter. Eight of *King George V*'s large guns were in two quadruple turrets, with two in a twin turret. The larger turrets with all their equipment each weighed 1,500 tons, nearly one thousand times more than a complete 25-pounder. A battleship carried the same quantity of armour plate as many thousands of tanks.

It had been intended to produce a second class of battleship to follow those above, the Lion class. Only one further battleship, the *Vanguard*, would be ordered – it was launched in 1944 but not completed till 1946. An extraordinary mixture of old and new, it used 15in guns and mountings made in the Great War. Battleships were replaced as 'capital ships' by the aircraft carriers. Here Britain was strong too: by the end of 1941 it had built more carriers and of a greater tonnage than anyone else; in Europe it faced opponents that had no carriers at all.[66]

But the key feature of the wartime navy was not the shift from battleship to aircraft carrier, but rather the shift to small ships with small guns. This was one of the few great production changes of the beginning of the war, in this case in 1939. As a result the really large increases in production were in destroyers, submarines, corvettes and frigates, and minesweepers, trawlers and boom defence vessels.[67] The average size and power of a naval ship shrank as the Royal Navy became a force protecting convoys against submarines rather than a force designed to fight heavy naval forces. Much now consisted of corvettes and frigates with reciprocating steam engines, rather than the more complex and expensive turbines of the destroyers, cruisers, battleships and aircraft carriers. The Flower Class corvettes, and following ships, were derived from civilian ships, in this case whale-catchers, and were built to civilian standards. Here naval and civilian came together.

In the air force, the story was different, one of growth in the scale and power of aircraft. In the case of heavy bombers, they rose from 30 per cent of all aircraft production by weight of airframe in 1942 to 50 per cent in 1944.[68] The total mass of all the heavy bomber airframes produced in 1944 was about the same as the displacement of a battleship, 50,000 tons. Furthermore bombers grew in size and power, as did fighters.

The 30-ton, four-engine bomber was even more of an Anglo-American phenomenon than the battleship. The Axis powers had a few, but the numbers involved hardly compared with those in British and US air fleets. Just over 1,000 Heinkel 177s were made by the Germans, the Japanese made a handful of two types, while the Italians had tens of Piaggo P 108s. By contrast the Anglo-American air forces had tens of thousands of such bombers. They came in essentially three British types, the Lancaster, Halifax and Stirling, and two American, the B-24 (Liberator) and B-17 (Flying Fortress). Britain produced around one third of the combined production. It made over 7,000 Lancasters (another 400 were made in Canada), over 6,000 Halifaxes and just under 2,000 Stirlings. The US made 18,000 B-24s and over 12,000 B-17s. The Lancaster may well have been the best of the lot, and in 1943 was the only bomber which could carry the projected atomic bomb. The eventual atomic bomber, the B-29 (Super Fortress), was a 60-ton aircraft used to bomb Japan (nearly 4,000 were built through to 1946).

By coincidence the power of the bomber can be straightforwardly compared with a battleship. Two hundred Lancasters, roughly the establishment of an RAF group, cost about the same as a single battleship.[69] On a single sortie they could deliver roughly the same tonnage of ordnance, around 1,000 tons.[70] The total crew, around 1,500, was about the same, since each Lancaster had a crew of seven. The amount of fuel carried was about the same too. On other measures they were very different. The total power output of the 200 Lancasters was six times that of a battleship and the total weight around one sixth. The Lancasters could, obviously, get to targets much faster, and do so many more times. On the other hand, the Lancasters were not self-contained in the way a battleship was. They needed a huge ground infrastructure to return to after a few hours, while battleships kept going for weeks. The great majority of the staff of Bomber Command never left their base. At the end of the war 20 per cent of the RAF were aircrew (all volunteers), though the proportion actually on operations would have been lower. In the navy a much higher proportion went to sea. The battleship gunnery officer was at as much risk of death as the stoker, the cook or the medical orderly. In the air force aircrew were at great risk, while the erks at base, the cooks and medics were generally quite safe.

More than half of the 6,500 Lancasters used in operations were lost, whereas just under half of all Bomber Command aircrew were killed.[71] At a rough average the maximum number of operational sorties for each crew was thirty, after which they were released to other duties; of course, the average was lower owing to losses. Unlike aircrew, aircraft were not retired after a maximum number of sorties but only thirty-four Lancasters survived to make more than 100 operational sorties. One, the 'Spirit of Russia', went on its first operation in June 1943, its last in February 1945. In 1943 alone it bombed Berlin seven times, Milan four times, Hamburg three times, Mannheim twice, Nuremberg twice, as well as Düsseldorf, Cologne, Gelsenkirchen, Turin, Remscheid, Peenemünde, Mönchengladbach, Modane, Frankfurt, Oberhausen and Leverkusen. Warships saw nothing like this intensity of action.

In its first year of operations 'Spirit of Russia' was flown eleven times by the same RAAF crew, five times by an RCAF crew and the remaining thirteen by various RAF crews.[72] This reflected the fact that the world of the bomber based in Britain was far from being a British one. Bomber Command (which operated only from the UK) suffered operational aircrew dead (to the nearest 500) of RAF, 33,000; RCAF, 8,000; RAAF, 3,500; RNZAF, 1,500; 1,000 Poles; and 500 other allies.[73] The US Eighth Air Force, consisting mostly of B-17 and B-24 bombers, was itself an enormous force based in Britain, peaking at around 200,000 personnel. It lost over 9,000 aircraft and about 44,500 men killed or originally reported missing, many of whom would have ended up as prisoners of war.[74]

TRAINING

New forms of work and the handling of new machines were the standard problem for the forces as for the arms industry. Over time, the effectiveness and efficiency in the use of weapons was to increase dramatically. Indeed, the efficiency of destruction increased much faster than the efficiency of production. Thus the rate at which German cities were destroyed by British bombers rose faster than the output of bombers from the factories.

The new world of the armed services involved a great deal of training. Most of the British army's activity until 1944 was at home, not fighting, but training. Many units had years of training in the UK before going into battle. Thus there were enough troops in Britain in autumn 1941 to have an exercise involving twelve divisions and three independent brigades and the RAF, followed by another in early 1943 of ten divisions and five independent brigades.[75] We may estimate the total number of troops involved at around 200,000, very crudely. One type of tank, the Covenanter, was produced in vast numbers but only ever used in training.

Although the air force was in constant operation, the effort devoted to training was enormous. Indeed, one of the important reasons for using the heavy bomber, with many times the carrying capacity of the medium bombers, was that it economized on pilots. There was even an RAF feature documentary on training, written by Flight Lieutenant Terence Rattigan and directed by Flight Lieutenant John Boulting, though this missed out two important late stages of training.[76] Training involved travel abroad and more flying hours than a bomber crew would fly in operations. The bomber aircrew received the 'most expensive education in the world', £10,000 for each man, enough to send ten men to Oxford or Cambridge for a full three-year degree course, according to Sir Arthur Harris.[77] Perhaps eighteen months could elapse between arrival at Aircrew Reception Centre and allocation to a fighting unit, omitting vacations, roughly the same as a university degree in wartime. Most of the early training took place alongside the training of aircrew from the Commonwealth abroad, largely in Canada, through the Empire Air Training Scheme. But training continued at home. One third of the heavy bombers, 700, were used for the very last stage of training in Heavy Conversion Units and Finishing Schools, in addition to 1,200 medium bombers in the Operational Training Units.[78] Half the airfields of Bomber Command were for training. Though training was clearly very important, 'learning by doing' was even more so. The air force thus invested a great deal in realistic training with aircrew retired from the front line.

In the modern army the proportion of men actually fighting was falling rapidly, especially in the British army. The army had its 'teeth' but also an increasingly long 'tail' of supply formations, labour battalions and repair teams. A whole new corps was formed in 1942, the Royal Electrical and Mechanical Engineers (REME), to repair equipment – the lower ranks were not 'fusiliers' or 'riflemen' or 'gunners' but 'craftsmen'. By the end of the war REME had 8,000 officers, 150,000 soldiers and 100,000 civilians in its ranks. Its officers were engineers, like Donald Stokes, educated at public school, an engineering apprenticeship at Leyland Motors in Lancashire, and the Harris Institute of Technology in Preston (now the University of Central Lancashire). Stokes became a Territorial in 1938, was mobilized in 1939 in the Royal Army Ordnance Corps and transferred to the REME in 1942, and ended the war as a lieutenant-colonel.[79] Everywhere there was demand for specialist technical officers as well as skilled men from the forces, to a degree which had not been expected at the beginning of the war. During the war the number of merchant navy radio officers (some very young) trebled; there were now three for every ship of any size.[80]

In the summer of 1940 a shortage of all types of radio and radio experts became clear. Lord Hankey was put on to the task of increasing supply, and a special 'radio syllabus' for physics and maths students at university was offered. In 1941 a bursary scheme of full fees, and a maintenance grant irrespective of parental income, was available to future students of radio, engineering and chemistry, with an expectation that half would be in radio (which included radar). The proportions were later fixed at 50 per cent radio (including radar), 40 per cent engineering and 10 per cent chemistry. Two thousand bursaries were offered for the first entry in October 1941, and around 6,000 in total. The state also supported engineers with ONDs by letting them do an HND course full-time with maintenance grants (around 4,000 over the war), and then engineering cadetships for those going to be engineer officers in the services, around 4,000 also.[81] With this, higher technical education was transformed, and indeed university education too. There was a much stronger tendency among students to see the university system as a national one and to study away from home.

Active service in war was a young man's occupation. Liability for service increased to the age of forty by June 1941, but not all those men up to forty went into the forces. Apart from the medically unfit, hardship cases and conscientious objectors, those above certain 'reservation ages' in 'reserved occupations' were not called up. The reservation age, in some cases, depended not only on the occupation but the enterprise worked in: in one firm, working on armaments, the reservation age might be eighteen, in another, thirty-five.[82] In addition, individual young men could get deferral on application by their employers. From 1942 there was no reservation under the age of twenty-five, though there were many deferments.[83]

The power to call up was used in a very discriminatory way when it came to young men in different training regimes.[84] Male arts students were generally called up after a very brief period of study.[85] The number of male arts students fell drastically – they went to become officers and completed their studies after the war. Those on degree courses in approved areas of science and engineering were deferred first to their final exams, and then to a maximum of two years and three months (two years and nine months in Scotland) to achieve parity with apprentices, leading to the compression of courses. Furthermore, to ensure parity between state bursars, the maximum that could be spent on a course was to be the same.[86] Medical students were deferred until they qualified. The net result was that British universities, as far as male students were concerned, became much more scientific, technological and medical than they had been; indeed, while the number of male students in 'pure science' decreased a little, numbers in 'technology' went up. During the war the number of full-time male university students studying 'technology' – largely engineering – overtook the number studying 'pure science' – that is, subjects like chemistry and physics – for the first time.[87] Furthermore students were expected to serve in the Officer Training Corps and knew that failure in exams meant call-up. Universities became grim militarized places. Just before graduation, at least in the last years of the war, officers from the Ministry of Labour and National Services' Central Register would interview students, and 'after explaining to them their liabilities under the National Service Acts provisionally allocated them to the Services, Government departments or industry'.[88] All other sorts of careers into which scientists, especially, went, notably teaching, were closed off. Large numbers went of course into technical arms of the forces. Only a small proportion went to research and/or development. For example, George Porter, later President of the Royal Society, graduated in chemistry from Leeds in 1941 and joined the navy as a uniformed radar officer, serving on board ship.[89]

THE DISTINCTIVE BRITISH ARMIES

By nearly every standard the British army was much better equipped than the German army from the beginning to the end of the war.[90] Even more surprising given the relentless impressions to the contrary is the fact that the British army had more tanks per soldier than the German army and the US forces in Europe. There was even a period when the much smaller British army had an absolutely greater number of tanks. In the latter part of the war, most of the tanks used by British forces came from the USA, but Britain's own tanks, and adaptations of the standard US tank, were probably superior to US tanks. Some of these British tanks had the edge on German tanks too. Even if the Germans did achieve a certain qualitative superiority in some respects, they never achieved both qualitative and quantitative superiority.

As we have seen, in 1940 British tank production was approaching that of Germany and in 1941 and 1942 it was decisively ahead (see Table 3.2). Thereafter German production was higher, but by then the British army was getting large numbers of US-made tanks: in 1943 the British imperial army was supplied with more tanks just from the USA than German industry supplied to the entire German army. German production would surge in 1944, but even so the British army got roughly the same number of tanks (from British and US sources) as went to the larger German army, which was in any case fighting much more intensively (see Table 7.2). In 1944 British domestic production of medium and heavy tanks was one quarter of German production, including that of self-propelled guns on tank chassis (which accounted for half German production). On a tank-only basis, British production was half that of Germany for an army a fraction of the size, and one also supplied with tanks from the USA (Table 7.2).

Table 7.2 Supplies of armoured fighting vehicles to British forces and German production

	UK production	Supplied from overseas	Total British forces	German production
1939	969	—	969	n/a
1940	1,399	—	1,399	1,643
1941	4,841	1,390	6,231	3,806
1942	8,611	9,253	17,864	6,174
1943	7,476	15,933	23,409	11,961
1944/I-IIQ	2,474	6,670	9,144	8,929

Sources: CSO, *Statistical Digest of the War*, Table 126; USSBS, *Tank Industry Report*, 2nd edn (January 1947), p. 14, and Exhibit A, available at http://www. angelfire.com/super/ussbs/tankrep.html. British and German production includes self-propelled guns, etc. Excludes armoured carriers, armoured cars, etc. German production also includes self-propelled guns on older tank chassis (38t, Panzer III and half the Panzer IV – these accounted for nearly half of production in 1944). The British production was almost entirely of tanks. The proportion of self-propelled guns, etc., from overseas is not known. The German figures include production in new parts of the Reich, including Austria and Bohemia–Moravia. The Germans supplied their allies with armoured vehicles.

For German production in 1940 and 1941 the source used here gives slightly different numbers from that used in Table 3.2.

Although it was widely believed from 1941–2 that British tanks in North Africa were inferior to German in quantity and quality, this view was shown to be incorrect by an official history published in the 1950s, by Sir Basil Liddell Hart's history of the Royal Tank Regiment published in 1959, and in his later work.[91] Throughout the North African campaign between 1940 and early 1943, British forces had a two- to threefold tank advantage in medium and heavy tanks over the German forces, a figure which would be reduced somewhat by including Italian tanks, but was still overwhelming.[92] For the November 1941 offensive (Operation Crusader) British and imperial forces in

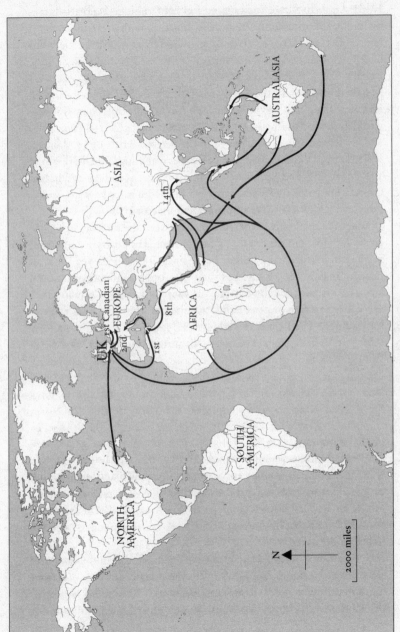

British and British Imperial Army movements, 1939–45

North Africa had over 300 cruisers (mostly Crusaders) and 200 infantry tanks (Matildas and Valentines), together with large numbers of light tanks, and large reserves of all types. By contrast the Germans had fewer than 174 of the comparable Panzer IIIs and IVs, and the Italians 146 13/40s, with no reserves.[93] Crusader was a success.

Yet by early 1942 the Germans and Italians had counterattacked and were deep into Egypt. Their tank forces were still comparatively weak: at the end of August 1942 the Germans had only 166 Panzer IIIs and 37 Panzer IVs.[94] On the eve of the Second Battle of El Alamein, which started the definitive rout of the Germans and Italians, they had 172 Panzer IIIs and 38 Panzer IVs fit for battle, as well as 278 Italian tanks. This compared with over 900 fit medium tanks with British forces, of which 488 were of British manufacture.[95]

What the qualitative difference was between the tanks is very difficult indeed to establish, but the picture presented by critics in Parliament in 1942 in particular, and echoed ever since, is deeply flawed. That all British tanks in the Middle East had the 40mm 2-pounder gun until late 1942 is broadly correct, but the idea that many German tanks carried a much more powerful 75mm gun from around 1940 is misleading.[96] The Panzer IV did have a 75mm gun, but it was a low-velocity gun firing high explosive. This turned out to be a very useful feature for a tank gun, which the British and Americans adopted by using a dual-purpose 75mm gun (as mounted in the Sherman and later in other tanks). But in regard to tank-mounted anti-tank guns the Panzers were no better equipped than British tanks. In North Africa up to May 1942 Panzer IIIs, which in contrast to the Panzer IVs did have anti-tank guns, had 50mm (short) anti-tank guns, which were no better than the 40mm 2-pounder.[97] The other side of the equation was the armour. One study suggests that in 1941 all the main British tanks (Valentines, Matildas, Crusaders) and the US Grant were superior to any available German type in North Africa.[98] Up-armoured Panzer IIIs capable of resisting British tank guns only started arriving in December 1941.

Where German forces had an early advantage was in the non-tank long 50mm anti-tank gun and in the very small numbers of 88mm anti-aircraft guns used as anti-tank weapons.[99] Britain did not introduce a dual-purpose anti-tank/anti-aircraft gun of the 88mm sort but it had plenty of 94mm (3.7in) anti-aircraft guns. By May 1942 there were 100 6-pounder 57mm anti-tank guns with British forces in the Middle East.[100] This was a successful anti-tank gun, which following its production in the United States became the standard anti-tank gun for US forces too.

Superior German tanks appeared in North Africa in the spring of 1942, but in small numbers. In May 1942 the new long-barrelled 50mm gun was first deployed on German tanks in Africa; and the long-barrelled 75mm gun in June 1942. In August 1942 there were 73 of the former on Panzer IIIs and 27 of the latter on the IVs.[101] At the Second Battle of El Alamein there were 88 Panzer IIIs with the long 50mm and 30 Panzer IVs with the long 75mm.[102] Improved British tanks were not far behind and arrived in quantity. In June 1942 the first 57mm 6-pounder Crusader arrived, of which 78 were fit in late October 1942.[103] This gun was more effective as an anti-tank weapon than the US 75mm dual-purpose gun and, at least in later versions, as effective as the German long 75mm gun as well and would be used in tanks to the end of the war.[104] The summer of 1942 also saw the arrival of very large numbers of Sherman tanks from the USA, armed with a 75mm long anti-tank/HE gun. Three hundred had been offered to Churchill in June 1942, and were put on seven ships in July, one of which was lost, but the tanks were replaced. By 11 September, 318 had arrived.[105] In other words, if there was a German superiority in tank gun and armour quality it was very small and very short-lived indeed.

The old view espoused by Basil Liddell Hart that the British army was reluctant to embrace the tank has been decisively criticized by historians who now suggest the British army overemphasized the use of the tank, especially independently of infantry.[106] British armoured divisions were, as of 1940, very tank-heavy, and based on the idea of fighting tanks with tanks equipped with anti-tank guns. Between April 1940 and May 1942 British armoured divisions had two armoured brigades, artillery and engineers, and very little infantry. Their medium-tank strength was 300, three times that of a contemporary panzer division.[107] From May 1942 until the end of the war one armoured brigade (150-plus medium tanks) was replaced with a whole infantry brigade, and the artillery increased.[108] However, the armoured division was still thought of primarily as a weapon for fighting tank formations, with the infantry there to help the tanks. Only in 1942 and 1943, with experience from North Africa, was the role of the tank downgraded. This led to the recognition of the need to have tanks designed to attack things other than tanks, expressed in the introduction of the Sherman and its dual-purpose gun.

The Allies enjoyed a 3:1 superiority in tanks deployed in Normandy, and the vast majority of German tanks were of the same quality as British and US tanks.[109] The powerful German Panthers, Tigers and King Tigers, some armed with improved 75mm and 88mm guns, were available only in small numbers, though the Panther was nearly as common as the Panzer IV, the most common tank. The Germans deployed only some 650 Panthers and 120–30 Tigers in Normandy.[110] The Allied forces had over one armoured formation to two infantry, whereas for the Germans it was 1:4; for the British the ratio was about 1:1.

The million-strong British-commanded 21st Army Group that fought from Normandy to Germany (excluding attached US forces) was an extraordinarily armoured force. It had seven armoured divisions (four British, two Canadian, one Polish) and six armoured/tank brigades (three British, two Canadian and one Czech), as well as eleven infantry divisions.[111] In terms of gun-power the German tanks were matched by the Sherman Firefly, a British adaptation of the Sherman to take the new British 17-pounder 76mm anti-tank gun. The 17-pounder was apparently superior to all German guns[112] and with the new but rare discarding sabot round could destroy even the heaviest German tanks.[113] The 21st Army Group had around 300 Fireflies at the beginning of the Normandy campaign, one for every three Shermans.[114] The proportion of Fireflies increased to 50:50 as new HE ammunition for the 17-pounder became available. It was not until the very end of the war that Britain produced the Comet tank, armed with a gun related to the 17-pounder, firing discarding sabot shot if necessary, which was close to the Panther in overall performance. This is not to say that British tanks before the Comet were necessarily inferior in armour: the Mark VII Churchill was more heavily armoured than the Panther or Tiger.[115]

Just as surprising is that British tank forces were better equipped than US ones. The US forces also acquired a Sherman with a 76mm gun, but this was inferior to the 17-pounder as an anti-tank gun.[116]

The Cromwells and the Churchills of the Anglo-Canadian forces were regarded as equal or superior to the Sherman. Except when they carried the 6-pounder these tanks had essentially the same gun as the Sherman; the Cromwell was just a little bit lighter than the Sherman, but faster and similarly armoured. The Churchill was considerably heavier, better armoured but slower.[117] When mounted with the 6-pounder 57mm gun, the British tanks had a more effective anti-tank (though single-purpose) gun than the 75mm.[118]

There was one other distinctively British aspect of the Normandy landings and the campaign in North-West Europe. Churchill appointed the pioneering tank officer Percy Hobart to command a special very large armoured division (the 79th) to hit the beaches of Normandy, and then proceed, with a whole series of wonderfully adapted tanks. Among them were swimming 'Duplex Drive' or 'DD' Sherman tanks, which could make it ashore on their own power; Sherman Crabs, with anti-mine flails attached; AVREs, Churchills armed with a spigot mortar, firing a 'Flying Dustbin' of explosive a short range to remove concrete obstacles; Crocodiles, Churchills armed with flamethrowers; and Grant CDL tanks, which had powerful searchlights to dazzle the enemy.[119] Some of these machines were clearly useful in particular contexts, but whether they warranted the investment involved is not at all clear. The scale of the 79th Division was extraordinary. It had – at D-Day – three brigades: one of engineers with AVREs, one of Sherman Crabs and one CDL, its DD Sherman brigade having been attached elsewhere for D-Day. It would later have five brigades: one of AVREs, one of Sherman Crabs, one of Sherman DD swimming tanks, one mainly of Crocodiles and one of US-designed and US-built tracked landing craft. It is striking that the much larger US forces, while they adopted the DD tank for the D-Day landing, did not take up most of the other devices. As in the case, possibly, of Mulberry, and certainly of PLUTO, it is evidence of both the plenitude of resources available to British forces, and their tendency to go for complicated devices which the Americans avoided.

Despite this, some officers in Normandy claimed that British tanks were once again much inferior to German, and once again questions were raised in the House of Commons by Richard Stokes. Once again, historians followed, condemning British tanks and holding their low quality responsible for setbacks suffered by British forces.[120] More generally, historians have overplayed the differences in quality between British and German equipment, systematically neglecting the good, sometimes superior qualities of the former. To an even greater extent they have underplayed the comparatively lavish scale of supply of weapons to the British army.[121] It was not just a matter of tanks. The British forces had ample supplies of transport of all kinds, from lorries to jeeps to universal carriers, Weasels, Kangaroos, Buffaloes and Terrapins. The Germans relied on horses and carts. Although German infantry units had a notionally higher proportion of machine guns and submachine guns, in 1942 and 1943 Britain alone out-produced Germany in these weapons by a factor of four to five and both were producing about the same in 1944, without even taking account of overseas British production.[122] In Britain, submachine gun (Sten) production was in peak years much greater than rifle production; German submachine gun production was never more than about a third of rifle production.[123] Sten guns were not just used by British forces, with some 600,000 sent to resistance units by the SOE, but it seems highly unlikely that this explains the apparently low proportion in British units.[124] The Sten was similar, though perhaps slightly inferior, to the German MP40,[125] and both were copied by the Americans in their M3, introduced late in the war to replace the Thompson (or 'tommy gun').

On top of this British soldiers had much better medical care than did the Germans. In North Africa British troops had a much lower rate of sickness than the Germans, and better treatment for the wounded.[126] The army was effective in adopting new techniques and medicines, such as the sulphonamides, Treuta's methods of encasing wounds in plaster, and tetanus immunization; the army blood transfusion service was 'the envy of all combatant nations' in the war. Britain was the only nation to go to war with a fully working blood transfusion system.[127] British troops landing on D-Day had DDT impregnated shirts, and sprays and so on were used to control flies and other insects. The German forces had rates of louse infestation thousands of times greater than those of the British.[128] On top of all this, the British soldier was rich by European, if not US, standards. 'A British private, wretchedly paid as he is, earns more than a foreman at the Navale Meccanica, while an American private – who can shower cigarettes, sweets, and even silk stockings in all directions – has a higher income than any Italian employee in Naples,' noted the eye-witness.[129]

BUREAUCRACIES

One very obvious feature of the war was the growth of bureaucracy, a machine (to use the language of the time) which was needed to organize the vast enterprises, military and civil, which the state had created, and to control much of the rest. One index of its growth was the huge and increasing amount of paper consumed by government in a world where paper supplies were much reduced. While supplies for newspapers and books were cut, government (including HMSO, the government publisher) consumption of paper surged from 3 per cent of use for printing to one third, quadrupling in quantity.[130] By far the biggest users were the service ministries (headed by the War Office, way out in the lead with 25,000 tons in 1942).[131]

Great controlling ministries were built up to run arms production, based in pre-war corporate headquarters. The Ministry of Aircraft Production moved into ICI's headquarters in Millbank; the Ministry of Supply into Shell-Mex House on the Strand, which was the centre also for the control of oil supplies. Some parts of ministries went out of London: some Admiralty offices went to Bath and much of the Ministry of Food was to operate from the Welsh seaside resort of Colwyn Bay. The bureaucracy spread abroad. Cairo and Washington, and many other places, sprouted military and civil British bureaucracies to coordinate and control the most mundane materials, as well as decide on high policy.[132] In 1943 Lord Louis Mountbatten was sent to New Delhi as head of South-East Asia Command; the HQ was moved to Kandy, in the centre of Ceylon, and further from the front, in early 1944. Mountbatten installed himself in the King's Pavilion within the grounds of the old royal palace of the kings of Ceylon, on the shore of Lake Kandy, and installed his HQ in the beautiful Botanical Gardens in nearby Peradiniya. A new airfield was built, as well as headquarters for army and air force. The total staff was an incredible 10,000. Headquarters life was a 'byword for elegance and luxury' and was indeed 'an expertly contrived theatrical entertainment'.[133] The British government even had ministers stationed abroad. There were Ministers Resident in the Middle East, West Africa and Washington from 1942 to the end of the war.

What kind of men were at the top of the machine which acquired the weapons of modern war? It was certainly not politicians or technocrats of the left. At the very top were, as we have seen, Churchill and his production ministry cronies, and Cherwell, with a tiny leavening from the left. In the bureaucratic machine an old elite, drawn from the state, industry and the military-industrial complex dominated. Their control was challenged, but not very strongly. Before the war the armed services had been in charge of supply and research for their services. However, in a powerful expression of the view that there was a radical difference between the worlds of production, science and engineering, on the one hand, and the military users on the other, there was a move to put production and research under civilian ministries. A Ministry of Supply was created before the war, to supply the army, and Churchill created the Ministry of Aircraft Production in May 1940. The navy kept control of its procurement and research and even added merchant shipbuilding to its responsibilities. Except in the naval case, it seemed, the logic of modern industrial war had been institutionalized in the new arrangements, the power of civilian ingenuity and efficiency unleashed from the dead hand of the backward military. Indeed, stories used to be told of how Lord Beaverbrook transformed aircraft production by taking it away from the air marshals and bringing in the push and go of the industrialist.[134]

In fact, what had happened was not the creation of new institutions, but rather that the procurement and research parts of the air force and army were renamed ministries, and ministers put in on top. At official levels senior personnel remained much the same, or were supplemented primarily with men from industry. The Chief Executive of the Ministry of Aircraft Production between 1942 and 1945 was the brilliant Air Marshal Sir Wilfrid Freeman, who had run production and development at the Air Ministry from the late 1930s.[135] The Ministry of Supply in its early days had its production departments under Engineer Vice-Admiral Sir Harold Brown, who had been in charge of munitions production at the War Office since 1936. He was, according to Harold Macmillan, who served in the Ministry of Supply, a 'man of quite outstanding quality' who was 'always cheerful, always smiling, never rattled, combining efficiency with charm, loved and trusted by all'. Macmillan made the point that as an engineer admiral he had 'great technical knowledge and experience, and was personally known to nearly all our main contractors'. Sir Harold Brown, wrote Macmillan, 'was soon one of my heroes and has remained so'.[136]

At one stage of the war the three key procurement officers all knew each other from their days as naval officers, and lunched every week at the Carlton Hotel, where they were joined by others, including junior ministers for supply and aircraft production. They were Admiral Sir Harold Brown, Commander Sir Charles Craven, who had long been the key figure in Vickers (and the great merchant of death of the 1930s) and was now at the Ministry of Aircraft Production, and Admiral Sir Bruce Fraser, the only one still in uniform, who had been a gunnery officer and was Controller of the Navy. The group had been founded by Sir James Lithgow, the shipbuilder and the man responsible for merchant shipbuilding at the Admiralty. They 'were known throughout Whitehall as the "boilermakers"'. Many crucial disputes between their ministries were solved in a 'genial manner in private' by these 'old and intimate friends'.[137] While Fraser was the Controller of the Navy, Brown and Craven were the Controllers-General of their ministries.

It would be tedious to list all the businessmen and senior servicemen central to this world of military-industrial control, and especially to try and give their positions and when they held them, but some examples are needed to make the point. Let us take some of the later 'boilermakers': they included at other times Lieutenant-General Sir Ronald Weeks of the War Office (where he was in charge of supply – he was a Pilkington employee), Sir Robert Sinclair (of the War Office, and later the Ministry of Production, previously Imperial Tobacco), Sir Graham Cunningham (Ministry of Supply, previously Triplex Safety Glass), Sir George Turner (civil servant, Supply) and Sir Cyril Hurcomb (Ministry of War Transport).[138] Other senior officials included Alexander Dunbar of Vickers, another Controller-General at the Ministry of Aircraft Production, and at the Ministry of Supply, Percy Mills of W. T. Avery Weighing Machines, Alexander Roger of Birmingham

Small Arms, Oliver Lucas of Joseph Lucas (the electrical-component maker who supplied the car and aircraft industries), William Rootes of the eponymous car firm, and Hugh Weeks from Cadbury's. Hugh Beaver, a partner in the civil engineering firm Alexander Gibb and Partners, Engineers, was Director-General and then Controller-General of the Ministry of Works.[139] In 1942 there were two ICI directors, five chairmen/directors of groups and thirty-three other staff seconded to ministries, including the Ministry of Supply. The absence of trade unionists, and indeed of academic scientists, should be noted.[140] As well as serving officers and businessmen the higher reaches of the ministries were occupied by scientific civil servants, who ran many of the important laboratories of the ministries. The directors of technical development and of engine development at the Ministry of Aircraft Production were long-standing technical civil servants, as was the man who ran the jet engine project. So were the heads of great laboratories like Farnborough and the Telecommunications Research Establishment and the heads of the Porton Down Station.

There was a wartime revolution in Whitehall, but it was not a general one. It concerned mainly the armed services and defence supply departments, bodies which even before the war did not correspond to the usual images of Whitehall. These were places where serving officers and technical experts had very great power even at the top of their ministries. These bodies grew and grew, changing the character of the state as a whole. Within them, technical experts certainly prospered as against serving officers, and outsiders (mainly from the military-industrial complex) were indeed brought in. But they reflected a world of the armed services, industry and technical experts in arms.

Academics did enter this world. Universities lost a good proportion of their most talented, ambitious and well-connected dons, for universities were not places where one could have a good war. They were seen as sources of the kind of men who might be called on to help the state in senior administrative and advisory roles. It was to dons, for example, that the administrative civil service looked for recruits to enter at the rank of principal, and it was to the universities too that the state often looked for senior professionals and for expert advice, and for senior recruits to the various expert services of the state. Many of them, especially those who rose to the top, were Conservatives, often long connected to the services: men like Charles Goodeve, who rose to be in charge of naval research and development, and John Lennard Jones, who went to Fort Halstead to run armament research. The academic economists concerned with the planning of production where it mattered – in the supply ministries – were often personally hostile to economic planning. A key case was John Jewkes, a practitioner of 'realistic economics' from the University of Manchester.[141] Jewkes went on to run the planning section of the Ministry of Aircraft Production and then went to the Ministry of Reconstruction. He was hostile to Keynesianism too.

Of course, many less doctrinaire liberals, and indeed the odd socialist intellectual and expert, were to be found in the wartime state – from Maynard Keynes and William Beveridge to the economist and politician Hugh Dalton. Dalton went to the Ministry of Economic Warfare in 1940 and brought in the economists Hugh Gaitskell and Noel Hall, formerly of University College.[142] But Dalton and his Labour-sympathizing economists were later to go to the Board of Trade, the old industry ministry which ran non-arms industries during the war. Similarly, the scientific intellectuals in favour of planning were generally engaged in the rather special realm of scientific advice and operational research.

The intellectual left did, however, have a disproportionate influence on how the role of scientists and intellectuals was perceived. Angus Calder's *The People's War* stated that 'the war was fought with the willing brains and hearts of the most vigorous elements in the community, the educated, the skilled, the bold ... who worked more and more consciously towards a transformed post-war world' while arguing they were betrayed; 'the brilliant scientist willingly devoted his creative energy to military purposes' but was, it was realized too late, 'working, not for the people, but for those few who would control the atomic bomb'.[143] But this view is to deny what might have been the most obvious features of the war – that the immensely powerful and creative warfare state, its scientists, its technicians, were generally not working for a transformed world; they were not betrayed. It was the old world which had created the new worlds of war – from the airfields to the bombers, the battleships to the tanks, and the research laboratories and new weapons.

The Second World War brought the classes together, but neither side liked what it saw. Something similar happened when academics and intellectuals stepped down from the ivory tower and faced the real world. Faced for the first time en masse with experts from all sorts of backgrounds they learnt a great deal, and also worked with these 'practical men'. But they were often appalled at what they saw – the want of education, vision, sophistication. Thus Lord Keynes, easily the most important academic intellectual in the state machine during the war, railed against the bone-headedness of the cotton masters; Hugh Dalton, another academic economist hardly used to the world of business, was less than impressed by them and by many other industrialists. Academic scientists from Cambridge sneered at the poor devils in the scientific civil service. Academic historians were aghast at the reactionary witlessness of the clubland adventurers, soldiers and former colonial policemen of the secret services. The academic outsiders had a particular advantage – they kept diaries, wrote books, spoke to the influential, wrote the histories, and not surprisingly they came out best. However, most of the war effort was run by men they often despised.

8

Boffins

In March 1942 it was claimed that 'The one great glory of this country in the last twenty years has been our scientists. It was the greatest country for physical sciences in the world.'[1] The speaker was a middle-ranking civil servant and personnel officer for scientific civil service officers addressing a meeting of the Parliamentary and Scientific Committee. Continuing in this vein of bombastic overstatement C. P. Snow claimed that 'At one point in this war nothing on earth could have saved us had it not been for Radiolocation', going on to say that 'perhaps it was not known, but at the start of this war, Radiolocation was entirely financed from the petty cash of Cavendish laboratory'. This bizarre claim was made in the presence of Sir Henry Tizard, then scientific adviser to the Ministry of Aircraft Production and member of the Air Council, who, to judge from the record of the meeting, did not demur.[2] Yet it was utterly false. Radiolocation, or radar, was the product not of the academy, but of the scientific civil service; the project could have funded the entire Cavendish out of its petty cash. There has been, as in the case of Snow, a tendency not only to claim too much for British devices, but to attribute British success in the technical war to academic science. Such stories are bolstered by the conceit that the Second World War was a 'physicists' war' (whereas the Great War had been a 'chemists' war').

Even before the war C. P. Snow had emerged as a minor scientific intellectual.[3] Yet his reputation was to become ever larger, and he was to shape perhaps more powerfully than anyone else how British science and indeed expertise were thought about. Although to attack Snow today is to attack a straw man, for decades he spoke what seemed the obvious and scandalous truth. He consistently identified science with academic physics (as well as making the equally silly identification between the 'traditional culture' and the novel) and he moralized about science in a naive way, casting out scientists who didn't fit his story, especially Frederick Lindemann. Yet his biggest contribution was his seemingly authoritative echoing of the claim that the scientific expert was not listened to by the powers that be. Such stories linked themselves to different ones in which inventors were not listened to. Indeed, after the war, Barnes Wallis and Frank Whittle became, contrary to all reasonable interpretations, prime examples of inventors who were insufficiently supported by government before and during the war. It was said that Britain was good at inventing but bad at developing, as their cases supposedly exemplified.

However, the usual examples cited were of inventions which Britain did in fact develop during and/or after the war – jet engines, penicillin and atomic power. They were also bad examples in that two were not in any event unambiguously British inventions. It is rarely if ever said that Britain was moderately good at inventing, but very good at developing, yet many cases fit that pattern well. Neither radar nor television were recent ideas in the 1930s; nor were they by any stretch of the imagination exclusively British. Yet by 1936 Britain had the first public television service and in 1939 still had the largest number of receiving sets of any country in the world. It also had the largest radar system of the late 1930s.

Wartime Britain saw a quite extraordinary cult of invention and the inventor, whose high priest was the Prime Minister himself. Gadget factories of all sorts flourished under his leadership, driving others to exasperation. Inventors of many different kinds were indulged to an extraordinary extent, so much so that, as we have also seen, some academic scientists and advisers were concerned that resources were being wasted on a large scale.

The degree to which wartime Britain invented, developed and refined machines would be scandalous if Britain had been as poor and weak as used to be suggested. In fact it was no surprise that a rich country not be especially affected by the war, and one committed to novel ways of fighting, should do exactly this. Far from being a matter of academic petty cash, invention was the province of industry, government establishments, the entrepreneurial inventor, and, indeed, the charlatan and conman. There were many worlds of invention, sometimes in conflict with each other. Cobblers did not always stick to their lasts.

Invention in war and in peace is necessarily a story mostly of failure. Many inventions are bound not to work as well as expected, and even fewer are likely to end up being used. In a country as inventive as Britain it was simply impossible to take up but a small fraction of workable inventions. We tend to know only (some of) the successes, which often come with stories of the superhuman effort of the inventor against the conservatism of the authorities. We need to recognize the equally superhuman efforts which led to failure, and to acknowledge that in the war the authorities were not defending old machines against attacks from the novel, but rather choosing between any number of novel machines.

We should also note that even the most successful inventors, like Whittle and Wallis, were to tell stories of frustration, delay and indifference from government. From these and many other accounts we could easily adduce a story of wartime invention utterly at variance with what happened.

THE BIG PROGRAMMES

When thinking about large-scale inventive effort during the Second World War it is as well to start with aero-engines and airframes. They had, from the interwar years, consumed perhaps half of what would now be called 'research and development' expenditure. We can get a good picture of the state of play for the period May 1940 to spring 1941 from a brief memorandum on research and development produced by Lord Beaverbrook for his Cabinet colleagues, in which he spoke of production as the 'business' and development as the 'passion' of his ministry. This period saw the appearance in service of three new fighters and three new bombers, trials on a new fighter, and the first flight of four prototype fighters and three bombers. By contrast only five new engines either came into service or came close to it, while the first airborne test of the 'Whittle Jet Propulsion Engine' was a few weeks away. Beaverbrook claimed that 'From the first day the Ministry of Aircraft Production was formed, enquiry and experiment, research and development, has been pursued with restless zeal. There has been no restriction or limitation.'[4] This is not, however, the view of the official historians, who see a nine-month interruption in some development, though this account is open to challenge, as this chapter will show.[5]

But what was the fate of these introductions into service and innovations? Of the engines, the Rolls-Royce Vulture was a disappointment which led to the redesign of the Manchester to become the Lancaster with Merlin engines (one of the aircraft to have its first flight). The Napier Sabre (with twice the power of the Merlin) proved very problematic in early production and was in effective service only at the end of the war, in the Typhoon (being tested in this period) and its development, the Tempest. The Hercules VIII was a turbo-supercharged engine for one of the new aircraft, the stratosphere bomber Wellington V (not to be confused with the Victory Bomber, with the embryonic Barnes Wallis project discussed below) – this project did not proceed beyond prototypes. The other two engines were refined Merlins: the Merlin XX, which was to power the Lancaster and Halifax bombers and later-model Hurricanes, and the Merlin XLV, which would power the Spitfire Mk V. Of the airframes, the Westland Whirlwind twin-engined fighter with a Rolls-Royce Peregrine engine was never as successful as was hoped as a successor to the Spitfire and was withdrawn in 1943. The Fairey Fulmar, a two-seater fighter for aircraft carriers, was similar to the Fairey Battle bomber; also equipped with a Merlin, it was to be replaced early in the war with the naval Spitfire (the Seafire) and US aircraft. The Beaufighter I and II would both be successful minor aircraft, especially as night-fighters. The Mosquito photo-reconnaissance aircraft would turn out to be an extraordinary success in that role and as a bomber. The Miles fighter – like the Mosquito made of wood – powered with the Merlin XX, did not make it past the prototype stage. And the Hawker Tornado failed because of its Vulture engine. As far as the bombers were concerned, the Stirling, Halifax, Manchester and Lancaster all went into service. The Wellington V was not developed, and the Fairey Barracuda torpedo bomber, originally designed for the Rolls-Royce Exe, was converted to Merlins but did not enter service with the Fleet Air Arm until 1943.

Problems with new engines were critical. The high-power Napier Sabre and the Bristol Centaurus were seriously delayed by technical problems. Many other pre-war engines were cancelled or became of minor importance, such as the Rolls-Royce Vulture, Exe, Peregrine and Crecy, and many airframe projects failed without them.

The complexities and uncertainties of development and production didn't go away because of the emergency. There were, to be sure, serious attempts to speed things up, but they didn't work, not least because machines got more complex. One method was 'ordering off the drawing board', that is to say, putting in production orders before the machine had been made, so that firms could prepare for production. It had an appalling record. Thus the Hawker Tornado was ordered off the drawing board but never put into production; the Whittle jet engine order kept precious factory space unused for years; the Churchill tank took years to get right.

WALLIS AND WHITTLE

Barnes Wallis and Frank Whittle, probably the best-known British inventors of the twentieth century, worked in the air-industrial complex. Both were supported very strongly by it because they came up with new ideas which were felt to be worth trying; their revolutionary inventions were supported because they were so novel, because they promised new offensive capabilities, which promised early victory.

In the summer of 1940 Wallis hatched an incredible war-winning scheme, 'a monster bomber to smash the Germans'. By mid-July 1940 he had come up with a plan for a 'High Altitude Stratosphere Bomber' (soon to be called the Victory Bomber). This would be a gigantic thing, with a fully laden weight of more than 50 tons, much more than the 30-ton future Lancaster bomber. It would carry 10 tons of bombs but at nearly twice the altitude of the future Lancaster, 40,000 feet, and for a considerably longer range.[6] Wallis claimed to his superior in Vickers that 'the new machine is going to be the instrument which will enable us to bring the war to a quick conclusion'. He explained that given the altitude they would operate at, they could fly 'at their leisure and in daylight'. Large numbers were unnecessary: 'irreparable damage could be inflicted on the strategic communications of the German Empire by ... ten or twenty machines within the course of a few weeks'. He was confident that a new gyroscopic bombsight would give bombing accuracies of 150 yards from 40,000 feet.[7] Such a machine had another advantage: Britain would have 'the machine and the experience to enable us to step right into the forefront of trans-oceanic civil aviation directly war ceases'. It could fly direct to New York in twelve hours.[8]

Vickers put this plan for a Victory bomber forward to Lord Beaverbrook, the Minister of Aircraft Production, in November 1940: the aircraft, pressurized and built to Wallis's geodesic principle, which was successfully used in the Wellington bomber, would have six engines. A map showed that it could bomb Moscow from London. The idea was to attack coal mines, underground petrol and oil storage, oilfields and hydroelectric dams, to strike at these 'sources of power' rather than humble power stations.[9] Wallis's concept was one of 'anti-civil engineering bombing', as he called it; he wanted to destroy great works.[10] The project was taken up by the Ministry of Aircraft Production, but work was dropped in 1942.

Two crucial aspects of the scheme did live on. The first was the idea of attacking civil engineering works, in this case dams. In 1943 Lancasters carrying the 'bouncing bomb' designed by Wallis destroyed, temporarily, some dams in the Ruhr region. The RAF did not bomb the repair work; nor did it repeat the use of the bouncing bomb – it was far too expensive in terms of crews lost. In fact the remaining stock of bouncing bombs was dumped in the sea at the end of the war.[11] The second was the idea, already in play in the Victory bomber, of bombs much larger than conventional ones, which would destroy large and strong structures. Great ingenuity and resources went into developing this concept. Wallis planned that the Victory bomber should drop new 10-ton bombs, many times bigger than current ones. Wallis designed the 6-ton Tallboy and then the 10-ton Grand Slam bombs, which were dropped from 1944 and 1945 respectively, but from Lancasters. Tallboys were used with some success against great structures like U-boat pens, the *Tirpitz*, which was sunk in late 1944 as a result of the third attack with these bombs; the Grand Slams were used to destroy viaducts and bridges, including the Bielefeld railway viaduct. In fact the viaduct had been under continuous attack since June 1944 by US and then British airforces, the latter using Tallboys. It was damaged but was quickly repaired, and a bypass was built making the viaduct significantly less important. The bypass had to be used for years after the war, as the viaduct was destroyed in March 1945 by the first, and much celebrated, Gland Slam raid.[12] Wallis's 'bouncing bomb' was of course the subject of *The Dambusters* film of 1955. As told in the film, bizzarely yet influentially, Wallis battled with a bureaucracy disinclined to accept new ideas from engineers. Yet Wallis received massive support, to produce a weapon which was in the event used on only one raid, one whose results were not as great as expected.

After Wallis, perhaps the most famous wartime inventor was Frank Whittle, in British eyes *the* inventor of the jet engine. Whittle was an air force officer who had been sent to Cambridge to study engineering in the 1930s; like others, he was given the opportunity to stay on and he developed his idea of the jet engine. With air force support he was seconded to a private company set up to develop his ideas, and this company, with state contracts, was to carry the main but not the only early development effort. All this happened before the war: by 1939–40 the project was well under way. By 1939 plans were being made to fit it to an aircraft; by early 1940 the jet engine was on a list of 'war winners' and by late 1940 was ordered into production.[13]

There was a brief hiatus in mid-1940, but Lindemann's interest in jets was 'an important factor in the remarkable renaissance in jet propulsion' in autumn 1940.[14] During the war the number of jet engine projects multiplied, with the Whittle group one of many, in what was a state-orchestrated collaborative programme, involving many large firms in turbine engineering and aero-engines – essentially Metropolitan-Vickers from October 1940, De Havilland, who got an order in May 1941, Rolls-Royce from June 1941 (though they had employed a jet specialist, A. A. Griffiths, from 1939) and lastly Armstrong-Siddeley.[15] Bristol too were involved. Yet belief in the power of the inventor meant Whittle, without question the key pioneer, was given a lot of scope despite the fact that many and in many respects more powerful players were in the game. His Power Jets company continued to be supported. The first British jet engine into service was a Rover/Rolls-Royce development of the Whittle W2, the W2B/23, called the Welland by Rolls-Royce. The second was the Halford H-1 (later called the Goblin), designed by the piston-engine designer Major Halford for the De Havilland company. Halford was one the big three engine designers of the war alongside Roy Fedden, chief engineer at Bristol until 1942, and Ernest Hives, general manager of Rolls-Royce.[16] The Goblin was considerably more powerful than the first Whittle engines but was, like Whittle's, a centrifugal compressor design.

The Gloster Meteor, a fighter equipped with two Wellands, went into partial service in June 1944 and was used against V-1 doodle-bugs. However, it was no match for the new fighters powered by the Sabre and Centaurus engines. The De Havilland Vampire, with a single De Havilland Goblin, only came into service after the war, as did a new Meteor powered with Rolls-Royce engines. Such late deployment was not intended. Production orders had been given 'off the drawing board' to the Rover car company in 1940. In early 1941 it was envisaged that British jet fighters would be operational in the winter of 1942–3. A jet engine factory at Barnoldswick in Lancashire costing £1.5m and employing 1,600 workers was ready before the engine, resulting in a serious waste of resources in 1942.[17]

Engine development was in effect to be taken out of Whittle's hands late in the war. From 1943 it was intended to nationalize his Power Jets firm, which was supported by private interests, as indeed happened in 1944. Through a series of steps it became part of the civil service after the war. But as he became less important in the actual design, Whittle was celebrated as the inventor of the jet. In 1944 he became a public figure, garlanded with honours from the state and from engineering institutions. Though still a young man he was never to design a jet engine again. Development was now very firmly in the hands of great private firms committed to the next generation of jets. The surprising ending to this story should not obscure the level of support this RAF officer got from the ministry long before the war – remaining a career officer he was seconded to a private firm to develop his invention with government money. The authorities believed in the individual inventor; his special qualities were seen as essential to creativity.[18]

'We have a gift for improvisation,' claimed the Air Ministry's former head of research, Herbert Wimperis, endorsing the view that 'great men are more important to science than great laboratories'.[19] But it would be quite wrong to assume that Britain relied on a few great eccentric boffins, even those within great organizations like Vickers or the RAF. British technical development encouraged individual inventors, but also relied on a massive infrastructure not only of laboratories, but also workshops, research establishments, design offices, testing facilities and teams of anonymous experts.

The arms industry and the government's own specialized inventing institutions were the main source of new devices of war. The largest institutions predated the Great War – they were the Royal Aircraft Establishment at Farnborough and the Research Department at the Woolwich Arsenal. Among the second rank in scale were Porton Down, the centre for chemical warfare work, the various naval centres like the Anti-Submarine Experimental Establishment at Portland, the Admiralty Research Laboratory at Teddington, the Signals School at Portsmouth, and the army's Air Defence Experimental Establishment and Signals Experimental Establishment. The one new body of any size which emerged in the interwar years was the new radar establishment on the east coast, directed by Robert Watson Watt. This was to move repeatedly in the early years of the war, ending up in Malvern as the Telecommunications Research Establishment (TRE). It was run by a scientific civil servant, A. P. Rowe.

Government establishments were very productive of new devices before the war. The Bawdsey research station produced radar, devising the air defence and other systems in place early in the war. Farnborough developed the axial jet engine from the 1930s. Anti-submarine sonar (ASDIC) came from the anti-submarine establishments, and so on. The Sten gun was designed in 1940 by Major R. V. Shepherd, of the Design Department, and Mr Harold Turpin of Enfield. Their names yielded the first initials of the name, 'st', and 'en' for Enfield, echoing the Bren gun. The Enfield Royal Small Arms Factory designed and refined the Lee-Enfield rifle and it was from Enfield that drawings and experts were sent to other factories. Experts in small arms were generally anonymous civil service engineers, usually the products of local grammar schools who, while working as engineering apprentices at Enfield, also studied at London technical colleges.[20] New guns came from Woolwich, from Vickers and from arms makers abroad like Bofors, Oerlikon and Hispano-Suiza.

The first generation of wartime armoured vehicles were designed mostly by Vickers: the Bren gun carrier, the light tanks, and the A9, A10 cruisers and the Matilda I infantry tank (all used in France) were the work of Sir John Carden, 6th Baronet, of the Carden-Loyd light-tank firm bought by Vickers in 1928. Carden, the aristocratic self-taught engineer, was also a promoter of light aviation, especially the French sensation of the mid-1930s the 'Flying Flea', for which he provided an engine. He died in an air crash in 1935. He is practically unknown. His case may be contrasted with that of another Vickers employee, R. J. Mitchell, the designer of the Spitfire, who died two years later and had also come into Vickers in 1928, when the company bought the Supermarine firm.

Private firms come up with many products just as important to the war effort as weapons. The Liberty ship was the work of Robert Cyril Thompson, born in 1907, the son of a Sunderland shipbuilder, a public schoolboy who studied engineering at Cambridge and went on to an apprenticeship in the conventional manner.[21] He joined the family firm in 1930, joining the board a year later, at the age of twenty-four. 'Mr Cyril' led the development of a new kind of cheap ship for the Depression years; its new hull shape was tested at the National Physical Laboratory and it was equipped with a new and efficient reciprocating steam engine from the North-Eastern Marine Engineering Company. After a second visit to the USA in mid-1941, he found himself turned down by the navy, and echoing T. E. Lawrence joined the air force as an aircraftsman, second class, and trained to be a flight mechanic; he served in Italy and was commissioned by the end of the war.[22]

Before the war ICI had spent more on research than any other company in Britain, straddling a number of different areas. Its great project of the 1920s and early 1930s was in the high-pressure chemistry needed to make ammonia and convert coal to oil, techniques which as we have seen were greatly used during the war. But ICI's inventive activities went much further, and were to prove significant for the war. For example, it came up with a way of making what it called 'Perspex', a material also made in Germany and the USA which was to be widely used in aircraft; it invented 'polythene' on the eve of the war, a material which found a quite unexpected but hugely important role in airborne radar sets because of its particular dielectric properties. But that is just to scratch the surface of the mountain of inventions from ICI: it developed processes for the production of key wartime anti-malarials, Atebrin (also known as mepacrine) and Plasmoquin (also known as pamaquin). Neither had been previously made in Britain – they were inventions of the great German firm IG Farben – and ICI was to produce both on a large scale (50 million and 32 million tablets respectively in 1942, later much increased).[23] ICI also developed a new anti-malarial, which came into use after the war – Paludrine.[24] It also worked on sulphonamides and on penicillin. In addition ICI was to get major research and development contracts in the British atomic bomb programme and, indeed, wanted to take over the whole project. ICI Billingham needed and attracted some of the best chemists and engineers in Britain (often Oxford chemists and Cambridge engineers), many of whom would play crucial roles elsewhere. (Sir) Wallace Akers, who graduated in chemistry from Christ Church, Oxford, chairman of ICI's Billingham division until 1937, went to ICI HQ to direct expansion of war capacity and then ran the British atomic programme. In 1944 he became the first ICI main board director of research. He was succeeded at Billingham by (Baron) Alexander Fleck, a Glasgow-trained chemist, later a chairman of ICI. The key technician concerned with hydrogenation was Major Kenneth Gordon CBE, MC, educated at St John's College, Oxford. At Billingham from 1923, running hydrogenation there, he built and ran Heysham. A younger Oxford chemist was (Baron) C. F. Kearton, who went to Billingham in 1933, later working at the iso-octane plants there and at Heysham, and worked on the British and US bomb projects from 1941 to 1945, apart from a spell in 1943 back on high-octane fuel. His specialism was gaseous diffusion; he ran the uranium

hexafluoride diffusion plant at the Valley chemical warfare factory.[25]
Another important Billingham figure was the Cambridge-educated
chief engineer, (Sir) Frank Ewart Smith, who would recruit many like
him to ICI. He was to become the government's Chief Engineer,
Armament Design, in 1942. ICI's (Baron) Christopher Hinton was
seconded to work in wartime ordnance factories. Hinton came from
a family of schoolteachers in Wiltshire; he did a premium apprenticeship at the Great Western Railway works at Swindon, which charged
parents £100. He went on to Trinity College, Cambridge, thanks to a
grant administered by the Institution of Mechanical Engineers, where
he was, he recalled, the 'only man there with a Secondary School
background'. He joined ICI and rose quickly, despite not being part
of the cadre of ICI managers and scientists on the fringe of Cheshire
county society, a problem compounded by his marrying a working-
class woman.[26]

As one might expect, every major British firm was involved in wartime research and development in some way or other. In the case of
radar the view of the official historian was that the government establishments were firmly in the technical lead (aided by young men from
universities), but that the main firms had an important role in radar in
'development for production' and in valve development; while in
radio they did initiate more.[27] The GEC laboratories at Wembley,
and especially their magnetron expert, E. S. Megaw, played a very
important role in the cavity magnetron story (see below). As GEC
themselves put it: 'In collaboration with Birmingham University the
multiple circuit copper block magnetron was developed as a practical
generator.'[28] GEC also developed Group Captain Helmore's 'Helmore
light' (a powerful searchlight in the nose of night-fighters), his scheme
for a gigantic searchlight illuminating the sky and then his gigantic
'Helmover' (sic) torpedo. This was a 5-ton monster launched from a
bomber, powered by a Meteor engine on the surface, and an electric
motor underwater.[29] Carrying a ton of explosive it could travel fifty
miles on the surface, breathing through a mast. It would attack sub-
merged, for up to three miles. It was controlled, by radio, from an
aircraft flying overhead. It was ordered in 1945 for use against the
Japanese, but it turned out to be too late.[30] Helmore was elected to
Parliament in 1943.

One can give only the tiniest flavour of the inventive work of the other radio and electronic industries. EMI, and in particular Alan Blumlein, made important contributions to the H2S radar.[31] British Tabulating Machines, the British subsidiary of IBM, made and helped design the 'bombes' used to break the Enigma codes. The Post Office Research Station at Dollis Hill was involved in the design of the Robinson decrypting machine, and it is possible that the Colossus computer developed from it was entirely the work of a Post Office team led by Thomas Flowers. At the end of the war Bletchley Park had three Robinsons and ten much larger Colossi.[32] Marconi worked on quartz for frequency stabilization, voltage stabilizers and cavity magnetrons, among other things.[33] Pye designed radio sets which were adopted instead of radios designed by civil servants. In the summer of 1939 the Ministry of Supply put out tenders to manufacture their newly designed No. 8 set, a portable device for the infantry. Pye got the contract, but used the money not to make the set but to develop an alternative, which was much lighter and consumed less battery power. It won a trial against the No. 8 in January 1940 and was put into production. From June 1940 it designed a new tank radio, which beat three official designs in trials. The No. 19 radio became the standard tank radio, manufactured in large numbers in the USA and Canada, as well as in Britain.[34] Some of the No. 19 radios made in the USA were supplied to the USSR. Pye had a research staff of around thirty in the war, most of whom started life as apprentices; there were a handful of graduates.[35] Pye engineers who worked on radar 'felt gagged by official secrecy all their lives' and thus did not get the credit they felt their due.[36] Such was the politics of invention.

TEAMS

In 1942 Viscount Halifax, British ambassador to the US, told the Carnegie Institute of Technology that it was 'difficult to exaggerate the importance of the chemist, the engineer and the inventor', these were the 'back room boys', the often unknown workers in both 'laboratories and drawing offices'.[37] By 1945 'the expression "back room boys" had been much publicised', claimed the *Bristol Evening World*.[38]

But this chummy language, like the wartime expression 'boffin', did little to illuminate the public about the sheer scale and nature of inventive activity, and how it was changing. Such terms hardly did justice to what happened in laboratories – in say ICI or Metropolitan-Vickers – in the drawing offices, design offices and workshops of the engineering, aircraft and similar industries, or the government research establishments. The war brought them under a single head, 'research and development', which included the design of a new aeroplane as much as the discovery of a new drug.

The period also saw a clearer emergence of a distinction between the scientist and engineer in general and the research and development scientist and engineer. The latter image has proved so powerful that the researcher, creator, developer, now stands (quite misleadingly) for all scientists and engineers. Before the war the Ministry of Labour had in fact asked the Royal Society to establish a register of research scientists and engineers, in addition to its own registers of engineers, chemists, etc. By November 1939 the Royal Society register had 6,484 researchers, of whom the largest category were the chemists at 1,487, the second category being bacteriology/pathology with 950 and physics a close third with 947. The list included 521 experts in research in 'engineering sciences'.[39] Research, and also development, managers of the interwar years, and certainly of the war period, were keenly aware that different sorts of skills were involved, but that not everyone appreciated this.[40] Thus Charles Goodeve complained that in the navy 'it was not sufficiently known that the qualifications for development work are entirely different from those for routine design'.[41] One particular problem in the army and navy appears to have been that the officers assigned to development work, and to command research and development units, were technical officers, but not researchers. The main wartime President of the Ordnance Board, a gunner and former captain of HMS *Hood* and a protégé of Churchill and Lindemann, Admiral Pridham, raged in his autobiography against the 'Blokes': the Design Department at Woolwich, he complained, 'included no true designers, being composed of "passed-over" Naval Gunnery Officers, and Gunners who had not been considered suitable officers for the Royal Regiment of Artillery. Under these supposed experts worked a number of Draughtsmen brought up and trained in the close confines of Woolwich Arsenal.' The Ordnance Board itself, he further recalled, was a comfortable 'pre-mortem' appointment for inspection officers, who as a result got promoted to Captain RN, Colonel or Group Captain; in addition many 'blokes' were to be found in the Ministry of Supply.[42]

Furthermore there were many criticisms about the quality of the researchers in government establishments. Pridham claimed 'only second rate scientists were to be found in the Research Department, and these insisted on working behind a jealously guarded veil of secrecy, and refused to consult anyone not directly concerned with the Woolwich Arsenal'.[43] Secrecy is here a tool in the hands of scientists, not something imposed upon them. The bright young men from elite universities could be scathing about the men they found in the scientific civil service. Poor B. S. Smith, the chief scientist at HMS *Osprey*, the anti-submarine establishment (which was at Portland, but moved to Fairlie in Scotland), was described by the wealthy young Cambridge geophysicist Edward Bullard as a 'first class obstructionist' who 'insists on work on quite hopeless schemes being carried out'; 'he was only interested in demonstrating what a fine fellow he was and what fools I and the mining experts are' – 'the fundamental difficulty is that Smith runs the place for his own glorification and not for the sinking of submarines'.[44] C. S. Wright, the Director of Scientific Research, was in Bullard's view a 'most unsuitable war time DSR as he just lets things drift and can only be persuaded to take action when it appears that a worse row will occur if he does nothing'.[45] The war took elite academics into jobs they would not have considered suitable in peacetime. There were important cultural differences between a naval laboratory and an elite university. They were probably much greater than those between different types of elite culture. Take, for example, the case of Richard Keynes, who in 1940 went to work as a naval scientist, first in ASDIC and then radar. He had a very low opinion of his first boss, the unfortunate B. S. Smith, and kicked up quite a stink in mid-1942 (that period again) about how it was run.[46] His father, Geoffrey Keynes, a surgeon, was Rupert Brooke's literary executor and an authority on William Blake; his mother was Elizabeth Darwin, granddaughter of

Charles. Among his uncles were the economist John Maynard Keynes and the physiologist A. V. Hill. Or take Andrew Huxley, just slightly older – grandson of T. H. Huxley and half-brother to Julian and Aldous Huxley – who worked in operational research; and his collaborator Alan Hodgkin, from a family of historians, who spent most of the war at TRE. All were physiologists, all from Trinity.

Many of the military and naval research establishments were run by military and naval officers. The Research Department at Woolwich was under a military or naval Chief Superintendent, with the rank of brigadier or naval captain. This was held essential 'to secure what was called "User" interest', recalled the first civilian and scientific director, John Lennard-Jones.[47] But for some the users should have been men in the field, not uniformed technical officers. Admiral Pridham set out to represent the user interest against the production and inventor interests. He was interested in quality and looked to independent scientists to help. The Admiral had a touching admiration for the senior scientists of the Royal Society: 'I have never elsewhere met such sincere humility; the greater their knowledge the more humble they become,' he recorded.[48] He often consulted A. V. Hill, and had as Resident Advisers on the board the Oxford mathematician and astrophysicist E. A. Milne and the UCL statistician E. S. Pearson; both were concerned with examination of trials and of statistics.[49]

The war was to see major changes in the direction and staffing of the research and development establishments of the Ministry of Supply. They happened in the crucial year 1942. In particular, the Ministry's Woolwich Research Department and the Design Department were both civilianized, thus throwing out the military and naval old guard. New civilian staff were also brought in from both the academy and industry. Frank Ewart Smith, the chief engineer at ICI Billingham, became Chief Engineer, Armament Design and head of the Design Department. He had played a key role in the development of the Blacker Bombard mortar and PIAT (Projector Infantry Anti-Tank gun) at ICI.[50] He brought with him many talented engineers from ICI and elsewhere, including Dr Richard Beeching from Mond Nickel. Also civilianized was the Research Department. Its first civilian director was John Lennard-Jones, Professor of Theoretical Chemistry from Cambridge, and the director of the Cambridge Mathematical Laboratory. He found it 'a somewhat bewildering experience to be put in charge of an army when you have previously been in charge of a platoon'.[51] Of all the academics recruited into government research Lennard-Jones was running by far the largest research organization – the other big ones were in the hands of career scientific civil servants. Lennard-Jones established his headquarters in the deserted former Projectile Development Establishment in Fort Halstead, near Sevenoaks, and recruited senior staff largely from the universities, many like himself fellows of the Royal Society.[52] In the last years of the war the Research Department was a vast, widely scattered organization, manned by thirteen higher scientific and technical officers with titles such as superintendent and the like, twenty-one principal scientific officers and 426 scientific and experimental officers of various grades, together with 303 in assistant grades, a total of 763.[53] To them one needed to add large numbers of military personnel, non-industrial and industrial workers, taking the total staff up to more than 3,000 in 1944.[54] When Professor John Lennard-Jones was appointed, one of the first things he was told was that the scientists in his care felt neglected and insufficiently recognized, both in the world of science and through the honours system.[55] One of the tasks he set himself was to 'seek for recognition of staff by political and scientific honours'.[56]

Lennard-Jones was to single out two technical achievements of the research department. The first was a new drive in anti-tank weapons which led to the discarding sabot; the second the development of new aluminized explosives for the RAF.[57] By 1942 the powerful 17-pounder anti-tank gun had been developed; attention switched to the development of high-velocity shot, and expert teams containing mathematicians, physicists and engineers were assembled in the Armament Research Department and Armament Design Department. After a few months they concentrated on developing the discarding sabot shot, for both the 6- and 17-pounder guns. The idea was to use a small shot with a tungsten carbide core encased in a 'sabot' which was discarded the moment it left the gun. The result was a relatively light but very strong shot fired at exceptionally high speed. The Research Department did not invent the discarding sabot, but rather it made it work effectively.

The new shot for the 6-pounder was demonstrated some weeks before D-Day but 'much "user research" was necessary' before the army could use the ammunition effectively. Shot for the 17-pounder was also produced. As the post-war PR put it: 'With the new invention, an advance through quality had overcome the German mania for the huge and monstrous.'[58] As we have seen, there was something in this, since the 17-pounder with the discarding sabot was indeed capable of destroying the heaviest German tanks and was the most effective anti-tank gun of the war. British tanks and anti-tank units in Normandy thus had very powerful new weapons.

The story of aluminized bombs is also in some respects one of catching up with the Germans. It is remarkable for the simplicity of the innovation and also for its impact. It is also a powerful instance of the value of a discerning adviser and critic, in this case Lord Cherwell. In a discussion of the effects of possible German rockets in autumn 1943, Cherwell had noted that it had been assumed that German explosives were twice as effective as British. This led to an investigation which established that this was indeed the case, because the Germans were adding aluminium powder to explosive. The British had, back in 1941, got the effect of adding powdered aluminium wrong. According to Churchill, Cherwell's 'roving eye' led to immediate efforts to aluminize, and to huge increases in power.[59] In mid-1943 bombs with the powerful RDX/TNT mixture had appeared. The aluminized Torpex (TNT/RDX/aluminium) was used in torpedoes and the Tallboy and Gland Slam bombs. Bombs appeared filled with Tritonal (TNT/aluminium) and Minol (TNT/ammonium nitrate/aluminium).[60] It seems that the US followed the British in this, preferring Tritonal to Minol.[61]

ACADEMIC INVENTORS

Universities were not, as we have noted, inventive organizations. Yet there were close connections to industry, and some examples of university inventions before the war. The still-existing rodent control firm Rentokil was created by the Professor of Entomology at Imperial College, who had devised a poison for woodworm; he was killed in

14. Potatoes as a substitute for imported wheat and flour

15. A merchant ship in wartime grey unloads flour from the USA and Canada into lighters, 1943

16. Food imports: Bacon from the USA replaces Danish bacon, 1941

Meat shipped to Britain actually increased during the war, including corned beef from Uruguay

18. Domestic food production: A refugee Belgian fisherman working out of Brixham, Devon

19. Domestic food production: Sugar beet harvesting with women workers and the US-designed Fordson tractor, made in Dagenham

20. Fuel: Land Army women training to saw larch poles to lengths for pit props, previously imported and essential for coal-mining

21. Fuel: One of the greatest new wartime factories – the synthetic aviation spirit plant at Heysham, Lancashire, which relied on imported base-oil

22. Fuel: The largest British aviation spirit plant – the Anglo-Iranian refinery at Abadan. Much of its aviation spirit went to the Soviet Union

23. The recently completed HMS *Howe*, 1942. It cost as much as a large factory

24. Inspecting Merlin aero-engines at Rolls-Royce Hillington, near Glasgow, 1942

25. A munitions factory in Beverley, Yorkshire, 1944, painted by Frederick William Elwell

26. The Rolls-Royce Hillington factory from the air, 1940

27. Forging a big gun, Sheffield, 1941, painted by Sir Henry Rushbury

28. Avro Lancaster bombers nearing completion at the A V Roe & Co Ltd factory, Woodford, Cheshire, 1943

29. The most senior of the many scientifically trained ministers in wartime: Sir John Anderson, member of the War Cabinet 1940–45 and photographed here as Chancellor of the Exchequer, 1944

30. Asymmetric warfare: the tonnage of bombs dropped on German Europe compared with German bombing of the UK

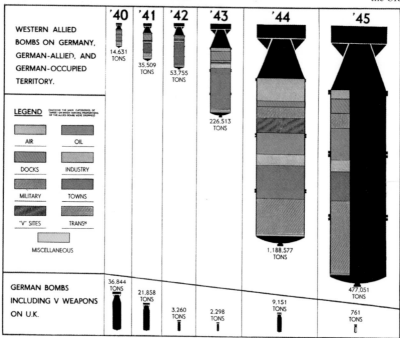

	'40	'41	'42	'43	'44	'45
WESTERN ALLIED BOMBS ON GERMANY, GERMAN-ALLIED, AND GERMAN-OCCUPIED TERRITORY.	14,631 TONS	35,509 TONS	53,755 TONS	226,513 TONS	1,188,577 TONS	
GERMAN BOMBS INCLUDING V WEAPONS ON U.K.	36,844 TONS	21,858 TONS	3,260 TONS	2,298 TONS	9,151 TONS	761 TONS

LEGEND [showing the main categories of target on which varying proportions of the allied bombs were dropped]

AIR · OIL · DOCKS · INDUSTRY · MILITARY · TOWNS · "V" SITES · TRANS^N · MISCELLANEOUS

(477,051 TONS)

1925 in an accident with the chemical warfare agent Lewisite.[62] In the interwar years Lindemann had designed a photo-electric cell, which was produced by a company he set up called Oxford Instruments (not related to the existing firm, a later spin-off from Oxford University) with some success; they were, for example, made use of by the Cambridge Scientific Instrument Company.[63] He was also a consultant for Baird Television.[64] Tizard, while Reader in Chemical Thermodynamics at Oxford, worked with Shell on aviation spirit. Alexander Todd, Professor of Organic Chemistry at Manchester, doubtless among other consultancy arrangements, was paid £400 per annum by ICI Dyestuffs as an adviser on medicinal chemistry. He claimed that 'nowadays, much academic research in chemistry has almost immediate commercial value and is liable to be exploited abroad to the detriment of our own industry'.[65] J. R. Baker of Bristol University (the inventor of the Morrison shelter) worked closely with the Department of Scientific and Industrial Research on steel structures and had his own consultancy practice. But these were exceptional cases, as their wartime work suggests, and did not reflect either the nature of the university or what was expected from them.

In 1939 the expectation had been that the universities would go through the war as essentially teaching institutions, perhaps with reduced numbers of students. Thus the head of the physics department at the University of Manchester, Patrick Blackett, estimated that he could manage with two thirds of his staff if student numbers were steady 'by working the staff harder and letting them drop their research'; halving the number of students would mean that he could manage with four people, just over one third of his staff. Blackett contemplated the possibility that students specializing in physics might disappear, but that the department would have more medical students to teach.[66] In fact undergraduate numbers in the sciences increased, and the number of staff fell somewhat, with a heavy fall too in the number of graduate students, and in research more generally. The ten full-time teaching staff had been reduced to seven, and the number of graduate students, who had helped in teaching, fell from eight to just one.[67] The Bristol physics department, one of the best endowed in the country, also saw an increase in undergraduate teaching, a collapse in graduate teaching, and a fall in expenditure on salaries, wages and so on, after a continuous increase in the 1920s and 1930s. Buildings were taken over by the Admiralty.[68]

Universities became places concerned above all to turn out graduates for the war effort. The most energetic and creative dons were often gone, leaving major laboratories and institutions in the hands of those who had nowhere to go. The famous Cavendish Laboratory's importance to the war effort may be gauged by the fact that it was run by Alexander Wood, a Scottish muscular Christian who ran the hardline pacificist organization the Peace Pledge Union. In 1940 he and other leaders were unsuccessfully prosecuted under a law outlawing the disaffection of the soldiery; they were acquitted, in what was seen by Vera Brittain as a powerful upholding of civil liberties.[69] Unlike the case in the US, wartime universities were not important centres of government-funded research.

Most British universities were quite small and were dominated by medicine, science and engineering. Bristol, for example, had only twenty-nine professors in 1939, thirteen of whom were in medicine; the remainder consisted of three engineers, three chemists, two physicists, five in all arts subjects, plus one each in law, geography and economics.[70] At the beginning of the war, they had lost only a professor of engineering (John Baker) and the professors of French and obstetrics to full-time state service, and two chemistry professors to part-time service.[71] Bristol had a particularly strong physics department, and two key members of staff, Neville Mott and Herbert Skinner, were away for most of the war. But the senior professor, Arthur Tyndall, stayed, as did C. F. Powell, the communist nuclear physicist, who continued quietly with his nuclear emulsion work, undisturbed it seems by the atomic bomb project. Some laboratory space was taken up with academic physicists evacuated from London; some was taken over by the government.

The University of Manchester, which was considerably bigger than Bristol, still only had just over 200 academic staff in total just before the war. Like Bristol it lost only a minority to government service. But they included some its most famous names. The historian Lewis Namier went first to Naval Intelligence and then the Jewish Agency for Palestine (run by a former Manchester academic, Chaim Weizmann).

The physics department came to be headed by a non-researching senior lecturer, as its two professors, Patrick Blackett and Douglas Hartree, were off on war work, as was one other member of staff. This is not to say that all the researching professors went.[72]

The government did not, in general, look to universities to do research or develop new ideas. Indeed, the amount of research done within universities almost certainly fell during the war. Only in some subjects and some places was research kept up or increased, and that where government departments placed contracts. It seems many chemistry departments fell into this category. Todd certainly turned over part of his Manchester laboratories to work on chemical weapons, among other projects, with a quarter of Todd's time, at the beginning of the war, given over to supervising a team of researchers devoted to chemical warfare research.[73] Prof. Garner of Bristol spent most of the week in London but had an expanding team of researchers working on the chemistry of explosives for the Ministry of Supply.[74] The contrast with physics may well have been a fairly general one – the senior physicists leaving for government labs, while the senior chemists directed war research in their own labs. This was probably due to the smaller scale of chemical research. Consultancy might have been an issue, in that senior chemists would have lost a great deal more than physicists by going into government service. In some aspects of physics, however, university departments were used. Hartree's differential analyser at Manchester was used by a group of people, working in two teams, one on the theory of the magnetron, the other on atomic bomb work.[75] The Cambridge differential analyser was run by Lennard-Jones in his Mathematical Laboratory. Both were taken over by the Ministry of Supply. Lennard-Jones built up a research team working there on ballistics of guns and rockets, theories of sound-ranging and other subjects.[76]

Perhaps the most notable developments to come out of British universities were the atomic bomb, the cavity magnetron, penicillin and British Anti-Lewisite. The story of the development of the early stages of the bomb within universities was not, however, an indication of the strength of British atomic research. The idea of the bomb had come from outside, and in any case Britain's leading atomic physicists were engaged in other work. Harold Nicolson's *Public Faces* (1932) and Eric Ambler's *Dark Frontier* (1936) are both about atomic bombs, which were changing the world in fiction long before the Manhattan Project was thought of. The early bomb project made use of former aliens in universities who could not be brought into government service.

The improvement of the resonant cavity magnetron was the most significant academic invention of the war, since it produced stable higher-power short-wave radio waves used for what was called 'centrimetric radar', that is, radar using radio waves with wavelengths measured in a few centimetres. The magnetron was a well-known device, but its capacity was quite radically changed. As a high-power, high-frequency device producing waves of a few centimetres wavelength it produced radar images which could pick out small details, like submarine periscopes, or provide a map-like picture of a target from the air. The new magnetron first operated at the physics department of the University of Birmingham on 21 February 1940. It was under Admiralty contract to work on high-frequency radar valves, among them the klystron. Just over a year later the three men responsible, John Randall, Henry Boot and S. M. Duke, wrote a report on it, with a preface by the head of the department, Prof. Mark Oliphant. Rather than boasting that this was an academic invention Oliphant went out of his way to stress the industrial connection:

> This is a good example of how effort in physics may be transferred from very different subjects to produce good results in another. Dr Randall and Mr Duke are both ex-members of the staff of the Research Laboratories of the G.E.C. and their 'valve-sense' was developed in that atmosphere. Dr Paterson, and Mr Megaw, have co-operated whole-heartedly in the development work.[77]

Even less well-known is the developing, also in Birmingham, of the 'strapping' of the magnetron, by James Sayers, in 1941, which made it stable and more powerful.

Penicillin was different in that government was not, at first, involved. At the Dunn School of Pathology of the University of Oxford Howard Florey got a grant from the Rockefeller Foundation in late 1939 to fund work by Ernst Chain on anti-bacterial proteins. One aspect of the work involved the complex isolation of penicillin from its mould. Very quickly recognized as potentially as important as sulphonamides, its development and production were spun out to a government laboratory in the USA – before the US entered the war – and to British firms.[78] In 1943 production levels were similar, but in 1944 and 1945 the USA surged ahead; very soon afterwards, however, British production was on a per capita par with the USA – leads and lags changed rapidly.[79]

At the beginning of the war the Oxford department of biochemistry was put on to looking for antidotes to chemical weapons by the Ministry of Supply. The head of department, Rudolph Peters, had worked at Porton Down in the Great War. The work did not yield an antidote for mustard gas, but one for another chemical weapon, Lewisite. Lewisite was an American poison gas based on arsenic. Perhaps because it was never used in action, Lewisite acquired a fearsome reputation during the interwar years, when it was widely tested and produced. British Anti-Lewisite, or BAL, was secretly developed and produced during the war, one perverse effect being to reduce the perceived effectiveness of Lewisite as a weapon since it was assumed (wrongly as far as is known) the Germans would also have found an antidote.[80]

Such developments within the university were rare. During the war it was government laboratories which particularly expanded, bringing in many important university researchers, but, perhaps most importantly of all, large numbers of very young researchers. Old and young, senior and junior, they did not bring inventions of their own to government. Instead, they were inducted into existing research and development programmes.

One particularly interesting case is the making of the Mark XIV bombsight. It was designed by Patrick Blackett, working at Farnborough between September 1939 and September 1940, and taken further by another academic, Dr H. J. J. Braddick.[81] Blackett was to become a noted opponent of strategic bombing, and perhaps for this reason this aspect of his war work is not usually stressed, despite its extraordinary success. His became the standard bombsight of Bomber Command, both in its British-made version and in the US-made T-1 version. The Mark XIV bombsight had the great advantage that it needed training on the target for a short period only. It could also be used in non-level flight, and it was thus ideal for taking evasive action. It was to become, from 1942, the sight of choice for area bombing. The more accurate SABS came in later and was used only for specialized operations.

Or take Dudley Newitt of Imperial College, chemical engineer, and soldier in the Great War, recruited in 1941 to run the scientific and research side of the Special Operations Executive. He recruited at least forty civilian officers, mainly in Station IX, a country house called The Frythe. It was an invention factory coming up with folding bikes, silencers, incendiaries, and disguised this and that.[82] After the war the buildings were used by ICI as a central research laboratory.

Radar was, contrary to the image given by C. P. Snow, a pre-eminent example of the recruitment of academics to work on an existing state project. As it happens, this was made very clear in the 1946 radar feature film *School for Secrets*.[83] It was concerned with 'the story of a handful of boffins', all academics, except for the man from the 'Eureka radio corporation', based loosely on Alan Blumlein. They are brought to radar in 1939, they 'go back to school' and they invent, with the process being invisible, new uses for radar, mostly offensive. The film deals with airborne interception radar, and something like the Oboe and H2S navigation systems, as well as the Bruneval raid. Links with the air force are stressed, with all the boffins flying, either in action or otherwise. The boffins are seen getting into uniform towards the end of the film, something which did not happen. Otherwise the story is broadly correct. In the early 1930s radar was exclusively the work of civil service scientists (except as advisers), and only when it was up and running did the academics get involved. It is also the case that the main radar work during the war was around offensive bombing operations. Academic boffins rose to be very important. Although the main radar laboratory, the Telecommunications Research Establishment, was run by a scientific civil servant, A. P. Rowe, his deputy, W. B. Lewis, was a Cambridge physicist. The army radar laboratory was run by a Cambridge physicist, Professor Cockcroft. Below them there were very large numbers of usually junior academics, many of whom were brilliantly successful not as creators but as developers.

GADGETS

At the beginning of the war inventors were a decided nuisance. In the first few weeks, the Admiralty was getting more than 1,000 letters a day suggesting war-winning inventions; some 25,000 were received on the magnetic mine alone.[84] Some serious time-wasting resulted: in 1940 a Lancashire inventor managed to get the editor of the *Manchester Guardian* to write to the Prime Minister about a television device of his invention. Churchill passed the letter on to Neville Chamberlain, who as Lord President of the Council was in charge of much of government's civil scientific research, and he passed it down to Sir Edward Appleton, Secretary of the DSIR. Extraordinarily, Appleton went with the head of the Radio Division of the National Physical Laboratory to visit the inventor in the small Lancashire town from where, with his invention, he claimed to be able to see Snowdon, the tallest mountain in Wales. It was, one need hardly add, not worth the journey.[85]

At the end of the war the Inventions, Patents and Security Section of the Admiralty had some serious fun dissecting the types that sent in inventions, in the pages of the restricted and recently launched *Journal of the Royal Naval Scientific Service*. First was the 'little man who only hopes to be helpful and claims no knowledge of ships or science'. Second came 'the ignoramus with grandiose ideas', among whom the 'death-ray specialists' were prominent. They were 'indignant in advance' that their theories would, as they knew, be rejected by the professional scientists – their schemes had, however, 'a magnificent plausibility'. Third was the personal friend of Sir John Voluble, MP; he 'Can (and often does) Get Questions Asked'. He often prefers writing to 'the King or to the Prime Minister or the First Lord in person'. Some will only reveal the invention in private or after having 'received a small consideration – say £100,000 – which as they rightly point out, is a negligible price to pay for a quick end to the war'. Lastly came the 'charlatan', the 'confidence trickster', whose object, secured with the help of a letter of introduction from, say, a Cabinet minister, is to get the Admiralty to write that 'trials of his scheme have been witnessed with interest' in order to get money from a 'promoters' syndicate' for 'non-existent or demonstrably impossible inventions'.[86]

There were, however, some more serious individual inventors – men of varying character and positions – who were an important source of ideas for new weapons. Consider three rather special cases of inventors who had been or became Members of Parliament. Noel Pemberton Billing (who had had success in the past) turned to designing fighters and bombers which took off under the power of an aircraft sitting above them.[87] He got nowhere with them. By contrast Sir Dennistoun Burney (the inventor in the Great War of the mine-sweeping paravane) got support for many of his ideas. The Toraplane was a set of detachable wings and tail fitted to a torpedo which could then be released at a very great distance from its target, gliding most of the way; two versions were tested, and though both failed the project nevertheless ran for three years.[88] The wings and tail were detached when a small paravane hit the water in advance of the torpedo. The essential problem appeared to be a profound lack of accuracy.[89] He was also the inventor of the recoil-less Burney gun, as well as its High Explosive Squashed Head (HESH), an anti-concrete and later anti-tank round; they were used after the war.[90] A 1944 prototype of his recoil-less Burney gun is in the Firepower Royal Artillery Museum at Woolwich.[91] Or consider the Conservative MP for Watford from 1943, Group Captain Helmore, who used contacts to get work going through unorthodox channels: while he was an MP, GEC and other companies were working on his gigantic radio-controlled torpedo, the Helmover (see above). As we have seen, Churchill and Amery also dabbled in invention.

Churchill created his own little invention factory, which was dubbed 'Winston Churchill's Toyshop'. It was headed by Major, then Colonel (later Major-General Sir), Millis R. Jefferis, an army engineer. Of Jefferis a colleague recalled that 'with a leathery looking face, a barrel-like torso, and arms that reached nearly to the ground he looked a bit like a gorilla. But it was at once obvious that he had a brain like lightning.'[92] It was known by the rather mysterious abbreviation MD1 and was created first under the War Office and then the Ministry of Supply, but outside its normal routines. Churchill himself recorded that in 1940:

in order to secure quick action, free from departmental processes, upon any bright idea or gadget, I decided to keep under my own hand as Minister of Defence the experimental establishment formed by Major Jefferis at Whitchurch. While engaged upon the fluvial mines in 1939 I had had useful contacts with this brilliant officer, whose ingenious, inventive mind proved, as will be seen, fruitful during the whole war. Lindemann was in close touch with him and me. I used their brains and my power.[93]

This was June 1940, and one of the first products of the outfit, which Churchill pushed vigorously, was the 'sticky bomb' – a primitive anti-tank weapon.[94] Churchill constantly chased progress: in May 1942, for example, he demanded to know why none of MD1's 'puffballs' (anti-tank bombs dropped from aircraft) were available in the Middle East.[95]

Despite coming under Cherwell's control[96] Jefferis had to fight to keep free of the Ministry of Supply. He was remarkably forthright: 'My contention', he told the car maker Billy Rootes, then a senior figure in the Ministry of Supply, 'is that I have a better brain for the purposes of producing warlike weapons than the D.S.R. [Director of Scientific Research], the D. of A. [Director of Artillery] or the Ordnance Board'. He warned him that 'if I am left alone, I might produce something, otherwise there is no hope'.[97]

Jefferis and his team did indeed produce an awful lot: as of 27 May 1942 they claimed twenty-one devices, of which thirteen were in production, two had been discontinued in production and four had been discontinued in development.[98] A list of October 1945 was far longer, including production data: 30,000 29mm spigot mortars (usually known as the Blacker Bombard); the Hedgehog, which claimed thirty-seven submarines; a million PIATs (infantry anti-tank weapons); a new fuze for the PIAT; 1.5 million sticky bombs; many limpet mines, and 2.5 million of the smaller clam mines; 2.5 million 'L' time delay switches for demolition, as well as other delays, fragmentation bombs and booby traps; one million puffballs; and 2 million anti-aircraft fragmentation bombs for balloon cables . . .'[99] Just how significant all this stuff was is another matter. The official history commented that they 'were sometimes merely ingenious'.[100]

Some of the most important developments MD1 claimed for itself came from Lieutenant-Colonel Stewart Blacker, who had a bruising encounter with the outfit. A man who spoke of 'the abominable no-men of Whitehall', he, like so many other inventors, managed to get his ideas into production, indeed on a large scale. He developed a spigot mortar anti-tank gun, under the name of 'Arbalest', which although it performed well in official tests did not get through what he called the 'zareba put up by the technical bureaucracy'.[101] In June 1940 he wound up in a unit run by Jefferis, a predecessor to MD1. There he resuscitated his spigot mortar and had prototypes made by Boosey & Hawkes, the musical-instrument maker. After it was suc-cessfully demonstrated to Churchill and the top brass at Chequers, Churchill ordered 5,000 there and then. As usually happened when orders were placed in this way, it took much longer to get things into production than anticipated. It was not till February 1941 that it was successfully tested and then drawings were passed to the Ministry of Supply, and after lots of debate the Blacker Bombard, as it became known, ended up in the hands of the Home Guard. The essence of the device was that the projectile, with an explosive in its base, was fired off a spigot, rather than out of a tube as in a normal mortar; it was in some respects an inverted gun, in that the barrel, linked to a projectile, flew off.

Blacker's spigot mortar was to find many varied uses, mostly but not all invented by himself. The spigot mortar was the basis of the anti-submarine Hedgehog device, which fired multiple bombs over submarines – it was developed by MD1 together with a naval unit discussed below. It was to be used on a whole class of convoy escorts, replacing the depth charge. Yet it was not until the second half of 1944 that it sank more submarines than the depth charge, and this was at a time when there were relatively few attacks on submarines.[102] The spigot mortar was also the basis of the PIAT anti-tank gun, which was to become the main infantry anti-tank weapon of British forces towards the end of the war. There were two original versions, one by Jefferis and one by Blacker, and Imperial Chemical Industries was to develop a version based on Blacker's.[103] In fact a press report in 1944 gave credit for the PIAT and the Bombard to Jefferis, to which Blacker took exception, suggesting to Jefferis an equal division of any post-war award, after his expenses had been deducted.[104] The Ministry of Supply had already paid Blacker £50,000 for his expenses in relation to the Bombard and PIAT.[105] Blacker went on to develop the Petard, a spigot mortar which fired a 'Flying Dustbin' of plastic explosive at concrete emplacements. It was fitted to the Churchill AVRE tanks, one of the key 'Hobart Funnies'.

Churchill himself was, as we have seen, deeply involved in the story of the Blacker Bombard, and indeed less directly the Petard. He also got involved in the PIAT case. Writing to the Secretary of State for War, in January 1943, he asked: 'Why should the name "Jefferis shoulder gun" be changed to P.I.A.T? Nobody objected to the Boys rifle, although it had a rather odd ring.'[106] The Boys rifle was the weapon the PIAT would replace. Churchill got embroiled in claims for credit for the gun, supporting Jefferis's claims, who he claimed had also invented the Blacker Bombard. 'It seems therefore a little hard to grudge having Jefferis's name attached to the shoulder gun, especially as everyone speaks of Mills grenades, Stokes mortars, Hawkins mines, Kerrison predictors, Northover projectors, etc.,' he asserted. He even maintained it was 'quite incorrect' that the PIAT was developed by ICI and incorporated 'points of design from both the Jefferis shoulder gun and from a similar weapon invented by Lieutenant-Colonel Blacker'.[107] Churchill did not get his way and Blacker won substantial sums from the Royal Commission on Awards to Inventors for the Bombard, Hedgehog, PIAT and Petard (see Table 8.1).

Churchill wasn't alone in having a crazy gang of inventors. The navy's 'Wheezers and Dodgers', officially the Department of Miscellaneous Weapons Development, was set up in 1940 to work on anti-aircraft measures.[108] The department was created by Professor Charles Goodeve of University College London, a Canadian chemist and reserve naval officer, 'a slim scientist, with greying hair over an intellectual forehead ... Beneath his placid exterior there lurked indomitable energy and considerable guile in naval politics.'[109] His unit had within it the aircraft maker (and novelist) Nevil Shute.[110] The group worked with MD1, for example on the Hedgehog. It was also involved in the invention of devices for landings on enemy-controlled coasts. One of these projects, the 'Great Panjandrum', has been immortalized by repeated television showings of the ridiculous progress of the contraption along a wartime beach, and a re-created version was central to an episode of *Dad's Army*. A giant rocket-powered pair of wheels separated by a cylinder meant to contain a ton of explosive; it was designed to be released from a landing craft, and to proceed up a beach to destroy concrete defences. A more significant invention was that by Robert Lochner, an electrical engineer who had been involved in degaussing and later served at sea. He invented the 'Bombardon' or floating breakwater, a key element of the Mulberry Harbour. But perhaps the most significant of all, and the simplest, was due to Edward Terrell, a barrister born to the French mistress of the unorthodox Victorian divorce lawyer and inventor Thomas Terrell (who as well as publishing *Terrell on the Law of Patents* in 1884, a work now in its 16th edition, also reported on the Dreyfus trial).

Terrell was not even a qualified scientist, yet invented, developed and put into place a remarkable invention known as 'plastic armour'. At the beginning of the war sandbags were used to protect vulnerable parts of merchant ships; in early 1940 the use of concrete panels started. Both were very ineffective, Terrell noted. But on 12 August 1940 a report caught his attention; it stated that a material used to waterproof decks had apparently reduced the effectiveness of a shell which had hit a ship during the evacuation of Dunkirk. He called in the manufacturers of the material – mastic asphalt mixed with cork – and, thinking that variants of the material would act as protection, ordered samples to be made up by the firm.[111] He took them to a new rifle range at the Road Research Station, where they signally failed to stop bullets. But with a steel plate backing, and at an angle, bullets were stopped more effectively than by concrete. That suggested to Terrell that replacing cork with hard granite, which would deflect bullets, might do the trick. Working with the government's Director of Road Research, Dr William Granville, he ordered from the same small firm their material with granite chips of different sizes added. All this was done within days, and back at the rifle range the new material showed enormous promise. That led to making materials without cork – just mastic asphalt and granite, with the crucial steel plate. Terrell called his invention 'plastic armour' – plastic, because the mastic asphalt was melted by heat, and armour, because that gave confidence in its strength. All this was done in ten days. The material was not only effective but cheap, as it could be made by otherwise redundant wartime road-building firms.

As needs to be the case in such stories, the authorities on armour, the naval constructors, were hostile, but by clever manoeuvring the material was sent for testing to the gunnery school HMS *Excellent*, where successful trials took place on 6 September 1940. By late October 1940 a directive had been issued that the bridge and gun posts of merchant ships should be protected with the new material, and in December it was authorized for naval ships. From then thousands of ships, British and later American, were equipped with this novel material.

This invention, like most others, would have many fathers – Dr William Granville, Director of Road Research, was included as an inventor in the initial patent, issued in 1941, though the Royal Commission on Awards to Inventors made plain the invention was Terrell's only. The company that had produced the mastic and cork mixture challenged the patent in 1946 and lost.[112] A naval scientist claimed in an article immediately after the war that it was 'a product of naval scientific ingenuity', which was about as accurate as saying it was the work of a lawyer.[113] Terrell was awarded nearly £10,000 in 1949.

Terrell went to work in the department of the First Sea Lord, coming up with and promoting inventions of all sorts. He promoted the dropping of hollow-charge (also known as shaped-charge) bombs rather than depth charges from aircraft – supported by Lindemann, his boss Admiral Usborne and Jefferis. But none the less the depth charge was retained.[114] In 1943 Terrell turned his attention to means to destroy U-boat pens and came up with the idea of a heavy bomb that would be given extra power of penetration with rocket motors. This rocket bomb was designed by the Chief Engineer of Armament Development at Fort Halstead. However, bombs came under the air force and the Ministry of Aircraft Production, and it took some fixing to get it past them. This involved: a minute from the First Lord to Churchill as chairman of the Anti-U-Boat Committee; a trip to Oxford by the project's champion, Edward Terrell, to convince Lindemann; a technical meeting to deal with Air Ministry technical objections; a meeting between the First Lord, Sir Stafford Cripps of the Ministry of Aircraft Production and air force chiefs on Christmas Eve 1943; and a large meeting convened by the Ordnance Board in January 1944, which approved design go-ahead. Continued objections by the air force led to the issue going to ministers in May 1944. The meeting was chaired by the Minister of Production and was attended by the Secretaries of State for War and for Air and the First Lord of the Admiralty, the Ministers of Supply and of Aircraft Production, Lord Cherwell, Dr Crow, two air force officers, the Deputy First Sea Lord and Edward Terrell. Lyttelton decided in favour of continuing with high priority, despite the opposition of the air force and the Ministry of Aircraft Production.[115] The rocket bombs, powered in their final phase by nineteen 3in rocket motors, went supersonic and could penetrate twenty feet of concrete.[116] They were not available for use until February 1945, which was too late.

But although individual inventors were important, it is vital to remember that they were not outsiders. They were men intimately familiar with the worlds they were inventing for, and they were typically products of that world. The cases of the individuals discussed above attest to the importance of the military – Jefferis, Blacker and Whittle were or had been career officers; Goodeve was a long-standing Naval Reserve officer; and Wallis was a designer of military aircraft and an employee of Britain's greatest arms firm. The great exception was Edward Terrell, a lawyer.

The assessment of novelty and import of some inventions was to occupy quite a lot of time of lawyers and technical experts after the war. Certain inventors, by no means all, could lodge demands for recompense from the government which might go to the Royal Commission on Awards to Inventors. The commission dealt only with special cases: those not agreed by departments, those not covered through normal contracts for weapons, and those not covered by the usual terms of service of state employment. State servants had to prove that they pushed these things in the face of opposition or independently (and some rather extraordinarily did, notably the radar and rocket teams). Medical and surgical inventions were not considered, on the grounds that it was established practice that rights were given up freely.[117]

The list of large awards (over £5,000) is given in Table 8.1. It is a motley collection from which nothing much can be deduced except that the awards, in most cases, involved a measure of impact on the war. Radar would be an example of a new technique which was very widely deployed very quickly and had a large effect; the cavity magnetron a case of an invention in wartime which was very quickly developed and used, indeed the only wartime invention with a significant impact on the war, claimed Sir Henry Tizard.[118] The fame of an invention clearly influenced the court. How else to explain the large awards for PLUTO, the bouncing bomb and the rockets? But the list tells us also of highly rewarded inventions of no resonance whatever – the Messier undercarriage for bombers, the obturator for naval guns and so on. It is worth noting that the only academics on the list are Randall, Boot and Sayers for the cavity magnetron and Goodeve for degaussing.

Table 8.1 Awards made by the Royal Commission on Awards to Inventors of over £5,000

Claim number and name(s) of claimant(s)	Description	Sum awarded
1. Major A. S. J. Du Toit	Anti-mine flail for tanks	£13,000 (other claimants later got a total of £3,800 for the same)
5. Sir Donald Bailey	Bailey Bridge	£12,000
8 & 9 Major-General Sochaczewski	Swift training rifle and accessories	£8,000
24 Lieutenant-Commander Stephens	Dome anti-aircraft trainer	£5,000

Table 8.1 – (*continued*)

Claim number and name(s) of claimant(s)	Description	Sum awarded
30 Air Commodore Sir Frank Whittle	Jet engine	£100,000 (not a claim but RC asked to come up with a figure)
31–8 Yvonne Levy and SIAM	Messier etc. undercarriage patents	£120,000
50. Vosper Ltd	Motor torpedo boats	£35,000
66–86 Caswick Ltd, Zbrojovka Brno Narodnik Podnik	Ceskoslovenska Zbrojovka Brno Akciova Spolecnost BESA machine gun and accessories	£5,000 (additional to earlier payments in other forms)
62 Mr A. C. Hartley	PLUTO	£9,000
63 Mr B. J. Ellis	PLUTO	£5,000 (plus £2,100 to various others)
89 Mr E. Terrell	Plastic armour	£9,500
98–100 Professor Randall, Dr Boot, Professor Sayers*	Cavity magnetron	£36,000
114 and 326 M. R. Grundlach and others. (Grunlach was a Pole who invented the periscope in the 1930s; he spent most of the war in France)	Periscopes for armoured vehicles	£17,000 (in addition to more than £63,000 previously granted plus £750 granted later)

Table 8.1 – (*continued*)

Claim number and name(s) of claimant(s)	Description	Sum awarded
125–6 Various, including A. C. Hartley	FIDO	£5,000
131–4 Wm Denny and Sons Ltd and Brown Bros Ltd	Ship's stabilizers	£27,500
142–7 Lt-Col. Blacker	Blacker Bombard, Hedgehog, Petard, PIAT	£17,000 (in addition to £25,000 already received)
149 P. M. Otway	Joints in concrete structures	£6,000
159 Wing Commander H. De V. Leigh	Leigh Light	£6,000
170 Mr B. N. Wallis	Mine	£10,000
171 Sir Robert Watson-Watt**	Radar	£50,000
173–7 A. F. Wilkins**	Radar	£12,000
178–87 E. G. Bowen**	Radar	£12,000
**175–276 Many people involved in radar before-1940	Radar	£20,375 (in total)
284–5 R. A. W. A. Lochner	Degaussing	£5,000
312–13 Prof. Goodeve	Degaussing	£7,500
286–303, 314–17 many people	Degaussing	£10,800 (in total)

Table 8.1 – (*continued*)

Claim number and name(s) of claimant(s)	Description	Sum awarded
323–4 Bruno Jablonsky	Propellers	£15,000 (plus additional £2,000 to others)
330–33 K. J. Sommerfeld and others	Sommerfeld track	£5,000 (plus more for US use)
338–9 C. H. and A. W. Wood	Flight training equipment	£7,000
344 Wing Commander D. Salisbury Green and Mr W. H. Parkin	Hydraulic remote control	£5,700
363 Mr A. C. Crossley	Obturators for breech-loading ordnance (a new form of device to seal the breech of a heavy gun)	£7,115 (in addition to £8,000; also royalties and US payment under related)
369–72 Various	Proximity fuzes	£15,000 (in total)
404–10 K. H. Nicholls	Various concerned with gunnery and servo motors	£9,000
412–14 Lt I. Porteous and others	Gyro rate unit stabilizers etc.	£8,500
433 Geigy companies	DDT	£6,000

Table 8.1 – (continued)

Claim number and name(s) of claimant(s)	Description	Sum awarded
434 Guy Motors	Welding of tanks	£5,000
435 Sir A. Crow, W. R. J. Cook and H. J. Poole†	Loose-charge rocket motors	£5,000

* One million cavity magnetrons were made, at a total cost of £5m: *Second Report of the Royal Commission on Awards to Inventors*, Cmd 7832 (London, 1952), para. 25.

** These awards were additional to some £200,000 awarded departmentally without reference to the commission. The commission dealt mostly with the claims from Watson-Watt and his team.

† Rocket motors worth £19m used though the degree of invention was small, the commission admitting that these men were charged with rocket development (and thus not eligible) and the inventive step was not great, but nevertheless awarded a 'heavily discounted' £5,000 given what they took to be the highly effective use of rockets at the end of the war: *Fourth and Final Report of the Royal Commission on Awards to Inventors (Awards to Inventors) 1955–56*, Cmd 9744 (London, 1956), para. 28.

Sources: *First Report of the Royal Commission on Awards to Investors 1949*, Cmd 7586 (London, 1949); *Second Report of the Royal Commission on Awards to Inventors 1949*, Cmd 7832 (London, 1949); *Third Report of Royal Commission on Awards to Inventors (Awards to Inventors) 1953*, Cmd 8743 (London, 1953); *Fourth and Final Report of the Royal Commission on Awards to Inventors (Awards to Inventors) 1955–56*, Cmd 9744 (London, 1956).

Wartime technical development took place, as we have seen, on a huge scale. It was partly the work of scientists, but it was also the result of the 'enterprise of engineers' and inventors.[119] There were profound continuities in institutions, from arms firms to government laboratories, in the development of the new machines of war. Yet there was a change in scale and character which was recognized by a change in language. Before the war people spoke of research as something to do with laboratories, and used 'science' very broadly, using it often where 'technology' would later be used. Otherwise lots of different terms were used, such as 'rapid developments in air and artillery tech-nique'.[120] The use of the term 'technology' was generally restricted to meaning the academic subject of engineering broadly defined.[121] New machines were described with terms like 'invention', 'appliance', 'appar-atus', 'engineering' and 'mechanical'. More specialist works might refer to 'scientific invention', 'technical progress', 'technological progress', 'scientific innovation'.[122] By the end of the war 'science and technology' and 'research and development' came to be central to talk of the national future. 'Technology' slowly came to mean what it means today, and to be used widely.[123]

CONCLUSION

In November 1945 Sir Edward Appleton, Secretary of the DSIR and the government's most senior scientific employee, addressed the Insti-tution of Mechanical Engineers on 'The Scientist in Wartime'. He claimed that before the war Britain had been 'somewhat slow in applying the discoveries of science to practical ends', but that that had changed during the war. People had awoken to the benefits of what he called 'scientific research', because, he said, 'in a period of fast-moving events the four processes of research, development, production and use, follow so swiftly on the heels of one another that even a short memory recognises cause and effect'.[124] Appleton, like many academic scientists, naturally wanted to highlight the importance of scientific research and created the impression that this was the main source of innovation. His prime examples were atomic bombs and radar. He also discussed radio, ionospheric forecasting and something he called the 'science of armaments'. He also gave a great deal of attention to operational research. Not even touched on in his talk were such things as aviation, the jet engine, gas or chemical warfare, or penicillin. The reason for the emphasis was that Sir Edward was presenting arguments for the importance of the academy and of his department, the DSIR. But for all his pay and status Appleton and the DSIR were minor players and had become relatively less important during the war. The atom bomb project had the DSIR merely as its postal address.[125] But the DSIR launched a serious post-war propaganda effort. It recruited

J. G. Crowther, a noted propagandist for the academic and scientific left, and the physicist Richard Whiddington to write a multi-volume 'popular account of the scientific war effort'. Only one volume, called *Science at War*, was ever published. Half dealt with radar and the rest with the atomic bomb, operational research and the war at sea.[126] Some 17,000 copies were printed, but it appeared late and was not wanted by the Central Office of Information.[127] It was to prove an influential choice of examples. The academics would continue to over-emphasize these cases and to ignore the work of the most important government departments and of industry. They would write as if sci-entists had worked on practical issues and in multidisciplinary teams for the first time, and as if it was the first time anyone had thought of applying statistics to the analysis of operations.[128] In doing so they painted an exceptionally partial and misleading picture of the com-plex world of wartime invention, research and development.

9
Machines and Modernities

During and after the war Britain changed and interpretations of the history of the war changed too. Some of the most important and interesting changes in both were not advertised; nor were they well correlated with standard political understandings of history. This is particularly the case for the topics covered by this book.

One of the most important and least discussed changes was that Britain became more economically and ideologically national after the war. This is not to ignore the crucial and often neglected point that the Empire was more important to British trade than ever before, that more troops were stationed in imperial garrisons than in 1939, and that more officials were engaged in imperial projects.[1] For within this imperial context the economy was much more nationally focused, and more committed to national self-sufficiency than before. This was reflected in politics, though not in the ways it is usually interpreted. In the Labour Party election manifesto of 1945, 'Socialism' appears once, whereas 'nation' and 'people' appear repeatedly, as did, though to a lesser extent, 'Britain' or 'British'; by contrast the 1935 manifesto hardly invoked 'Britain/British' or 'nation' at all. Tellingly, in an election newsreel in June 1945 Mr Attlee complained that the Conservatives impudently presented themselves as a national government; his response was that the Labour Party was the real national party.[2] He had a point. By 1945 the language of nation and nationalism permeated Labour's economic discourse. It is not surprising then that a National Coal Board would appear, a National Health Service and National Assistance, and a British Transport Commission and a British Electricity Authority.

This new nationalism affected how the war was interpreted. A 1945 government document called *What Britain Has Done, 1939–1945: A Selection of Outstanding Facts and Figures* is a powerful case in point. The imperial contribution to the war was radically downplayed, as was, less surprisingly, that of the USA and other allies. The stress was on British help to these allies.[3] A profoundly national bias is also very evident in the official *Statistical Digest of the War* published in 1951.[4] For example, only in sub-categorizations of trade did the Empire appear (as in trade with 'British countries' as opposed to 'foreign countries'). Otherwise the figures presented are for Great Britain and Northern Ireland. Thus the armed forces are the British ones (though a footnote recognizes there were some foreigners in them), while the very closely allied Commonwealth forces, not to mention colonial ones, or Allied ones, do not figure. There is no Indian Army, no Royal Canadian Air Force, no Polish Corps. The figures for arms production are those for production within the borders of the United Kingdom, not those for arms produced in the Empire and then supplied to either British or imperial forces, or those produced outside the Empire for British forces. The only movements of goods recorded (apart from Lend-Lease, and these inadequately) are those in and out of the United Kingdom, not the movements to the fighting British Empire, though munitions movements are often left out altogether. Thus the very statistical picture that was produced is shot through with a particular nationalist way of understanding, and it should be noted that this would become standard in nearly all the histories of war production and subsequent accounts of the economics of the war.[5] Some of the rare exceptions are incorporated in the tables and data in this book (for example, Table 9.1). As we have seen, taking a full global picture changes our understanding of the totality of British war production very considerably, even in the case of the standard infantry rifle. It also changes our understanding of what was produced and used in Britain itself.

MORE MOBILIZED

British historians and commentators in the nationalist post-war mood were to make much of the fact that Britain was supposedly the most mobilized nation during the war, without noting the critical importance

Table 9.1 Supplies of groups of certain war-stores from the United States and production in the United Kingdom and Empire, September 1939–August 1945

	Production in Britain and the Empire				United States supplies for War Office and Empire requirements, 000s (col. 5)	United States supplies as percentage of:	
	United Kingdom, 000s (col. 1)	Canada, 000s (col. 2)	Eastern Group, 000s (col. 3)	Total cols. 1, 2 and 3, 000s (col. 4)		United Kingdom production, % (i.e. col. 1)	Total supplies, % (i.e. col. 4 + col. 5)
Tanks*	24.8	3.6	—	28.4	25.6	104	47
Artillery, anti-aircraft equipment, tank and anti-tank guns	132.1	13.4	5.9	151.4	10.3	8	6
Small arms (for army only)	7,598.0	1,450.0	1,124.0	10,172.0	2,757.0**	36	21

Military vehicles†							
(a) Propelled vehicles††	682.0	582.0	1.6	1,265.6	264.0	39	17
(b) Tank transports (included under (a) above)	1.3	—	—	1.3	4.8	362	78
Aircraft							
(a) Combat aircraft‡	96.1	5.4	1.1	102.6	23.0##	24	18
(b) Transport (and Air–Sea Rescue) aircraft	1.8	—	—	1.8	2.7##	150	60
Landing craft and ships	4.3	—	—	4.3	2.6	60	38

Table 9.1 – (continued)

	Production in Britain and the Empire				United States supplies for War Office and Empire requirements, ooos (col. 5)	United States supplies as percentage of:	
	United Kingdom, ooos (col. 1)	Canada, ooos (col. 2)	Eastern Group, ooos (col. 3)	Total cols. 1, 2 and 3: ooos (col. 4)		United Kingdom production, % (i.e. col. 1 + col. 5)	Total supplies, % (i.e. col. 4 + col. 5)
Small arms ammunition, million rounds	11,094	4,520	3,790	19,404	7,453	67	28

Source: M. M. Postan, *British War Production* (London, 1952), Table 36, p. 247.

* To 30 June 1944 only, and excluding tank chassis for self-propelled artillery.

** Special release of rifles from stock in 1940–41 not included.

† For all armed services.

†† Excluding trailers and motorcycles.

‡ Heavy, medium and light bombers, general reconnaissance, naval types and fighters.

‡‡ Deliveries from North America.

to this of Britain's particular place in the world. 'Britain's war effort per head of population is greater than that of any other belligerent,' boasted *What Britain has Done*.[6] Churchill claimed in his history that Bevin and Anderson had developed a system which 'Enabled us to mobilize for war work or in the field a larger proportion of the men and women than any other country of the world in this or any previous war.'[7] The official historians of the war economy reproduced a graph of war expenditures as a proportion of national income for Australia, Canada, New Zealand, the United Kingdom, the United States and the USSR, which showed Britain top for all the full war years. They were careful to note, especially for the Soviet Union given its great losses and low income per head, that the figures could not be taken to mean commitment or effort, yet pointed to a British lead over the USA, especially noteworthy given that the US was richer, as they pointed out.[8] Later historians followed their lead with embellishments. Angus Calder noted Britain devoting a greater proportion of its resources to war than any of its 'western allies' and also Germany in 1943.[9] The claim is made to suggest a worthy commitment, a moral choice actualized and an argument for the effectiveness of British planning.[10] A. J. P. Taylor waxed lyrical about 'a revolution in British economic life, until in the end direction and control turned Great Britain into a country more fully socialist than anything achieved by the conscious planners of the Soviet Union'.[11]

One of the more extraordinary versions of this story was the suggestion that Britain mobilized more successfully than Nazi Germany. As we have seen, this is what British economists writing in 1939 and 1940 thought would happen. But the post-war analysis took high British mobilization not as an indicator of wealth, but rather of superior organization and commitment to modernity. A particular and important part of this story was the claim that the British mobilized more women than the Germans, who were held back by reactionary ideals of *Küche, Kirche und Kinder*. As Angus Calder put it, Britain went 'far further than Hitler's Germany' in conscripting women, 'where the idea that the role of the feminine sex was to breed and succour the master race' prevailed.[12] In fact the proportion of women employed in Germany was already higher at the beginning of the war and remained so. What Britain could do, but Germany could not, was to bring more women into the labour force.[13] Indeed, by British standards the German situation was extraordinary. In wartime Britain there were always many more men than women in the civilian workforce. By contrast, by the end of the war Germany had more German women than German men in the civilian labour force (that is, discounting slave and forced labourers). The male German civilian labour force fell by 10 million to 13.5 million, in other words by more than 40 per cent, while the female labour force remained constant at 14 million, nearly 6 million of them working on farms.[14] In Britain, male civil employment fell around 23 per cent, while women's employment went up by around 40 per cent, but there were still, by a considerable margin, more men than women in civil employment: 10.7 million men to 6.8 million women in 1943, the peak of women's employment.[15] The only measure by which Britain mobilized more women was the irrelevant one (in this case) of increasing the number of women in paid work. The extraordinary demands of the German army for men, and the extraordinary dependence on female labour, especially on the land, meant that Germany relied, to the tune of 7 million workers, on manpower from abroad, subjugated to varying degrees. The reliance on local agriculture was critical, and here Germany was indeed, by the usual pre-war standards of modernity, well behind that quite exceptional case, Britain. In this language Germany was economically not socially backward by the standard criteria, yet that categorization was inverted in post-war British accounts.

The comparison with the USA shows a greater British mobilization, despite the greater wealth of the USA. Britain had a higher proportion of the labour force in the armed services than did the USA and a higher proportion of the labour force in the forces or munitions production (55 per cent rather than 40 per cent); in terms of national income, defence expenditure was higher in Britain (for 1944 as for above figures, 53 per cent to 47 per cent).[16] These figures may in some respects downplay the comparative British effort: in Britain the proportion of the workforce in the military was higher than in the USA, but because military wages (by comparison with civilian) were lower in the UK, the proportion of UK national income expended on them was relatively lower. Furthermore the cost of munitions production was, relative to other production, higher in the USA.[17] These figures take no account of Lend-Lease. This added resources, particularly armaments, to the British war effort which if taken into account would raise the proportion of available resources mobilized.[18] At its peak, Lend-Lease net of British reciprocal aid gave Britain the equivalent of an additional 20 per cent of its own output (see Table 9.2). This huge addition to national resources did not figure in the national accounts, and for good accounting reasons: the materiel remained the property of the US and was never paid for by Britain. At this moment of triumph of national income accounting, the object they created, the national economy, was a peculiarly artificial construct.

As we have seen, the role of the US in supporting the British Empire in 1940–41 has been overrated. But the support at the end of the war is understated. In 1944 the Empire provided only 73 per cent of its arms, with the UK providing only 61 per cent of the total and Canada 9 per cent. Something like 20 per cent of British armaments came from the USA, mostly under Lend-Lease: it took 12 per cent of US

Table 9.2 British imports, Lend-Lease, reciprocal aid and GNP, 1940–45, £m

	Lend-Lease*	Reciprocal aid*	Imports** (cash) of goods	GNP factor cost[†]
1940			1,000	6,878
1941	300		1,100	7,921
1942	1,200	100	800	8,540
1943	2,100	700	800	9,080
1944	2,600	800	900	9,140
1945	1,300	500	700	8,754

*Sayers, *Financial Policy*, quoted in Charles Feinstein, *National Income, Expenditure and Output of the United Kingdom, 1855–1965* (Cambridge, 1972), pp. 112–13.
** Feinstein, *National Income*, Table 15, p. T39. Excludes Lend-Lease. See also pp. 112–13.
[†] Ibid., Table 2, p. T9. Excludes Lend-Lease. See pp. 77 and 112–13.

munitions production, 10 per cent in Lend-Lease, 2 per cent for cash.[19] Types of equipment supplied in large quantities included transport aircraft, above all the DC3/C47, or Dakota in British nomenclature (a veritable Liberty ship of the air), naval aircraft and Liberty ships. One particularly clear example of US dominance was in the control of the latest machines of war. In 1941 Britain had the largest atomic programme in the world; by 1945 the US led and financed the building of a bomb over which Britain had no control. While in 1942 the US navy relied on British cryptographers for naval intelligence in the Atlantic, by the end of the war the USN was doing the majority of decrypting (something not advertised in the popular accounts).[20] In 1944 the US supplied and controlled the use of the vital proximity fuze and produced the best gun-laying radar and fire control equipment; Britain relied on this equipment to defeat the V-1. As we have seen, the great bulk of British oil, in its many varieties, came from the USA. The principles of division of labour and of comparative advantage were applied and the British Empire gave up independent supply and action in the name of increasing the overall efficiency of the war effort (see Table 9.1).

External support was vital to high British mobilization. Indeed Lend-Lease was designed precisely to achieve this: 'It was the deliberate and combined policy to maintain and to increase British war mobilization, which was never quite equalled by the US even in the last year of war. Lend-Lease aid to the British . . . was one of the means whereby this policy was implemented.'[21] Surprisingly perhaps, this remained the case even after the USA entered the war.[22] Lend-Lease kept US workers in factories and on farms to supply British servicemen and servicewomen and also civilians. That was hardly a wholly un-sordid act. Nor was it wholly original. Britain had since the eighteenth century armed allies with the products of its factories and fields, thus keeping its own forces out of the front line. In fact Britain continued to do this with lesser powers, even sometimes in effect subcontracting Lend-Lease. Britain paid for and equipped the great Indian Army and organized, for example, Czech and Polish units. These were equipped on the British pattern, including Lend-Lease equipment.

The USA was far from being the only supplier to Britain of goods without the need to export to pay for them. Much of the rest of the world continued to supply Britain with food and raw materials, and its vast forces overseas with local supplies, as we have seen. Britain paid not in goods, or in money which could buy products from elsewhere, but in sterling which accumulated in London. Take the case of Argentina. Its imports fell very drastically during the war, from fuel to machinery, and as a result investment fell drastically too; both fell more than in the Great Depression. However, exports increased substantially.[23] For Uruguay exports were roughly constant while imports fell very greatly. The direction of exports changed: Europe other than Britain disappeared as a market, with the result that exports shifted to Britain and to the USA.[24] The 'sterling balances' Argentina, Uruguay and other countries accrued were a measure of the (temporary) privations Britain imposed by being unable to export to them. These balances have been misunderstood as British generosity, especially in the context of India. Thus A. J. P. Taylor thought that paying for the defence of India through sterling balances was an outburst of 'Imperial obstinacy' which 'sprang from habit'.[25] The implication was that India should have been left to fend for itself. But sterling balances were promises to pay, not payment, and what was being defended was the Empire not India.

The sterling balances increased enormously during the war. The figures for 31 December 1945 were as follows:

Dominions	£481m
India, Burma and the Middle East	£1,929m
Other sterling areas	£561m
North and South America	£519m
Europe and Dependencies	£273m
Rest of the world	£62m

The total came to £3,825m, whereas in December 1939 it had been £556m.[26] The increased sterling balance of, say, £3bn, compared with an estimate of Lend-Lease support net of reciprocal aid of £5bn to the UK alone (see Table 9.2). There was a big difference: Lend-Lease would not have to be paid back, the sterling balances belonged to others. But in terms of supporting the war effort both were of comparable importance, though the latter have been much less visible, and less well understood.

The mobilization celebrated as a particular national genius depended in fact on Britain's unique ability to rely on overseas supplies without reciprocating with exports, and support from the USA, a major asymmetry of immense benefit to Britain,[27] for this allowed Britain to devote most of that productive capacity and shipping capacity which went to exports to warlike purposes; furthermore it could avoid adding too much labour and equipment to domestic production of food and raw materials; it also allowed much larger armed forces to be equipped than would otherwise have been the case. Indeed, especially in the light of complaints from the 1980s in particular that Britain's production record in war was poor, it is as well to remember that this external support allowed factories in the United Kingdom to out-produce Germany in many classes of weapons. In absolute terms the United Kingdom alone out-built Germany in aircraft, warships, merchant ships, tanks in the early years, light armoured vehicles, bombs, heavy wheeled vehicles, guns under 75mm and infantry machine guns. Germany was ahead only in guns of 75mm and over, rifles, ammunition of all sorts, and tanks in 1943 and 1944.[28] This British record is remarkable in itself, but particularly so when one remembers that it had a smaller population than Germany (especially the greater Reich in wartime), had fewer people in its armed services, that its armed services were much less likely to be in battle, and that it was supplied with additional weapons from overseas. Britain's new international order, and then the order of the 'United Nations', proved more successful than Hitler's continental new order. The British were also much more successful imperialists than the Germans, mobilizing a huge imperial force, a large part of it effectively mercenary, which did most of the fighting in the East.[29]

The other side of the coin of British access to the world was that Axis trade was destroyed. Germany and Italy and Japan had traded with the rest of the world but were now cut out of anywhere other than territories they had conquered. They, not Britain, were blockaded with extraordinary intensity and were left indeed even without the means to trade: their merchant fleets went to the bottom of the sea. British naval power, and indeed military power, was never merely defensive, sustaining a lifeline to Britain. It actively bottled up the German and Italian economies in continental Europe, with overwhelming offensive naval power in every sea except the Baltic. This involved taking over extra-European territories of the Axis and their allies, and of states which might support and supply them. Hence the campaign of conquest from 1940 to 1943 of North Africa and the Middle East from the Italians and Vichy France, and the effective takeover of Egypt, Iraq and Persia.[30] Blockade imposed hardship not only on Germany and Italy, but on the peoples of all the nations conquered by the Axis. It led to some protest in Britain, which led to the creation of Oxfam (the Oxford Committee for Famine Relief, created in 1942). 'Economic sanctions', a mild term from the 1930s for a brutal policy, were of huge but under-appreciated importance during the war.[31] British ships, and British and imperial land and air forces, not the sea, cut off Nazi Europe from the world.

BOMBING

Strategic bombing was without question the most important large-scale novelty in the mode of warfare conducted by Britain and the USA. As we have seen, strategic bombing played a crucial part in the British war effort, and depended to a great extent on access to the rest of the world. Ideologically and practically strategic bombing was an internationalist project. It depended on training in Canada and elsewhere, on aircrew from around the Commonwealth and on aviation spirit from the USA. Indeed, the offensive was a combined one with USAAF units stationed in Britain and elsewhere in Europe. For some, let us recall, this added up to an international air police ready to extirpate militarism from the world.

Some intellectuals criticized bombing on moral grounds. Vera Brittain wrote a powerful pamphlet called *Seed of Chaos*, published in April 1944. It had little response in Britain, though the USA version caused a storm.[32] She argued against bombing both on moral grounds and for the practical reason that Britain would need Germany as a market for its goods after the war. She saw obliteration bombing having origins, not least through Arthur Harris, in interwar imperial air control. It was in her view the result of moral lapses rather than a continuous and conscious policy of the state.[33]

The bombing of civilians and industry was central to British warlike practice from 1940, and to policy long before that. It did not offend British values in warfare; it exemplified them. It was seen as an extension of the idea of blockade. In 1943, for example, the *New Statesman* argued that:

> In fact it would be fair to sum up the advantages of bombing Germany by saying that it is to be judged as part of the long-term policy of blockade. Battleships and submarines prevent supplies reaching Germany; bombing operations destroy the dwindling supply of consumers' goods ... [Russians] think not in terms of bombing duels and victory by blockade, but of the swift defeat of the German army.[34]

There was another very important element of this kind of thinking which is worth bringing out. It is that in modern parlance such methods of warfare meant that Britain could 'punch above its weight'. John Strachey argued that:

> in the coming age of air power we need never attempt to build a huge, mass army, a job at which the great Continental nations with their huge populations will always beat us. But the air is a different matter. There our highly developed industry, technical ability, unsurpassed science and our air crews' flying skill will always enable us to hold our own and more.[35]

'Is it really so?' enquired the air propagandist J. M. Spaight in 1948, asking whether Britain had become a secondary power:

> Must we not always be a Great Power so long as we maintain, besides a small spearhead Army, a powerful Navy and a powerful Air Force? They will count for more than masses of men in the new age. They will close and hold the gates of civilization as they did in 1940–41, and when co-operative defence is effectively organized they will enable us to do our share, and more than our share probably, in helping to frustrate the knavish tricks of any breaker of the peace ... The tragedy of the German cities will not have been in vain.[36]

Even during the war Spaight claimed that war would not be tolerated in a 'Beveridged world' and the bomber would abolish war: 'It has been to Britain that the main credit is due for the bringing of that change ... It has been the British way of using air power which has revolutionized war.'[37]

The Anglo-German confrontation was deeply asymmetric here too. By the end of the war Bomber Command could, in twenty-four hours, drop more bombs on Germany than the tonnage launched against Britain by all the flying-bomb attacks of 1944, and more too than was dropped on London during all the months of the Blitz.[38] The total deaths in London from V-1 and V-2 weapons in late 1944 and early 1945 (a few thousand) were what the RAF was then dishing out to Germany on a single bad night for the bombers. The British killed somewhere between 150,000 and 300,000 German civilians by bombing – the figures are highly uncertain – three to six times as many as British civilians killed by the Germans.[39] The Germans killed many more British soldiers, sailors and airmen than British civilians. Some 122,000 British soldiers were killed fighting the Germans and Italians. The navy, the merchant navy, the RAF and British civilians each suffered deaths in the mid-tens of thousands each.[40] To put the issue another way, the losses of RAF Bomber Command and the merchant navy were each of the same order as those resulting from a single very heavy RAF raid on a Germany city, say Hamburg (1943) or Dresden (1945).

How effective was this bombing? In a 1947 memoir Sir Arthur Harris claimed that he had not been allowed to try strategic bombing properly. Area bombing, pioneered by the British in Europe, had led to the defeat of Japan (he had no truck with the notion that the atomic bomb was responsible – it caused a mere 3 per cent of the devastation, he noted). He held that if he had been allowed to area-bomb Germany as much as he could have, Germany could have been bombed into submission, perhaps in 1943.[41] Britain had instead poured money into 'less powerful and sometimes wholly obsolete weapons',[42] among other things 'battleships – the most expensive and the most utterly useless weapons employed in the whole of the last war'.[43] He defended area bombing as against precision bombing. It compensated, in his view, for lack of accuracy and lack of knowledge of how the German economy really worked. In Harris's mind the economic experts suggested targets (which he called 'panacea targets') which they alleged were so important that knocking them out would destroy Germany's capacity to produce. However, not only could they not be knocked out by precision bombing, but their destruction would not of itself bring the German economy down. It was better, in effect, to destroy generally than to go for specific targets of unknown real significance. Harris had a point – for despite interception and decryption of messages, the investigation of captured equipment and photographic reconnaissance on a huge scale, how the German war economy operated was far from being an open book to the Allies. Even with full information, the means of understanding how an economy works would still have been elusive. Indeed, it still is.

In a book-length official despatch written in 1945 Harris argued that nevertheless 'the heavy bomber did more than any other single weapon to win this War'.[44] Yet while his statistics showed huge numbers of bombs dropped and vast acreages of German cities destroyed, his evidence on the impact of bombing on the Germany economy was weak. Harris claimed the slow-down in German production growth in 1943 was due to area bombing, but was more circumspect about what happened in 1944. Harris was clear in the view that between April and September 1944, when Bomber Command was, in his account, distracted from 'its proper strategic role', Germany was able to reorganize war production and increase the supply of weapons, particularly of new weapons.[45] The bombers had been redirected to attack communications in particular. In this period German war production did increase quickly again, peaking in July 1944 at a time when area bombing was negligible. When area bombing resumed, soon reaching levels three times greater than in 1943, production started falling very quickly. But there are two difficulties. The first is the assumption that the decline in bombing in the spring and summer was what allowed German production to increase. Many other factors were in play, not least a determination of the German leadership to increase the exploitation of the workforce, particularly slave labourers, to an ever greater degree.[46] The second difficulty lies in ascribing the post-July 1944 collapse to British area bombing, about a quarter of the total Anglo-American bombing, one half roughly of British bombing.[47] Not only was Germany collapsing under military assault, but its communications and oil production were under large-scale attack.[48] The official post-war British bombing survey by Solly Zuckerman, an advocate of transportation bombing, criticized area bombing and focused on transportation and oil, and, while recognizing the effectiveness of both campaigns, came down in favour of transportation as the more decisive.[49] The United States Strategic Bombing Survey came down against area bombing and highlighted the successes of bombing both transportation and oil.[50] The British official history of the bombing campaign, written by the historians Charles Webster and Noble Frankland and published in October 1961, was also negatively disposed towards area bombing.[51] In subsequent work Frankland's view was that the 1944 and 1945 bombing could have been much more effective had it been less directed at cities and more at transportation and oil.[52] As he put it in his memoirs:

> Maximum concentration upon oil or transportation could probably, it seemed to me, have finished the war before the end of 1944 ... Even so, by April 1945, bombing alone had sealed the fate of Germany. But this was too late to stand as a clear-cut victory and all that the bombers could legitimately claim was that their efforts had greatly facilitated the success of the allied armies.[53]

The debate on bombing has concentrated on 1944 and 1945 with good reason. In 1943 British bombing of Germany was one tenth of total 1944 levels, and area bombing at one third of late-1944 levels.[54] However, the effects of this low 1943 level of bombing have been underestimated, in both its direct and indirect effects. The largely area attacks on the Ruhr specifically between the spring and summer of 1943 (of which the Dambuster raid was a small part) had important effects on production of coking coal, steel and 'intermediate components of all kinds'.[55] The raids 'stopped Speer's armaments miracle in its tracks'; indeed, 'the Ruhr was the choke point and in 1943 it was within the RAF's grip. The failure to maintain that hold and to tighten it was a tragic operational error.'[56] Secondly, 1943 levels of bombing were sufficient to divert significant German resources to air defence. As A. V. Hill had suggested in 1942, forcing the enemy to take defensive measures did not necessarily require a lot of bombing. Speer in his memoirs made clear that the great diversion of anti-aircraft gun production towards air defence was occurring between 1941 and 1943.[57] All the figures support this account.[58] In 1944 only some 13 per cent of Germany's heavy-gun production (over 75mm) were anti-aircraft guns, whereas the proportion was significantly higher in 1943 and earlier years.[59] This contradicts the strong impression given by Harris and others since that the diversion took place in 1944, and was the result of the high levels of bombing of that year and affected the war in that crucial year. In his despatch Harris claimed there were 2 million Germans working in air defence and dealing with the consequences of bombing, and that, in 1944, 30 per cent of gun output was anti-aircraft guns and that the German army was thus much weakened in the 'final battles'.[60]

The 1944 and 1945 levels of bombing should not be defended by invoking the diversion effect. In 1944, despite the increased bombing, the Germans began to withdraw guns from air defence. Remarkably, by November 1944 'fully 45 percent of the Luftwaffe's 88-mm flak guns were located in the occupied Western territories, in Italy, or on the Eastern Front', that is, outside Germany, 'with a great number of these weapons being used for ground combat support instead of anti-aircraft protection'. In December 1944 100 heavy flak batteries were moved to support the Ardennes offensive; in January and February 1945 over 300 heavy flak batteries were diverted to the Eastern Front, primarily as anti-tank and artillery weapons.[61] During the last week of January 1945 the Luftwaffe transferred 110 heavy and 58 medium and light flak batteries to the German army defence against the Soviet offensive on Berlin. Transfers stripped entire areas of flak defences.[62] So desperate was the German position that the diversion was from anti-aircraft to anti-tank, not the other way round.

Of course, anti-aircraft guns were not the only German reactions to bombing. One was the development of the V-weapons, the V-1 and the V-2. The other was the escalation of fighter production in 1944, which was at the heart of the expansion of German aircraft production in that year. Again it is not at all clear that these effects were generated by 1944 levels of bombing, as opposed to 1943 levels. The offensive had as a secondary effect the engagement with and destruction of much of the defending air force.

In analysing the effects of bombing, listing a series of its positive effects, from falls in industrial production to the diversion of resources, is not enough.[63] The effects need quantification if the argument is to convince: just how much production was lost; what proportion was diverted? There is a second crucial issue, which is hardly addressed at all. What was the effect of bombing compared with the use of the resources to support other ways of waging war? Was the British strategic bombing campaign a worthwhile use of, at the very least, 7 per cent of the war effort, and perhaps nearer twice that?[64] What other uses might the particular skills of the best-educated young men in the forces, large numbers of creative academics and the most refined parts of the engineering industries have been put to? For some points in time, the answer is clear. For example, given the utter ineffectiveness of the bombing of Germany in 1940, Britain would clearly have been better off with the resources devoted to Bomber Command going instead to a larger army. Indeed, one military intellectual, writing in 1943, claimed that the policy of bombing had brought 'calamities upon the nation and Empire'.[65] But we need to look not just at what economists call static efficiency, the efficiency at one point, but efficiency over time – dynamic efficiency. Without a Bomber Command in 1940 there would have been no Bomber Command in 1943. Could it have been usefully dispensed with to create, say, another British army or two to operate in North-West Europe? Would such a force have secured an earlier German surrender? Or, conversely, would Britain have been better off putting more resources into the bombing? Imagine even greater British bombing (and, to extend the argument, US bombing too) in 1943, 1944 and 1945. Would the war have ended earlier? Surely much heavier or redirected conventional bombing, followed if necessary by a *coup de grâce* from an atomic bomb, would have won the war, without British or American troops in Europe? Indeed, an early victory by bombing, say in 1943, would have made Britain, as it hoped to continue to be, the greatest of the great powers: if things had turned out as Churchill hoped and believed in late 1941,[66] Britain 'would not have become so junior a partner in the Anglo-American effort'.[67] For in 1942-3 the USA was still catching up with Britain in this crucial machine, and the Soviet Union was nowhere. British air power would have dominated the world. But it didn't, not just because bombing wasn't as decisive as had been hoped: the nature of the continental war was quite different from the theories of modern war advocated by British analysts.

INVENTIONS, SCIENCE AND MACHINES

The debate about the impact of the heavy bomber on the war is the most extensive, long-standing and long-drawn-out discussion of the impact of a machine in war there has ever been. Perhaps no other machine, in any field, has been subject to so much assessment of its effectiveness. And yet assessment remains partial and inconclusive. That in itself points to the need for extreme caution in making claims for the effectiveness or efficiency of any other machine.

The question of the success or otherwise of British science and machines cannot be separated from that of the bombing campaign, which was central to scientific and technical endeavours, from the aviation spirit programme to most of the advances in radar and radio after the beginning of the war. Great effort went into the development of bombsights and new kinds of bombs and explosives, not to mention the atomic bomb. Furthermore Bomber Command had one of the very largest of the operational research sections. Of all the British high commanders Harris stands out for his commitment to new machines and systematic examination of operations – he was the very model of a modern warrior.

And yet, for a significant and influential part of the British scientific community, the bomber offensive was seen as a non-scientific way of warfare. Scientists of the left criticized area bombing, suggesting it flew in the face of the advice of operational researchers. J. D. Bernal claimed that 'Mass bombing of cities proved its futility as a means of winning the Second World War.'[68] Patrick Blackett was also strongly critical of the bombing in a remarkable critique of atomic weapons published in 1948, stressing its small contribution independently of achieving air superiority and the key contribution of the Red Army to victory.[69] Both were alluding not just to the debates of 1942, but to Solly Zuckerman's advocacy of a particular bombing policy in 1944.

Zuckerman wanted attacks on transportation, focused not on road and rail bridges, but on marshalling yards and repair depots. The former were difficult to destroy but remained so, while the latter were easily hit but were quickly repairable.[70] Yet this was not a position based on arguments which convinced all experts. The US economists and economic historians who had argued for oil plant bombing claimed that the transportation plan was based on very weak evidence from the Italian campaign. They later argued that what curtailed military transport in Normandy was not the attacks on marshalling yards/repair centres but rather on the bridges over the Seine.[71] It is not at all clear that Zuckerman was right in his assessment of the bombing effort it would take to destroy marshalling yards, another key issue at stake.[72] Experts were once again fighting experts, but one side was seen as right, the other hopelessly wrong and morally deluded. This kind of analysis would reach its absurd peak in C. P. Snow's account of the clash between Tizard and Lindemann over bombing in 1942.[73]

Distancing science from an association with bombing and promoting operational research was important for many scientific intellectuals. Indeed operational research became disproportionately important in the influential accounts of science in the war from the left.[74] Its reputation benefited from the argument that it was novel, British and not concerned directly with killing.[75] Consider J. D. Bernal's immediate post-war view that not only had 'the scientists of this country offered their services' but that they 'were more fully employed and with greater success than in any other combatant country'. During the war it 'was found that the value of scientists was as great, if not greater, in examining and advising on many and varied problems of warfare and supply, and particularly the evolution of the methods of operational research', and called for this to be used to raise productivity in peacetime.[76] In another account he claimed that 'at the beginning of the war the scientist was conceived of as a person who produced and developed scientific weapons', by which he meant, rather oddly, weapons which you need to know some science in order to operate – he had in mind the aeroplane, 'almost scientific through and through', and radar, in contrast to the rifle or the machine gun. In this period, he claimed, 'the whole emphasis of science in war was science in relation to scientific gadgets'. But, as the war progressed, what was important was 'problems raised in the factory, production problems, planning problems in the general preparation of war weapons, and finally, towards the middle and end of the war, the problems of actual operations'.[77]

The idea that British science was more effectively mobilized than German science was widespread after the war, and was articulated by, for example, Tizard, who claimed Germany suffered from Hitler's emphasis on mad inventions.[78] Churchill too celebrated British science, claiming that 'Unless British science had proved superior to German, and unless its strange, sinister resources had been effectively brought to bear on the struggle for survival, we might well have been defeated.'[79]

Churchill, as we have seen, was to write essentially about machines he was connected with personally. Indeed, there were lots of different stories of machines and victory after the war, though not in the talk of the scientific left. The Britishness of inventions was important. Radar and penicillin were regarded as British.[80] FIDO, a system which burned 100,000 tons of precious oil during the war to clear fog from airfields, was another celebrated 'all-British achievement'.[81] And when the atomic bomb exploded over Japan, the British newspapers made the British contribution disproportionately clear. 'British and American scientists,' intoned the *Manchester Guardian*, had 'achieved what the Germans were unable to do and have won the "greatest scientific gamble in history"'.[82] The 1945 Labour manifesto waxed lyrical about the 'genius of British scientists and technicians who have produced radio-location, jet propulsion, penicillin, and the Mulberry Harbours'.[83] Newsreels were full of Grand Slam bombs, Mulberries, PLUTO, jet engines, bouncing bombs. A model of the Mulberry Harbours was seen by millions in Britain and travelled to Europe.[84] After this orgy of techno-nationalist excess it became hard to believe that Britons were not the only, or indeed the main, inventors of the jet engine and radar.[85] Yet it is remarkable that what the most celebrated examples – the jet engine, the Mulberry Harbour, PLUTO, the bouncing bomb, the Grand Slam – had in common was that their contribution to the war was of doubtful value. The best that can be said is that Germany almost certainly wasted more on what turned out to be economically and militarily irrational new weapons.

At the great London victory parade of 1946 there were two columns – a marching column and a mechanized or motorized column in which a selection of overwhelmingly British machines paraded with name plates attached: Cromwell, Churchill and Sherman (perhaps a Firefly) tanks, Tallboy bombs, radar sets, fire engines and tractors, scout cars, London buses, artillery pieces, railway engines, cranes, motorbikes ...[86] That eclectic mix makes the point that a huge combination of machines was involved in the war. Indeed, unconstrained by the need to put machines on the road (and there was a huge fly-past, and many naval ships were moored in London), one could add many more. Recall some of the ones highlighted in this book: the aero-engine, the Liberty ship, the cold store and the reefer, the alkylation process for aviation spirit, aluminized explosives, machine tools from the United States, aerial mines, decoys, plastic armour, the APDS shot: so many are involved, and are so interconnected with each other, that assessment is fraught with difficulty. But the overall point that needs to be made is precisely that the machines were so many, the change so rapid: everywhere there were new devices of war, large and small.

Yet the tradition of celebrating very particular machines and devices and granting them war-winning status has proved enduring. Radar and the Spitfire won the Battle of Britain. According to Sir Arthur Harris: 'the Lancaster was the greatest single factor in winning the war'.[87] The two monster passenger liners, the *Queen Elizabeth* and the *Queen Mary*, allegedly shortened the war by a year through their ability to transport huge numbers of troops at high speed.[88] British cryptographers are supposed to have shortened the war by two years.[89] Indeed, if one adds up the times by which the war was shortened or won by particular machines, it was clearly won many times over, or would otherwise have gone on for many years after 1945.[90]

THE WELFARE STATE

One of the more long-lasting claims for national development during the war concerned the creation of a new kind of welfare state. A new conception of the British people, it is argued, set the country on the path to a universalistic welfare state. This interpretation of the war was to become central especially from the 1960s onwards – thus A. J. P. Taylor ended his *English History* with the assertion that in the war 'the British people came of age. This was a people's war ... Imperial greatness was on the way out; the welfare state was on the way in.'[91] In the Britain 'which faced defeat between 1940 and 1942', wrote Angus Calder, were the 'seeds of a new democracy'; the rulers had to ensure the cooperation of all society by making 'concessions in the direction of a higher standard of living for the poor, greater social equality and improved welfare services'.[92]

The key source for the contentions on welfare was a book by the pioneering sociologist of welfare Richard Titmuss, his 1950 official history covering *Problems of Social Policy*. Titmuss's book, limited in detailed treatment to the social issues directly concerned with the war (evacuation, the emergency medical service, rehousing of the bombed-out and such like), in fact makes minimal claims for wartime extensions of social services: increase in the number of school meals, their provision beyond the really needy, the national milk scheme for nursing mothers and children, and improvements to the old age pension.[93] It suggests that the Blitz was crucial in broadening social services to a wider population than the destitute, and made subtle and important points about the principles underlying welfare. Overall, however, he shows that the war was bad for social services; so bad, indeed, that this in itself promoted post-war reform. It was the dire treatment of the old, for example, that pricked people's consciences.[94] As Titmuss put it: 'Somebody had to pay the price of war by going without, waiting longer, getting less or being pushed about to make room for others', noting that 'among those who suffered most were the poorest, the most helpless and the "useless" members of the community'.[95] Far from being universal there was a hierarchy in provision of hospital treatment, with the aged and chronic sick at the bottom, followed by the ordinary sick and expectant mothers in need of beds, air raid casualties, evacuees and, at the top, 'Service patients, the most favoured group of all, who got the lion's share of hospital care throughout the war'.[96] The war 'deprived the civilian population of a large part of its pre-war medical resources', not only in hospitals, but in school medical and dental services, maternity and child welfare clinics, and public health. GP services were badly affected: the number of GPs decreased by over one third, and of those left many were old.[97] In 1943, in Britain, there were five times more doctors per soldier than per civilian.[98]

This was not all. Other studies showed what one would expect would happen to social services. 'The standard of educational facilities fell sharply', with the number of teachers in elementary and secondary schools falling by about 10 per cent and of male teachers by a third.[99] While there were additional civil expenditures out of wartime votes of credit, they appear to have been limited to measures specific to the war, such as the Emergency Medical Service, the Ministry of Food, evacuation, etc. There was certainly no general improvement in welfare services.[100]

Even after the war the welfare state had not much more to spend than before the war. Indeed, the war saw a collapse in investment in these areas, with slow recuperation in the late 1940s, typically not reaching the levels of the 1930s. It was a case of reconstruction from wartime cutbacks, rather than raising the level above the supposedly disastrous 1930s. No hospital was built in the late 1940s, and what expansion in capacity there was during and after the war may not have been enough to raise bed provision per capita.[101] In housing, too, it was not until the 1950s that 1930s levels of building were reached. In welfare the 1940s taken as a whole were an era of disinvestment. In many areas, reconstruction did not begin in 1940; it started painfully slowly in 1945, and levels of social investment took until the 1950s to return to the levels of the 1930s.

CONSEQUENCES

One of the great themes in studies of the Second World War is that even if Britain was not saved by the USA, it was certainly eclipsed by it. This process, however, should not be taken to imply British weakness; nor should it be attributed wholly or mainly to US superiority. US growth, not British failure, was the main cause of the spectacular British relative decline between 1941 and 1945. Much of the extraordinary rise of the US was a reflection not of its modernity and power but the previous under-use of its resources. Into 1940 the USA was not the land of opportunity but of mass unemployment: the recovery from the Depression, for all the activity associated with the New Deal, was notably weaker than in Britain, let alone Nazi Germany. In 1940 the US had a reserve army of unemployed man-power of 7 million, which would directly and indirectly provide fully one third of all US service personnel and arms workers in the coming years. During the war 20 million Americans entered the labour force, giving a remarkable 50 per cent increase, divided equally between the armed services and civilian employment.[102] Britain increased its work-force by only 10 per cent compared with 1940, and while it put 4.5 million into the forces, it lost one million from the civilian workforce. Put very crudely, the US labour force, including the military, was twice as big as the British in 1940, but three times as big at the wartime peak. Its coal production was twice as big as Britain's in 1940, and three times as big in 1945. In shipbuilding it was a minor player in 1940, and the biggest in the world by far in 1945. In 1942 UK forces were larger than those of the US and into early 1942 Britain produced more arms; by 1943 the US was ahead of the whole Empire in mobi-lized forces, having around twice the UK total; in arms production it now led four to one.[103] The US had people and resources to spare on a gigantic scale, and once got to work the US could supply itself and others on a stupendous scale. It could fight and win the war, making itself richer and more successful in the process. Even a growing, more powerful and more heavily armed Britain could only lose out rela-tively to such a state. Britain's stupendous relative decline in wartime was caused not by its decision to fight, but by that of the USA.

This casts doubt on the view that, in its most romantic version, Britain, led by Churchill, bankrupted itself to survive and to save the world from Hitler.[104] This view is in any case not as well-grounded as it seems. On the face of it some statistics support this view. British losses of capital have been estimated at about 25 per cent by the government in 1945, and in a recent revision at 18 per cent.[105] These estimates need breaking down, for by far the largest element in the loss is 'external disinvestment', the build-up of obligations to foreigners – that is, external debt – and to a much lesser extent the loss of foreign assets. Pre-war investment in the USA and Canada in 1938 amounted to £4,674m; the receipts from sales were £1,100m.[106] The biggest firm to be sold was the Courtauld-owned American Viscose, largest maker of rayon in USA.[107] Yet, surprisingly, investment income from abroad was not much lower in 1946 than before the war.[108] Major centres of British expatriate business, Rangoon, Singapore, Shanghai, Hong Kong, were lost along with great rural enterprises like Malayan rubber plantations, but were recovered, thanks in large measure to the USA, after the war, some only temporarily. Net losses due to UK physical destruction (including ships and cargoes) came to 6.9 per cent of capital stock.[109] That is a significant proportion, but hardly catastrophic.

It could be argued that Britain sacrificed its place in the world by reducing its exports, to maximize arms output and as a result of US pressure. However, although British exports fell during the war, so too did those of its key competitors, many of whom were completely eliminated from world markets. In the post-war years Britain was able to achieve, temporarily, a commanding position. Yet the key effect of the fall in exports was that they made Britain's position temporarily very difficult because it needed to import. This was so because the US suddenly and unexpectedly cut off Lend-Lease with victory in Japan, which came earlier than expected. This 'financial Dunkirk' of 1945 forced the British to take out a US loan to pay for imports. It was this loan which was paid off in 2006, though it was misleadingly implied that Britain had paid off Lend-Lease.[110] The loan had very strong and disliked strings attached. It is little wonder that large segments of the British elite felt very sore indeed about this.[111]

It was, however, only a financial and not a commercial or industrial Dunkirk. Indeed, contrary to what is often implied, the post-war balance-of-payments problem was only partially to do with the balance of trade in goods; it was also to a very considerable extent to do with military and relief expenditure abroad and a massive export of capital to the white dominions.[112] One may question, as historians have done, the wisdom of such expenditures compared with alternatives; for us the point is that Britain had the wherewithal to do all this.

Industrially Britain had been strengthened by the war. It is true that in vast swathes of the economy maintenance fell, investment fell. The great cotton-spinning mills and weaving sheds of Lancashire had seen practically no investment since the very early 1920s, and often much earlier, but Lancashire was now full of massive modern engineering and chemical works, from Heysham's aviation spirit plant to the new aero-engine and airframe plants. The same was true of much of the rest of the country.

Critics of Britain's position in 1945 also underestimate the great importance of victory and the cheapness of the victory. For most of the major combatants (the USSR, Germany, the USA, Japan, India, the last because of famine) the Second World War was worse or far worse than the First. The major combatants for whom the experience was better (France and Belgium) suffered occupation, deprivation, and the total dislocation and humiliation of their societies. For Britain and the white British Empire it was a considerably better experience, with lower casualties for the UK, Australia, Canada and New Zealand. Among the victorious it was better by far to be British than French in 1945, and better French than Soviet or Chinese. The price of victory was, for the Soviet Union, higher than the price of defeat paid by Germany. Britain, and above all the US, had cheap victories. There is a simple rule here – wealth meant low losses, poverty dire ones, whether victorious or not. Britain did not win *the* war – there were many wars, and Britain did not participate except marginally in by far the most important one, that between Nazi Germany and the USSR. As one military history put it: Britain came first in the Great War, but third in the Second World War.[113] But it won *its* war, and it won because it was rich and powerful, because it could depend on the rest of the world and on vast numbers of machines.

This is not to say that the war did not change Britain's position in the world. It did so drastically. Britain had lost large chunks of imperial territory and regained them only because of the victory in the East of the USA, its greatest reserve of manpower was headed for independence, and its Australasian dominions became more closely linked to the USA. Still, Britain emerged very obviously by far the most successful Western European power. Britain did not pay enough, could not pay enough, in either currency to become a world power again, but it paid enough to remain a very powerful player. Given the alternatives, it is hard to see that Britain was hard done by.

A NEW BRITAIN?

What was the new Britain that emerged from the war like? The standard themes of histories are of reconstruction, of the building of a new Britain promised and essayed during the war, a process characterized by a pall of continuing austerity. But it was not just a question of building a New Jerusalem, but also a new Sparta. Post-war Britain had very high defence spending (and much was spent abroad) and conscription, as well as levels of austerity often exceeding wartime ones. The economy was much more national than before the war in, for example, domestic food supply, and in such matters as oil refining. The economic language of post-war economics, of Keynesian demand management, inflation, deflation, nationalization, with a dash of planning and industrial policy, was not suited to highlighting what had been called protection, autarky, state trading. There is no doubt of the increased influence of social democratic thinking associated with the Labour Party, but that did not in itself capture the new nationalism or militarism either. Its main concern was welfare, together with nationalization in the narrow sense of transfer of local government and private industries to state ownership.[114] Yet other ideological changes made such developments seem normal, necessary and, indeed, in many respects beneficial, so much so that they elicited little adverse comment. After the war much intellectual culture was nationalistic and militaristic to a greater degree than before it.[115] These were implicit rather than explicit ideologies. There was no mid-twentieth-century British Friedrich List. Economic nationalism was a matter of lower-level politics, policy, national security, business lobbies, rather than the economics seminar.[116] Yet there were to be many critiques of the failure of the state to take a national approach, to blame it for putting international trade and finance before domestic industry. Political economists came to complain of the supposed lack of a developmental state, when the state was more powerful than ever. Scientific intellectuals lamented the lack of a more rigorous plan for a national development of science, technology and industry. There was no explicit language either for the new and historically unprecedented commitment to militarism. There was no British Clausewitz, yet the dominant tone of commentary on the military was that it was insufficiently supported, that it was too focused on the air and the sea and not enough on what counted, the large army. Military intellectuals came to see the orientation of British forces to the seas and the air as a distraction from what mattered – engagement on land with a Continental army. Such critiques, lambasting Britain for not being nationalistic, technocratic or militaristic enough, not *German* enough, profoundly affected how the history was written.

As we have seen, a very different country confidently declared war in 1939 and kept fighting in 1940 – a great power that was fighting not for survival but for victory. This was a country, an Empire, confident in its way of making war, and one which took its international connections and its liberal militarism to be a source of strength. The Britain of 1939 was by various measures the most powerful country in the world and not one which Germany could hope to beat easily (as became clear with the failure in 1940 of both the German navy and the German air force). Just as Hitler was to prove drastically mistaken about the nature of the war he had unleashed, so British policymakers would have been astounded by the sequence of events. Germany was not to be defeated as France was in 1940, nor as it had been in 1918. The Red Army had to take Berlin street by street, losing more dead in that campaign than the entire number killed in the British army through the whole war. We still hear that by its actions in 1940 Britain saved the world, but for all its strategic significance the defence of Britain didn't save Europe from Hitler's greatest crimes, from this most devastating war, which in 1940 was still in the future.

The cruel war on the Eastern Front, as historians now very clearly demonstrate, was something not to be understood in the usual schemes of modernity. It was not a war that military modernists anticipated. This was a land war, not a war in three dimensions; there was plenty of fire and steel, but it was also a matter of fodder and grain.

It involved huge numbers of tanks but also unprecedented artillery barrages and costly infantry assaults. And the radical brutalization of the front was reflected in the treatment of prisoners and civilians, and, indeed, on the home fronts too. Many more civilians – tens of millions – than combatants died, not through aerial bombing but from hunger, disease, small arms and improvised poison gases. That two powers so self-consciously presenting themselves as being in the van of modernity should fight by such means was a profound irony, one still difficult to fully grasp.

For the British the war was very different. This was a war fought in three dimensions, with radio waves, aeroplanes, tanks and ships; a large measure of fire and steel was visited on Britain's enemies. The expectations that such means would bring victory were not immediately borne out: in Britain the power of the economic and the mechanical in warfare were overestimated. British machines and logistic strength did not always, especially at the beginning of the war, translate into victory in the field.[117] In France and North Africa the German army made more effective use of its limited machines than did the lavishly equipped British army. In the East, Japanese soldiers fought on far beyond the point where British and imperial forces surrendered. Yet then, and since, British officers blamed shortages and poor quality of equipment. The obsession with equipment, with overwhelming firepower, perhaps led to failures in leadership, in resourcefulness, qualities which later accounts would celebrate as British complements to a supposed lack of equipment. Indeed, there was a tendency to celebrate quirky, inventive British devices, like Mulberry or PLUTO, as evidence of a peculiar technical resourcefulness in the face of material austerity. They were in fact compelling evidence of extravagant commitment to technical solutions, and of massive material and technical capacity.

But initial perceptions had substantially proved correct in some crucial respects: that Britain could successfully fight because of its tremendous wealth and assets and its huge range of military and civilian machines and expertise. It could produce a great deal, in factories with modern amenities, with well-fed workers who could be induced to work in new industries. Its wealth in weapons and resources and experts, supported by rich and poor peoples beyond the seas, allowed it to emerge victorious and relatively unscathed. It was fortunate that it was never alone and never had to fight a total, people's war of the sort that was fought on the Eastern Front. For that deliverance post-war Britain showed a remarkable lack of gratitude and an equally notable lack of appreciation of how it was achieved.

Notes

1. INTRODUCTION

1. The agenda-setting books on the Home Front were A. J. P. Taylor, *English History, 1914–1945* (Oxford, 1965); Angus Calder, *The People's War: Britain, 1939–1945* (London, 1969); Paul Addison, *The Road to 1945: British Politics and the Second World War* (London, 1975); and Correlli Barnett, *The Audit of War: The Illusion and Reality of Britain as a Great Nation* (London, 1986). Mark Donnelly, *Britain in the Second World War* (London, 1999), is a good summary of the literature. Henry Pelling, *Britain and the Second World War* (London, 1970), was refreshingly sceptical about many claims made for the war. The issue of national identity has dominated the academic literature on the war in recent years; see, for example, Angus Calder, *The Myth of the Blitz* (London, 1991); Sonya O. Rose, *Which People's War? National Identity and Citizenship in Wartime Britain 1939–1945* (Oxford, 2003); and Mark Connelly, *We Can Take It! Britain and the Memory of the Second World War* (London, 2004), but these rely on long-established accounts of the war and its impact. These domestic and national stories have been largely by the left of the left. There is a deep division between these accounts and the strategic and international histories of the war. Churchill's own *The Second World War* (6 vols., London, 1948–54), which now needs to be read with David Reynolds, *In Command of History: Churchill Fighting and Writing the Second World War* (London, 2004), set the agenda, one overwhelmingly developed by historians whose political sympathies have been Churchill's. British historians, like Churchill, have tended to write about the strategic and related aspects of the war as a whole, rather than specifically about how Britain fought. Max Hastings, *Finest Years: Churchill as Warlord 1940–1945* (London, 2009), is perhaps the exception which proves the rule, for which see, for example, Peter Calvocoressi and Guy Wint, *Total War: Causes and*

Courses of the Second World War (London, 1972); R. A. C. Parker, *The Second World War: A Short History*, rev. edn (Oxford, 1997); Norman Davies, *Europe at War 1939–1945: No Simple Victory* (London, 2006); Andrew Roberts, *The Storm of War: A New History of the Second World War* (London, 2009); and Michael Burleigh, *Moral Combat: A History of World War II* (London, 2010). The disconnection between foreign and domestic is now being seriously challenged, partly by some work on Empire (for a notable example, see John Darwin, *The Empire Project: The Rise and Fall of the British World-System, 1830–1970* (Cambridge, 2009)) and also more generally: see, for example, William Mulligan and Brendan Simms (eds.), *The Primacy of Foreign Policy in British History, 1660–2000: How Strategic Concerns Shaped Modern Britain* (London, 2010). This book is itself a contribution to the rethinking of the relations of domestic and international history.

2. David Edgerton, *The Shock of the Old: Technology and Global History since 1900* (London, 2007). For rich examples of how the material is treated, see Gerhard L. Weinberg, *A World at Arms: A Global History of World War II* (Cambridge, 1994), Chapter 10, 'Means of Warfare: Old and New'; Alan S. Milward, *War, Economy and Society 1939–1945* (London, 1977), Chapter 6, 'War, Technology and Economic Change'; Guy Hartcup, *The Effect of Science on the Second World War* (Houndsmills, 2000); Guy Hartcup, *Challenge of War: Scientific and Engineering Contributions to World War Two* (London, 1970).

3. The contributions of British historians of Germany to our understanding of the economics of the war as a whole, and of the British war economy, are striking. See above all Milward, *War, Economy and Society*; Richard Overy, *Why the Allies Won* (London, 1995) and Adam Tooze, *The Wages of Destruction: The Making and Breaking of the Nazi Economy* (London, 2006). See also the vital contributions of Mark Harrison (ed.), *The Economics of World War II: Six Powers in International Comparison* (Cambridge, 1998), and George Peden, *Arms, Economics and British Strategy: From Dreadnoughts to Hydrogen Bombs* (Cambridge, 2007), and the work of Peter Howlett, notably in Harrison (ed.), *The Economics of World War II*.

4. See my *England and the Aeroplane: An Essay on a Militant and Technological Nation* (London, 1991) and Jim Tomlinson, *The Politics of Decline: Understanding Post-War Britain* (London, 2000).

5. This was my definition of declinism in my *Science, Technology and the British Industrial 'Decline', 1870–1970* (Cambridge, 1996), p. 4.

6. David Edgerton, *Warfare State: Britain, 1920–1970* (Cambridge, 2005),

especially Chapter 7; Tooze, *Wages of Destruction*, pp. xxii–xxiii, 597–405.

7. This repositioning of Germany is used to brilliant effect in Tooze, *Wages of Destruction*.

8. Calder, *People's War*, is an instance of what was a general phenomenon. For an analysis, see Edgerton, *Warfare State*, Chapter 7.

9. For example, David French, *The British Way in Warfare 1688–2000* (London, 1990); Edgerton, *England and the Aeroplane*; David Edgerton, 'Liberal Militarism and the British State', *New Left Review* 185 (Jan–Feb 1991), pp. 138–69; and Peden, *Arms, Economics and British Strategy*, which has an exceptionally good and very substantial chapter on the Second World War. Peden's *British Rearmament and the Treasury 1932–1939* (Edinburgh, 1979) pioneered the serious historical treatment of the relations of economics and war for twentieth-century Britain.

10. See in particular the last two chapters of Edgerton, *Warfare State*.

11. For some suggestions as to how to approach this 'historiography from below' and how to deal with the anti-histories of the experts, see Edgerton, *Warfare State*, Chapters 5 and 8. See also Harold Perkin, *The Rise of Professional Society: England since 1880*, 2nd edn (London, 2002; first publ. 1989), especially pp. 117–18, 343–52, 398. For Perkin, the professional ideal – trained expertise and selection on merit – was critical in twentieth-century Britain, and the rise of this ideal was powerfully influenced by the war. Yet, revealing how often such stories are fitted too easily into standard historical narratives, Perkin sees only the welfare professions and the welfare state in this crucial period: medicine, education, town planning, social work. New work on British experts has been a feature of recent years, notably in relation to armed services in my own work, and to Empire: see Peder Anker, *Imperial Ecology: Environmental Order in the British Empire, 1895–1945* (Cambridge, Mass., 2001); Sabine Clarke, 'A Technocratic Imperial State? The Colonial Office and Scientific Research, 1940–1960', *Twentieth Century British History* 18 (2007), pp. 1–28; and Joseph Morgan Hodge, *The Triumph of the Expert: Agrarian Doctrines of Development and the Legacies of British Colonialism* (Athens, O., 2007).

12. See Edgerton, *Warfare State*, Chapter 7, for details of this argument.

13. Alan S. Milward, *The Economic Effects of Two World Wars on Britain*, 2nd edn (London, 1984), p. 27, is an early and very rare acknowledgement of this.

14. See especially Darwin, *The Empire Project*.

15. 'Cameron's historic blunder: Fury as PM says we were "junior partner"

to Americans in 1940': http://www.dailymail.co.uk/news/article-1296551/David-Cameron-describes-Britain-junior-partner-Americans-1940.html #ixzzov8uNaliT, *Daily Mail*, 22 July 2010.

16. http://www.bbc.co.uk/news/uk-politics-10719739, 30 July 2010.

17. Cameron, speaking on the *Today* programme, 28 July 2010.

2. THE ASSURANCE OF VICTORY

1. Stephen King-Hall, *King-Hall News-Letter* (13 January 1939).

2. Ibid. (24 March 1939).

3. *The Economist* (22 April 1939), p. 182.

4. 'War and Post-War Economics', *The Economist* (2 September 1939), p. 434.

5. Ibid.

6. Ibid., p. 436.

7. *The Economist* (25 November 1939), p. 275.

8. *The Times* (23 December 1939), quoting from Ministry of Information, *The Assurance of Victory*.

9. *The Times* (1 January 1940).

10. *The Engineer* CLXIX (1 March 1940), p. 209.

11. W. H. Bragg to Walter Moberley, 15 January 1939, The Papers of A. V. Hill, AVHL I 2/1, Churchill Archives Centre.

12. 'British and German Science', *The Engineer* CLXIX (22 March 1940), p. 281.

13. *The Engineer* CLXIX (5 January 1940), p. 15.

14. *Aircraft Production: The Journal of the Aircraft Manufacturing Industry* 2 (April 1940), p. 105.

15. Winston S. Churchill, *The Second World War*, Vol. I: *The Gathering Storm* (London, 1948), p. 320.

16. Ibid., p. 322.

17. Andrew Gamble, *Britain in Decline* (London, 1982). This use is quite different from that of Halford Mackinder, whose own 'world island' was Eurasia and Africa – which he took to be the great central mass of the world, to be contrasted with peripheral places like Britain, the Americas and Australasia.

18. Leslie Hannah, 'Logistics, Market Size, and Giant Plants in the Early Twentieth Century: A Global View', *Journal of Economic History* 68 (March 2008), pp. 46–79.

19. Major D. H. Cole, *Imperial Military Geography: General Characteristics of the Empire in Relation to Defence*, 7th edn (London, 1934), p. 64.

20. Alfred Zimmern, *Quo Vadimus?* (London, 1934), pp. 35–6.

21. On Empire and ideology, see the very useful survey of the literature by Stephen Howe, 'Empire and Ideology', in Sarah E. Stockwell (ed.), *The British Empire: Themes and Perspectives* (Oxford, 2008), pp. 157–76.

22. It is still common in the literature. For example, in Talbot Imlay's eyes: 'Separated physically from the Continent, the British were understandably more ambivalent about Europe – an ambivalence reinforced by the importance of imperial considerations. Put simply, the British were far more imperial-minded than the French': Talbot C. Imlay, *Facing the Second World War: Strategy, Politics, and Economics in Britain and France 1938–1940* (Oxford, 2003), p. 18, and *passim*.

23. *The Economist* (8 April 1939), p. 80.

24. L. Isserlis, 'Tramp Shipping Cargoes, and Freights', *Journal of the Royal Statistical Society* 101 (1938), pp. 53–146, on p. 86.

25. W. K. Hancock and M. M. Gowing, *British War Economy* (London, 1949), p. 183, note 1.

26. Heinz Schandl and Niels Schulz, *Using Material Flow Accounting to Operationalize the Concept of Society's Metabolism. A Preliminary MFA for the United Kingdom for the Period of 1937–1997*, ISER Working Papers, Number 2000-3 (2000), Institute for Social and Economic Research, University of Essex, UK.

27. 'Times of Attack on Merchant Ships Sunk by U-boat', Cherwell Papers, CSAC 80.4.81/F.334, Nuffield College Archives. See the endpapers of this book.

28. Ina Zweiniger-Bargielowska, *Austerity in Britain: Rationing, Controls, and Consumption, 1939–1955* (Oxford, 2000), p. 36.

29. Gerald Egerer, 'Protection and Imperial Preference in Britain: The Case of Wheat 1925–1960', *The Canadian Journal of Economics and Political Science/Revue Canadienne d'Economique et de Science Politique* 31 (August 1965), pp. 382–9.

30. F. Gerrard, *Meat Technology* (London, 1951), p. 288. The figures are for 1936.

31. Lewis C. Ord, *Secrets of Industry* (London, 1944), p. 85

32. Robert W. Steel, 'The Trade of the United Kingdom with Europe', *The Geographical Journal* 87 (June 1936), pp. 525–33.

33. Cole, *Imperial Military Geography*, p. 58. The text, however, shows evidence of the trade with Europe. A similar text, A. G. Boycott, *The Elements of Imperial Defence: A Study of the Geographical Features, Material Resources, Communications, and Organization of the British Empire* (Aldershot, n.d. but probably 1939), has a map showing

principal trade routes and commodities (between pp. 94 and 95) which does include Europe and its iron ore, timber, ham, bacon and eggs, etc., but since it is weighted by quantity it gives a misleading impression of its scale. Exports are not included.

34. C. B. A. Behrens, *Merchant Shipping and the Demands of War* (London, 1958), p. 5.

35. *The Economist* (16 March 1940), p. 485.

36. Isserlis, 'Tramp Shipping Cargoes, and Freights', p. 86.

37. Behrens, *Merchant Shipping and the Demands of War*, Appendix 1, p. 17.

38. Smaller versions carried bananas from the Caribbean, a trade dominated by the American giant United Fruit, and its British subsidiary, Fyffes.

39. The Royal Mail also had *Asturias* and *Alcantara* on the same run, plus others.

40. Richard P. de Kerbrech, *Harland & Wolff's Empire Food Ships 1934–1948: A Link with the Southern Dominions* (Isle of Wight, 1998).

41. *Report on Scientific and Industrial Research for the Year 1934–35*, Cmd 4787 (London, 1935), p. 18; *Report on Scientific and Industrial Research for the Year 1938–9*, Cmd 5927 (London, 1939), p. 10. There is evidence that the first shipments, on the same ship, went from Australia.

42. Nhamo Samasuwo, 'Food Production and War Supplies: Rhodesia's Beef Industry during the Second World War, 1939–1945', *Journal of Southern African Studies* 29 (June 2003), pp. 487–502.

43. D. J. Payton-Smith, *Oil: A Study of War-Time Policy and Administration* (London, 1971), p. 243.

44. The largest US fleet by far was that of Standard Oil (Esso); other fleets were owned by Socony-Vacuum Oil, Atlantic Refining, Gulf Oil, Sun Oil, Texaco, Union Oil, Standard of California and Standard of Indiana. The US fleet was primarily concerned with shipping refined products from the Gulf of Mexico to the north-eastern seaboard.

45. Roger W. Jordan, *The World's Merchant Fleets 1939: The Particulars and Wartime Fates of 6000 Ships* (London, 1999), pp. 88–9.

46. Ibid., pp. 103–4.

47. Ibid., pp. 87, 127.

48. That material history as defined here is different from economic history needs to be stressed. Interestingly, Cain and Hopkins, in their extraordinary recasting of the history of the British Empire, say they sought to write from the 'perspective of economic history and to underline the importance of material forces', recognizing that economic history

and indeed 'the productive forces' had become deeply unfashionable: P. J. Cain and A. G. Hopkins, *British Imperialism 1688–2000*, 2nd edn (Harlow, 2002), p. 2. And yet it is a peculiar sort of material force they focus on: their interest is directed at finance, at trade flows measured in money, the politics of sterling. These are vital topics, but hardly the same as the history of imperial agricultural and industrial production. It is more a financial and commercial than a fully economic history. Much of the material is also hidden behind abstract economic measures: an example would be histories of shipping that give no sense whatever of the size of capabilities of the ships that are aggregated into fleet tonnages. A good example of this is S. G. Sturmey, *British Shipping and World Competition* (London, 1962), which leaves one in complete ignorance of the nature of the ships discussed.

49. R. M. Kindersley, 'British Overseas Investments, 1937', *The Economic Journal* 48 (December 1938), pp. 609–34.

50. R. W. Ferrier and J. H. Bamberg, *The History of the British Petroleum Company*, Vol. II (Cambridge, 1994), p. 83. Employment figures from Table 3.4, p. 81.

51. J. R. Andrus, *Burmese Economic Life* (Stanford, Calif., 1947), p. 118.

52. Interview with Mr S. W. Johnson (12 May 2008) and 'Fray Bentos, Uruguay – 1920–1940', *The Bulletin* LII (May 2009), p. 13 (publ. Buenos Aires). Thanks to Joanna Richardson.

53. As reported by his son, J. G. Ballard, *Miracles of Life: Shanghai to Shepperton: An Autobiography* (London, 2008), pp. 45–8.

54. Gerrylyn Roberts and Anna E. Simmons, 'British Chemists Abroad 1887–1971: The Dynamics of Chemists' Careers', *Annals of Science* 66 (2009), pp. 103–28.

55. http://sites.scran.ac.uk/empire/EmpireFla.html.

56. Martin Daunton, 'Britain and Globalisation since 1850: II. The Rise of Insular Capitalism, 1914–1939', *Transactions of the Royal Historical Society* 17 (2007), pp. 1–33.

57. Henry Clay, 'The Economic Outlook of the United Kingdom', *The American Economic Review* 37 (1947), pp. 12–20, particularly p. 13.

58. Peter Scott, *Triumph of the South: A Regional Economic History of Britain during the Early Twentieth Century* (Aldershot, 2007), pp. xiv, 324.

59. *British Machine Tool Engineering* XXI (Sept–Dec 1939), pp. 224–30.

60. Imperial Chemical Industries, Ltd, *Imperial Chemical Industries, Ltd: A Short Account of the Activities of the Company* (London, 1929).

61. Part was taken over by the Air Ministry in 1940, the rest by the

Admiralty in 1941, and the whole worked very closely with the Admiralty Signals Establishment: G. D. Speake, 'The Marconi Research Centre – A Historical Perspective', *Physics in Technology* 16 (1985), pp. 275–81.

62. L. H. A. Pilkington, 'Research at Pilkington', *Physics in Technology* 19 (1988), pp. 114–20.

63. Sally Horrocks, 'Nutrition Science and the Food and Pharmaceutical Industries in Inter-War Britain', in David F. Smith (ed.), *Nutrition in Britain: Science, Scientists and Politics in the Twentieth Century* (London, 1997), pp. 53–74.

64. D. E. H. Edgerton and S. M. Horrocks, 'British Industrial Research and Development before 1945', *Economic History Review* 37 (1994), pp. 213–38; David Edgerton, *Science, Technology and the British Industrial 'Decline', 1870–1970* (Cambridge, 1996).

65. E. M. Forster, 'Havoc', in Clough Williams-Ellis (ed.), *Britain and the Beast: A Survey by Twenty-Six Authors* (London, 1937), pp. 45–6.

66. See my *Warfare State: Britain, 1920–1970* (Cambridge, 2005), Chapter 1, for the older historiography and my analysis of the industry in 1935.

67. A. J. P. Taylor, *English History, 1914–1945* (Oxford, 1965; Harmondsworth, 1975), p. 507.

68. W. Hornby, *Factories and Plant* (London, 1958), p. 91.

69. 'National Conservative General Election Manifesto', in Iain Dale (ed.), *Conservative Party General Election Manifestos, 1900–1997* (London, 2000), pp. 54, 57–8. See the Liberal and Labour manifestos in the companion volumes, published the same year.

70. Richard Grayson, 'Leo Amery's Imperialist Alternative to Appeasement in the 1930s', *Twentieth Century British History* 17 (2006), pp. 489–515. Given the argument of the paper, the title is misleading.

71. Ronald W. Clark, *The Life of Bertrand Russell* (Harmondsworth, 1978), pp. 580–81. Surprisingly, Alan Ryan, in his perceptive account of Russell's pacifism in passages of *Bertrand Russell: A Political Life* (London, 1988), pp. 144–50, does not mention the crucial Soviet connection.

72. 'England Awakes', *The Economist* (25 March 1939), pp. 601–2.

73. Winston S. Churchill, *Step by Step 1936–1939* (London, 1939), p. 144.

74. But for historians and other analysts writing after the Second World War such a policy was deeply misguided. As in the Great War, Britain could not hope to matter unless it developed a large, powerful army for continental European operations. To fight in Europe it had to become European in its approach to war is the argument. But it did not do so, it is suggested, because of an orientation to global, and in particular to

imperial, operations, centred on the navy and the air force. But the historians of the army complained of the smallness of the army without sufficiently acknowledging the crucial place the capital-intensive navy and air force had in British strategy for dealing with European and other great powers. For discussion, see Edgerton, *Warfare State*.

75. Elsewhere I have argued that we need to replace a notion of Britain as being anti- or non-militaristic with a notion of its specific approach to warfare, which I labelled liberal militarism, and highlighted its economic and technological aspects. While of course accepting that such doctrines represented tendencies, indeed ones which were challenged, and even overturned at times, there has been a powerful tendency to think and act in particular ways which are usefully labelled 'liberal militarism': David Edgerton, 'Liberal Militarism and the British State', *New Left Review* 185 (Jan–Feb 1991), pp. 138–69, and Edgerton, *Warfare State*.

76. Thomas Jones to Abraham Flexner (former senior figure in the Rockefeller Foundation and Director of the Center for Advanced Study, Princeton), 10 December 1939, in Thomas Jones, *A Diary with Letters, 1931–1950* (London, 1954), p. 443.

77. Ibid.

78. Britain had one battleship completed between 1920 and 1925, the USA five and Japan two; from the war years Britain had eleven, the USA seven and Japan six; from the pre-war period Britain had one, the USA three and Japan two.

79. S. Roskill, *The War at Sea*, 3 vols. (London, 1954–64), Vol. 1, p. 52.

80. 'Report of the First Lord of the Admiralty to the War Cabinet, No. 1', in Martin Gilbert (ed.), *The Churchill War Papers*, Vol. I: *At the Admiralty, September 1939–May 1940* (London, 1993), p. 107. The 'pocket battleships' were 'pocket' in the sense that they were about one third the size of a contemporary battleship, and 'battleship' in that they carried 11in guns, like the *Scharnhorst* and *Gneisenau*. These were, however, small compared to the 14in, 15in and 16in guns of British, Japanese and American battleships.

81. Roger Chesneau, *Conway's All the World's Fighting Ships, 1922–1946* (London, 1980), is my source. Norman Friedman, *British Carrier Aviation: The Evolution of Ships and Their Aircraft* (London, 1988), explains this phenomenon – essentially the Royal Navy chose to have all aircraft stowable below decks. Unusually, many British carriers had armoured decks.

82. B. McKercher, 'The Greatest Power on Earth: Britain in the 1920s', *International History Review* 13 (1991), pp. 661–880, particularly pp. 765, 766.

83. Personal communication, John Brooks, cited in Edgerton, *Warfare State*.

84. Richard Titmuss, *Problems of Social Policy* (London, 1950), p. 23, n. 1, points out that the statistical and logical basis of the argument was open to serious criticism.

85. Ibid., p. 14.

86. But Titmuss noted: 'It is not clear from the files whether this information was made available to all interested divisions in the Ministries concerned and, if so, what value was placed on it', ibid., Chapter 2, n. 3.

87. 'The general view which had emerged by 1938, and then became the most important single factor in determining the form of the wartime emergency services, contained the following basic features. At the outset (and perhaps without any declaration of war) London would be subjected to concentrated and intensive air attack by bombers operating from Germany. In the first twenty-four hours the Germans might attempt to drop as much as 3,500 tons. Subsequently, and for a period of weeks, the daily weight of attack might average 700 tons. A high degree of accuracy might be achieved by the enemy in bombing specific targets and areas. It was thought that high explosive would be employed to a greater extent than incendiary bombs, while the use of gas was considered possible': ibid., p. 14.

88. Joseph S. Meisel, 'Air Raid Shelter Policy and Its Critics in Britain before the Second World War', *Twentieth Century British History* 5 (1994), pp. 300–319.

89. This crucial point, obscured in most treatments, is made in David Gloster, 'Architecture and the Air Raid: Shelter Technologies and the British Government, 1938–1944', M.Sc. dissertation, Imperial College London (1997).

90. *The Economist* (21 January 1939), p. 140.

91. Angus Calder, *The Myth of the Blitz* (London, 1991), p. 83.

92. Titmuss, *Problems of Social Policy*, Appendices 7 and 8.

93. Sebastian Ritchie, *Industry and Air Power: The Expansion of British Aircraft Production, 1935–1941* (London, 1997).

94. F. R. (Rod) Banks, *I Kept No Diary: 60 years with Marine Diesels, Automobile and Aero Engines* (Shrewsbury, 1978), pp. 134–9.

95. Andrew Nahum, 'Two-Stroke or Turbine? The Aeronautical Research Committee and British Aero Engine Development in World War II', *Technology and Culture* 38 (April 1997), pp. 312–54.

96. Ibid.

97. Banks, *I Kept No Diary*, pp. 154–5.

98. See Richard N. Scarth, *Echoes from the Sky: A Story of Acoustic Defence* (Hythe, 1999).

99. Philip Judkins, 'Note on the costings of early UK radar', personal communication, 22 October 2009; Phillip Edward Judkins, 'Making Vision into Power: Britain's Acquisition of the World's First Radar-Based Integrated Air Defence System 1935–1941', unpubl. Ph.D. thesis, Cranfield University (2008).

100. Colin Dobinson, *AA Command: Britain's Anti-Aircraft Defences of the Second World War* (London, 2001), pp. 528, 555.

101. Five hundred sets were ordered early in 1939 from Metrovick and Cossor: M. M. Postan, D. Hay and J. D. Scott, *Design and Development of Weapons / Studies in Government and Industrial Organisation* (London, 1964), p. 383.

102. Dobinson, *AA Command*, pp. 115–16.

103. A. V. Hill to Sir Henry Tizard, 16 June 1936, HTT 67, quoted in Philip Chaston, 'Gentlemanly Professionals within the Civil Service: Scientists as Insiders during the Interwar Period', unpubl. Ph.D. thesis, University of Kent at Canterbury (2009), p. 254.

104. The remarkable discovery is due to Phil Judkins, who has also found remains of a site. See Judkins, 'Making Vision into Power'.

105. R. W. Burns, 'Early History of the Proximity Fuze (1937–1940)', *IEE Proceedings-A* 140 (May 1993), pp. 224–36.

106. Ibid.

107. Alwyn Crow, 'The Rocket as a Weapon of War in the British Forces', 34th Thomas Hawksley Lecture, November 1947, *Proceedings of the Institution of Mechanical Engineers* 158 (1948), pp. 15–21.

108. John Lennard-Jones, 'Some Wartime Experiences', Lecture to the University Chemical Society, Cambridge, 12 March 1947, The Papers of Sir John Lennard-Jones, LENJO 46, Churchill Archives Centre. In his memoirs Vice-Admiral Pridham mentioned 'upwards of a million pounds': Vice-Admiral Pridham, unpubl. memoir, p. 196, The Papers of Vice-Admiral Sir (Arthur) Francis Pridham, PRID 2, Churchill Archives Centre.

109. Crow, 'The Rocket as a Weapon of War in the British Forces', p. 20.

110. See Talbot Imlay, 'France, Britain and the Making of the Anglo-French Alliance, 1938–39'; Martin S. Alexander, 'Preparing to Feed Mars: Anglo-French Economic Co-ordination and the Coming of War, 1937–40'; William J. Philpott, 'The Benefit of Experience? The Supreme War Council and the Higher Management of Coalition War, 1939–40', in Martin S. Alexander and William J. Philpott (eds.), *Anglo-French Defence Relations between the Wars* (London, 2002), pp. 92–120, 186–208, 209–26.

111. For a useful survey, discussion and new evidence about the connection between the domestic and foreign influences, see Daniel Hucker, 'Franco-British Relations and the Question of Conscription in Britain, 1938–1939', *Contemporary European History* 17 (2008), pp. 437–56.

112. *The Economist* (25 November 1939), p. 275.

113. Ibid. (16 December 1939), pp. 403–4.

114. Lord Chatfield, Minister for the Coordination of Defence, HL Deb. (series 5), vol. 115, col. 386 (18 January 1940).

115. *Aircraft Production: The Journal of the Aircraft Manufacturing Industry* 2 (April 1940), p. 105.

116. Margaret Gowing, *Britain and Atomic Energy 1939–1945* (London, 1964), p. 43.

117. *The Economist* (30 March 1940), p. 565.

118. R. J. Overy, *The Origins of the Second World War*, 2nd edn (London, 1998), p. 53.

119. Adam Tooze, *The Wages of Destruction: The Making and Breaking of the Nazi Economy* (London, 2006), pp. 315–25, 664.

120. *The Engineer* CLXIX (19 January 1940), p. 66

121. Hancock and Gowing, *British War Economy*, p. 100.

122. Pathé *Review of the Year 1939* at http://www.britishpathe.com/record.php?id=24904, final shot.

123. Talbot Imlay argues that both had lost faith in the long-war strategy: Imlay, *Facing the Second World War.*

124. Talbot C. Imlay, 'A Reassessment of Anglo-French Strategy during the Phony War, 1939–1940', *English Historical Review* CXIX (2004), pp. 333–72.

125. Sir William Beveridge, 'Preparing the new war', June 1940, p. 12, TS in the A. V. Hill Papers, AVHL I 3/5, Churchill Archives Centre.

126. Churchill, *Second World War*, Vol. I: *The Gathering Storm*, p. 320.

3. NEVER ALONE

1. David Thomson, *England in the Twentieth Century* (Harmondsworth, 1965), pp. 190, 194, 210, makes clear that for a year it was 'the British Commonwealth' rather than Britain or even England which was alone (for example, pp. 190, 194, 210). Angus Calder, *The People's War: Britain 1939–1945* (London, 1969), hardly deals with 'alone'; it criticizes the notion of 'alone' for excluding millions of Asians (p. 19), yet remarks in inverted commas that Britain 'stood alone' (p. 110) while noting that Britain fought alone (p. 113). Calder excluded treatment of the Empire

because he did not believe that Empire affected domestic thinking (p. 19). Even economic historians have claimed that 'Britain stood virtually alone' (in this case from the summer of 1940 to December 1941): Sidney Pollard, *The Development of the British Economy 1914–1980*, 3rd edn (London, 1983), p. 193. The excellent general history of the war by R. A. C. Parker, *The Second World War: A Short History*, rev. edn (Oxford, 1997), even has a whole chapter called 'Britain Alone' which hardly refers to the Empire. 'Alone' is used and given explanatory weight in, among many works, Malcolm Smith, *Britain and 1940: History, Myth and Popular Memory* (London, 2000), and Mark Connelly, *We Can Take It! Britain and the Memory of the Second World War* (London, 2004). The issue of 'alone' is not problematized in studies which bring aspects of Empire into the national identity issue; for example, Sonya O. Rose, *Which People's War? National Identity and Citizenship in Wartime Britain 1939–1945* (Oxford, 2003), and Wendy Webster, *Englishness and Empire 1939–1965* (Oxford, 2005).

2. Winston S. Churchill, *Blood, Sweat, and Tears* (New York, 1941), p. 281.

3. Ibid., p. 345.

4. House of Commons and broadcast, 8 May 1945. The text was also read by Lord Woolton in the House of Lords. 'The End of the War in Europe, Broadcast, London and House of Commons', in Winston Churchill and Robert Rhodes James (ed.), *Winston S. Churchill: His Complete Speeches, 1897–1963*, Vol. 7 (London, 1974), pp. 7152–3, available at http://www.winstonchurchill.org/learn/speeches/speeches-of-winston-churchill/1941-1945-war-leader/95-end-of-the-war-in-europe; accessed 28 April 2010.

5. 'To V-E Day Crowds, May 8, London', in ibid.

6. Iain Dale (ed.), *Conservative Party General Election Manifestos, 1900–1997* (London, 2000), pp. 61, 62, and Iain Dale (ed.), *Liberal Party General Election Manifestos, 1900–1997* (London, 2000), p. 61.

7. 'Prime Minister for Two Years, May 10, Broadcast, London', in Churchill and James (ed.), *Churchill: His Complete Speeches*, Vol. 6, pp. 6629–34, available at http://www.winstonchurchill.org/learn/speeches/speeches-of-winston-churchill/1941-1945-war-leader/95-end-of-the-war-in-europe; accessed 28 April 2010. But see the very much more imperial terms he used in the speech at the Guildhall, 30 June 1943: Churchill and James (ed.), *Churchill: His Complete Speeches*, Vol. 7, p. 6792.

8. Winston S. Churchill, *The Second World War*, Vol. II: *Their Finest Hour* (London, 1949), p. 225. These words are from the opening sentences of

a chapter called 'At Bay', towards the end of Book I of the volume; the second of the two books of the volume is entitled 'Alone'. The 'theme of the volume' inscribed at the front was 'HOW THE BRITISH PEOPLE HELD THE FORT ALONE TILL THOSE WHO HITHERTO HAD BEEN HALF BLIND WERE HALF READY'. Around the same time, the military historian Cyril Falls wrote of the same period: 'Now, in all Europe, the United Kingdom stood alone' (*The Second World War: A Short History* (London, 1948), p. 61), though later noting that 'Britain . . . in maintaining the war alone or virtually alone was a vital element in victory, though it may be noted that she did not suffer heavy loss of life in that period' (p. 295).

9. *Britain at Bay* (1940), GPO film unit. Available at www.screenonline. org.uk.

10. Pathé *Review of the Year 1940*: http://www.britishpathe.com/record. php?id=11974, Pathé *Review of the Year 1941*: http://www.britishpathe. com/record.php?id=13123; accessed 29 April 2010.

11. *Winston Churchill and Allied Leaders Attend Tank Demonstration*, British Gaumont newsreel (24 February 1941) available at http://www. itnsource.com/shotlist//BHC_RTV/1941/02/24/BGU408070026/; accessed 29 April 2010.

12. Churchill and James (ed.), *Churchill: His Complete Speeches*, Vol. 6, p. 6423, available at http://www.winstonchurchill.org/learn/speeches/ speeches-of-winston-churchill/1941-1945-war-leader/95-end-of-the-war in-europe; accessed 28 April 2010. For newsreel, see *Prime Minister Winston Churchill Speaks to Allied Leaders at St. James's Palace*, British Gaumont newsreel (16 June 1941) available at www.itnsource.com/ shotlist//BHC_RTV/1941/06/16/BGU408110019/?s=churchill; accessed 29 April 2010.

13. Asa Briggs, *The History of Broadcasting in the United Kingdom*. Vol. III: *The War of Words* (Oxford, 1970; rev. edn 1995), pp. 186, 299, 354–6.

14. *Dawn Guard* (1941), produced and directed by John and Roy Boulting.

15. George Orwell, *The Lion and the Unicorn: Socialism and British Genius* (London, 1941).

16. *London Can Take It!* (1940), *Christmas Under Fire* (1941), *Battle for Freedom* (1942).

17. Joel Hurstfield, *The Control of Raw Materials* (London, 1953), pp. 155–60.

18. Ibid., p. 154.

19. Central Statistical Office, *Statistical Digest of the War* (London, 1951), Table 148.

20. Hurstfield, *The Control of Raw Materials*, p. 156.

21. Sir William Beveridge, 'Preparing the new war', June 1940, TS in The Papers of A. V. Hill, AVHL I 3/5, Churchill Archives Centre.

22. Philip Ollerenshaw, 'War, Industrial Mobilisation and Society in Northern Ireland, 1939–1945', *Contemporary European History* 16 (2007), pp. 169–97, particularly p. 173.

23. Ibid., p. 153.

24. C. B. A. Behrens, *Merchant Shipping and the Demands of War* (London, 1958), pp. 57–9.

25. Ibid., pp. 60–63.

26. Ibid., p. 64.

27. Ibid., p. 94.

28. Ibid., p. 96.

29. '"War of the Unknown Warriors", July 14, Broadcast, London', in Churchill and James (ed.), *Churchill: His Complete Speeches*, Vol. 6, pp. 6247–9, available at http://www.winstonchurchill.org/learn/speeches/speeches-of-winston-churchill/1940-finest-hour/126-war-of-the-unknown-warriors; accessed 29 April 2010.

30. W. K. Hancock and M. M. Gowing, *British War Economy* (London, 1949), p. 103.

31. Ibid., pp. 243f.

32. Ibid., pp. 239, 240.

33. Ibid., pp. 239–41. Alec Cairncross, *Living with the Century* (Fife, 1998), p. 79.

34. Keith Jeffery, 'The Second World War', in Judith M. Brown and Wm. Roger Louis (eds.), *The Oxford History of the British Empire: The Twentieth Century* (Oxford, 1999), pp. 306–28.

35. War Cabinet, memorandum by the Secretary of State for War (Anthony Eden), 'The Army Programme', 14 September 1940, CAB 66/11/50, The National Archives.

36. Chiefs of Staff Committee, 'Future Strategy', 4 September 1940, paragraph 50, CAB 66/11/42, The National Archives.

37. Sebastian Cox (ed.), *The Strategic Air War against Germany, 1939–1945: Report of the British Bombing Survey Unit* (London, 1998).

38. Most imperial territories did not conscript for overseas service for domestic political reasons. Even white settler communities were usually divided enough for conscription for overseas service to be politically impossible. Canada, for example, used volunteers until 1944, though only a very small number of conscripts were in fact sent abroad. Australian forces overseas were at first entirely volunteers – wartime

conscription was only for home defence, and from 1943 for South-East Asia south of the Equator only. Half a million Australians served overseas. White South Africa was deeply divided. The great exception is New Zealand, which conscripted for overseas service from July 1940. See Ashley Jackson, *The British Empire and Second World War* (London, 2006), *passim*.

39. See Adam Tooze's brilliant brief analysis *The Wages of Destruction: The Making and Breaking of the Nazi Economy* (London, 2006), pp. 368–80.

40. Ibid., pp. 374–5.

41. Parker, *The Second World War*, p. 23.

42. We need to be careful not to see in these ruminations a prescient analysis of the Nazi exterminist frenzy yet to come.

43. 'War against a Machine', *The Engineer* CLXIX (7 June 1940), p. 517.

44. Sir William Beveridge, 'Preparing the new war', June 1940, TS in the Hill Papers, AVHL I 3/5, Churchill Archives Centre.

45. Some recent (but hardly unique) cases: the British army in 1940 was 'virtually disarmed' (Malcolm Smith, *Britain and 1940*, p. 62), and the British, meaning the British army usually, had 'virtually nothing with which to resist an invasion' (Martin Kitchen, *Rommel's Desert War* (Cambridge, 2009), p. 2).

46. CSO, *Statistical Digest of the War*, Table 122, p. 144.

47. Ian Skennerton, *The Lee-Enfield: A Century of Lee-Metford & Lee-Enfield Rifles & Carbines* (Labrador, Qld, 2007), p. 196.

48. The Papers of 1st Viscount Weir, WEIR 20/13 and 20/28, Churchill Archives Centre.

49. M. M. Postan, *British War Production* (London, 1952), p. 110.

50. Oliver Lyttelton, Viscount Chandos, *The Memoirs of Lord Chandos* (London, 1962), p. 161.

51. These rifles were the product of an earlier Anglo-American alliance. They were British-designed Enfield Pattern 1914 rifles, produced for the US army in the Great War with the 0.30 calibre. The Home Guard also got US-made Enfield Pattern 1914 0.303 guns manufactured for the British along with First World War Canadian Ross rifles. Skennerton, *The Lee-Enfield*, p. 217.

52. Hancock and Gowing, *British War Economy*, p. 227.

53. Skennerton, *The Lee-Enfield*, pp. 307, 310.

54. D. P. Mellor, *Australia in the War of 1939–1945. Series 4 – Civil – Vol. V: The Role of Science and Industry* (Canberra, 1958), pp. 323–4. Production peaked in 1943 at well over 100,000.

55. Skennerton, *The Lee-Enfield*, pp. 340, 342, 367.

56. General Fedor von Bock, quoted in Karl-Heinz Frieser, *The Blitzkrieg Legend: The 1940 Campaign in the West* (Annapolis, Md, 2005), p. 302.

57. See in particular Frieser, *The Blitzkrieg Legend*. An early work to outline the argument was R. H. S. Stolfi, 'Equipment for Victory in France', *History* 55 (1970), pp. 1–20, noted in Eugenia C. Kiesling, 'Illuminating Strange Defeat and Pyrrhic Victory: The Historian Robert A. Doughty', *Journal of Military History* 71 (July 2007), pp. 875–88.

58. The conventional picture of the wartime army is very different, as for example expressed by Field Marshal Lord Carver: 'the BEF was not equipped for modern war' either in quantity or sometimes in quality (Michael Carver, *Britain's Army in the 20th Century* (London, 1999), p. 171).

59. David Pam, *The Royal Small Arms Factory, Enfield and Its Workers* (Enfield, 1998), p. 156.

60. National Army Museum, *'Against All Odds': The British Army of 1939–40* (London, 1990), pp. 17–21. 'By May, 1940, we had in France 400 light tanks armed with machine guns, 100 infantry tanks of which only 23 were equipped with 2-pounder guns, and 158 cruiser tanks armed with 2-pounders. All this comparatively small force of tanks was entirely destroyed or lost at the time of the French collapse', Viscount Cranborne, HL Deb. (series 5), vol. 123, col. 589 (1 July 1942).

61. Tooze, *Wages of Destruction*, p. 376.

62. Heinz Guderian, *Panzer Leader* (New York, 2001; first English trans. 1952), Appendix III, p. 472.

63. The figures given for British tanks in Frieser, *The Blitzkrieg Legend*, p. 40, mislead because as the text makes clear the number includes light tanks.

64. HC Deb. (series 5), vol. 381, col. 251 (1 July 1942).

65. John Ellis, *The World War II Databook: The Essential Facts and Figures for All the Combatants* (London 1993), p. 204.

66. Ibid., p. 150.

67. Postan, *British War Production*, p. 103.

68. The official historian of production, M. M. Postan, more upbeat about British tank production than later writers, wrote that 'The monthly output [1939 and 1940] was roughly equal to the German, but whereas the German figures are all Panzers Marks II, III and IV, i.e. medium and heavy tanks of infantry type, the bulk of English tanks at that time were made up of light and cruiser (Light Mark VI and Cruisers Marks I to IV) types, while the output of infantry tanks (the Matilda I and II) was

relatively small. Only sixty-three infantry tanks were produced in the last four months of 1939 and sixty-seven in the first four months of 1940': Postan, *British War Production*, p. 110. See also Table 3.2, on which he was commenting. However, he was not comparing like with like: the Panzer Mark II was a light tank, with a 20mm gun and weighing 7.2 tons, comparable to British light tanks of 5 tons which carried a 0.5in gun. The British cruisers were altogether different, weighing between 12 and 15 tons, and carrying a 40mm 2-pounder anti-tank gun. Postan's figures appear to exclude Germany's important Czech-designed tank, the 9.5-ton Panzer 38t. It by these biased comparisons that the image of British weakness was established and came to endure in the scholarly literature.

69. Viscount Cranborne, HL Deb. (series 5), vol. 123, col. 589 (1 July 1942). Lyttelton: 'We had, in this country at the time of Dunkirk, only 200 light tanks armed with machine guns and 50 infantry tanks' (HC Deb. (series 5), vol. 381, col. 251 (1 July 1942)).

70. Churchill, *Second World War*, Vol. II: *Their Finest Hour*, p. 128. An official history was to give the figures for 10 June 1940 of 'only 103 cruisers and 114 infantry tanks in the United Kingdom': J. R. M. Butler, *Grand Strategy*, Vol. II: *September 1939–June 1941* (London, 1957), p. 256.

71. Postan, *British War Production*, p. 103.

72. Harold Macmillan, *The Blast of War: 1939–1945* (London, 1967), p. 104.

73. Churchill, *Second World War*, Vol. II: *Their Finest Hour*, pp. 378–9. Some ministers like Harold Macmillan and Oliver Lyttelton, both concerned with the Middle East, lauded this decision in no uncertain terms: Lyttelton, *The Memoirs of Lord Chandos*, p. 175. The role of the Army and War Office and Churchill's hesitations about quantities are brought out in David Reynolds, *In Command of History: Churchill Fighting and Writing the Second World War* (London, 2004), pp. 191–3. A recent history reproduces the argument that this was one of the 'toughest decisions of the war ... virtually denuding Britain of tanks': Andrew Roberts, *The Storm of War: A New History of the Second World War* (London, 2009), p. 120.

74. I. S. O. Playfair, *The Mediterranean and the Middle East*, Vol. 1: *The Early Successes against Italy* (London, 1954), pp. 191–2.

75. E. R. Mayhew, *The Reconstruction of Warriors: Archibald McIndoe, the Royal Air Force and the Guinea Pig Club* (London, 2004).

76. Stephen Bungay, *The Most Dangerous Enemy: A History of the Battle of Britain* (London, 2000).

77. Vera Brittain, *England's Hour* (London, 2005; first publ. 1941), p. 61.

78. My *England and the Aeroplane: An Essay on a Militant and Techno-logical Nation* (London, 1991), pp. 62–3, got part of this story, with some errors, but the case is made with devastating clarity in H. W. Koch, 'The Strategic Air Offensive against Germany: The Early Phase, May–September 1940', *The Historical Journal* 34 (March 1991), pp. 117–41.

79. Postan, *British War Production*, p. 116.

80. Ibid., p. 164.

81. G. P. Bulman, *An Account of Partnership – Industry, Government and the Aero Engine: The Memoirs of George Purvis Bulman*, Rolls-Royce Heritage Trust, Historical Series No. 31 (Derby, 2002), pp. 269–73, 366.

82. Colin Dobinson, *Fields of Deception: Britain's Bombing Decoys of World War II* (London, 2000), pp. 211–13, 298.

83. Ibid., p. 108. As a boy I played on the site of the decoy, completely oblivious to the history of this patch of land. On 2 December 1940 Bristol suffered a very large raid, with 156 killed – and on the same night sixty-two bombs of unspecified size fell on the Stockwood Decoy. On 6 December an equally large and damaging raid on Bristol might have been seventy-three bombs heavier – they were dropped on the decoy. In both cases German operational reports noted the existence of the decoy. See John Penney, *Bristol at War* (Derby, 2002), pp. 94, 98, and for the operational reports John Penney (ed.), *German Air Operations against the Bristol Area, 1939–1945*, Vol. 4: *November 25th to December 11th 1940* (Bristol, 1991).

84. M. M. Postan, D. Hay and J. D. Scott, *Design and Development of Weapons: Studies in Government and Industrial Organisation* (London, 1964), p. 384.

85. Colin Dobinson, *A A Command: Britain's Anti-Aircraft Defences of the Second World War* (London, 2001), pp. 277–8.

86. Ibid., p. 279.

87. This is my reading of David Reynolds, 'Churchill and the British "Decision" to Fight on in 1940: Right Policy, Wrong Reasons' (first published 1985), in David Reynolds, *From World War to Cold War: Churchill, Roosevelt and the International History of the 1940s* (Oxford, 2006), pp. 75–98, and John Lukacs, *Five Days in London, May 1940* (London, 1999).

88. Hugh Dalton, reviewing W. N. Medlicott, *The Economic Blockade*, Vol. I: *1939–41* (London, 1952), in *The Economic Journal* 63 (September 1953), pp. 660–68.

89. Harold Nicolson to Vita Sackville-West, 10 July 1940, in Harold Nicolson, *Diaries and Letters, 1939–1945* (London, 1967), p. 97.

90. Chiefs of Staff Committee, 'Future Strategy', 4 September 1940, pp. 2–3, CAB 66/11/42, The National Archives.

91. Reynolds, 'Churchill and the British "Decision' to Fight on in 1940", in Reynolds, *From World War to Cold War*, pp. 75–98. This very important article, first published in 1985, does not analyse Britain's actual economic strength or Churchill's view that invasion would fail. See also Reynolds, *In Command of History*, Chapter 11. Reynolds's recognition that Churchill and the rest of government had reasons for believing in possible victory was a crucial point; my contribution is to suggest that the arguments were not as far-fetched as they now seem and were part of a general confidence in the power of modern weapons and Britain's ability to produce them, as well as a more positive, and justifiably so, view of Britain's economic position than has since seemed plausible, not least because of the greater role of the US from 1942.

92. Geoffrey Crowther, *Ways and Means of War* (Oxford, 1940).

93. Churchill, *Second World War*, Vol. II: *Their Finest Hour*, p. 130.

94. Macmillan, *The Blast of War*, p. 106.

95. Churchill, *Second World War*, Vol. II: *Their Finest Hour*, pp. 254–5.

96. Ibid., pp. 404–5.

97. Tooze, *Wages of Destruction*, p. 398.

98. William Foot, *Beaches, Fields, Streets, and Hills: The Anti-Invasion Landscapes of England, 1940* (York, 2006).

99. '"The Few", August 20, House of Commons', in Churchill and James (ed.), *Churchill: His Complete Speeches*, Vol. 6, pp. 6260–68, available at http://www.winstonchurchill.org/learn/speeches/speeches-of-winston-churchill/1941-1945-war-leader/95-end-of-the-war-in-europe; accessed 28 April 2010.

100. Memorandum by the Prime Minister, 'The Munitions Situation', 3 September 1940, in Churchill, *Second World War*, Vol. II: *Their Finest Hour*, pp. 405–7, particularly p. 407.

101. Ibid.

102. Prof. Lindemann to the Prime Minister, 8 August 1941, quoted in Lord Birkenhead, *The Prof in Two Worlds: The Official Life of Professor F. A. Lindemann, Viscount Cherwell* (London, 1961), p. 217.

103. 'Army Scales, Directive by the Minister of Defence [Churchill]', 6 March 1941, in Winston S. Churchill, *The Second World War*, Vol. III: *The Grand Alliance* (London, 1950), Appendix F, pp. 705–7.

104. War Cabinet, memorandum by the Secretary of State for War (Anthony Eden), 'The Army Programme', 14 September 1940, WP (40) 370, CAB 66/11/50, The National Archives.

105. David French, *Raising Churchill's Army: The British Army and the War against Germany 1919–1945* (Oxford, 2000), p. 107.

106. Reynolds, *In Command of History*, p. 268, very correctly notes that these plans were not prescient plans for D-Day, but something else entirely.

107. 'The Campaign in 1943', in Churchill, *Second World War*, Vol. III: *The Grand Alliance*, pp. 582–4.

108. '"Westward, Look, the Land is Bright", April 27, Broadcast, London', in Churchill and James (ed.), *Churchill: His Complete Speeches*, Vol. 6, pp. 6378–84, available at http://www.winstonchurchill.org/learn/speeches/speeches-of-winston-churchill/1941-1945-war-leader/95-end-of-the-war-in-europe; accessed 28 April 2010.

109. Butler, *Grand Strategy*, Vol. II: September 1939–June 1941, pp. 547–50.

110. Alexander Hill, 'British Lend-Lease Aid and the Soviet War Effort, June 1941–June 1942', *Journal of Military History* 71 (July 2007), pp. 773–808.

111. Richard Overy, *Why the Allies Won* (London, 1995), p. 198.

112. Sir Charles Kingsley Webster and Noble Frankland, *The Strategic Air Offensive against Germany 1939–1945*, Vol. IV: *Annexes and Appendices* (London, 1961), Appendix 49, Table iii, p. 469.

113. Churchill, *Second World War*, Vol. III: *The Grand Alliance*, p. 463.

114. But abandoned it on the advice of advisers: ibid., pp. 479, 489.

115. See Christopher Bayly and Tim Harper, *Forgotten Wars: The End of Britain's Asian Empire* (London, 2007). One of the few general texts which recognizes the significance of the fall of Britain's Asian empire was Henry Pelling, *Britain and the Second World War* (London, 1970), Chapter 6, 'The Crisis of the Empire'.

116. T. A. B. Corley, *A History of the Burmah Oil Company*, Vol. II: 1924–1966 (London, 1988), p. 94.

117. Churchill, opening speech, War Situation vote of confidence debate, HC Deb. (series 5), vol. 377, col. 601 (27 January 1942).

118. Ibid.

119. Churchill, *Second World War*, Vol. III: *The Grand Alliance*, p. 539.

120. A. J. P. Taylor, *English History, 1914–1945* (Oxford, 1965; Harmondsworth, 1975), p. 646. It is a general view that only in 1942, despite the travails, did it become clear that 'Britain was not going to lose; in fact she was going to win; not just yet, but eventually': Smith, *Britain and 1940*, p. 105.

121. David Reynolds, '1940: Fulcrum of the Twentieth Century?', in Reynolds, *From World War to Cold War*, p. 29.

122. Angus Calder, *The Myth of the Blitz* (London, 1991), p. 52.

123. Parker, *The Second World War*, p. 57.

124. P. J. Cain and A. G. Hopkins, *British Imperialism 1688–2000*, 2nd edn (Harlow, 2002), p. 620.

125. Gavin Bailey, 'The Narrow Margin of Criticality: The Question of the Supply of 100-Octane Fuel in the Battle of Britain', *English Historical Review* CXXIII (April 2008), pp. 394–411.

126. *Preliminary Agreement between the United States and the United Kingdom* (23 February 1942), available at http://avalon.law.yale.edu/20th_century/decade04.asp.

127. Reynolds, '1940: Fulcrum of the Twentieth Century?', p. 33.

128. Note for Minister, possibly by Lord Weir, 'Machine Tools from the United States', undated, but 1940 and almost certainly October or later, Weir Papers, WEIR 20/10, Churchill Archives Centre.

129. On explosives and propellants, see W. Hornby, *Factories and Plant* (London, 1958), p. 113.

130. R. F. Fowler (in Canada) to A. V. Hill (11 September 1940), Hill Papers, AVHL I 3/19, Churchill Archives Centre. A. V. Hill, a man with great experience of weapons, who had been in the USA and Canada earlier and had some experience in gunnery and air defence, agreed, although noting that Tizard, who would be leading a mission, and Fowler would 'be able to get effective help from them'. A. V. Hill to R. F. Fowler (in Canada, 4 October 1940), Hill Papers, AVHL I 3/19, Churchill Archives Centre.

131. David Zimmerman, *Top Secret Exchange: The Tizard Mission and the Scientific War* (Montreal, 1996).

132. H. Duncan Hall and C. C. Wrigley (with a chapter by J. D. Scott), *Studies of Overseas Supply* (London, 1956), pp. 92–102.

133. Weeks notes the main exceptions being the Merlin and the .303 rifle, and the compromise by which the Americans made the 6-pounder anti-tank gun as the 57mm: Hugh Weeks, 'Anglo-American Supply Relationships', in D. N. Chester (ed.), *Lessons of the British War Economy* (Cambridge, 1951), pp. 69–82, particularly p. 71.

134. http://www.lancaster-archive.com/bc_t1bombsight.htm; accessed 28 April 2010.

135. Ibid.

136. Mark Frankland, *Radio Man: The Remarkable Rise and Fall of C. O. Stanley* (London, 2002), pp. 112–22.

137. Margaret Gowing, *Britain and Atomic Energy 1939–1945* (London, 1964), p. 105. Hermione Giffard, 'The Development and Production of Turbojet Aero-Engines in Britain, Germany and the United States, 1936–1945', Ph.D. thesis, Imperial College London (2011).

138. A German cruise ship seized in 1945, it was used as a troopship and was

to become famous for bringing one of the first post-war consignments of immigrants from the Empire into Britain.

139. Peter Elphick, *Liberty: The Ships That Won the War* (London, 2001), p. 106.

140. This follows the authoritative account of Elphick, *Liberty*, p. 68, which differs from the standard ones.

141. She was lost in 1942, while *Empire Liberty* survived until 1960.

142. He was son of Sir George Hunter, of the Swan Hunter shipbuilding firm.

143. And a light weight (i.e. empty) of 4,000 tons, giving a total displacement of around 14,000 tons full. The gross 'tonnage' was around 7,000.

144. Elphick, *Liberty*, p. 47. The amount agreed was more than twice the £10m envisaged earlier.

145. Ibid., pp. 47–8.

146. H. Duncan Hall, *North American Supply* (London, 1955), p. 179.

147. Elphick, *Liberty*, pp. 47–52.

148. The losses were in absolute terms divided equally, roughly, between British Commonwealth, British allies and neutrals, and enemy fleets. Enemy fleets were reduced to practically nothing, making their loss relatively much more severe than the considerable British one.

149. Elphick, *Liberty*, p. 137.

150. S. G. Sturmey, *British Shipping and World Competition* (London, 1962), pp. 138–9. The US Liberties had the same engine as the British ships, but the boilers were oil-burning.

4. CRONIES AND TECHNOCRATS

1. An interesting exception is Victor Feske, *From Belloc to Churchill: Private Scholars, Public Culture and the Crisis of British Liberalism 1900–1939* (Chapel Hill, NC, 1996).

2. Geoffrey Best, *Churchill and War* (London, 2005).

3. Churchill Draft Memoirs, CHUR 4/76A, f. 13, quoted in Richard Toye, *Lloyd George and Churchill: Rivals for Greatness* (London, 2008), p. 256.

4. Oliver Lyttelton, Viscount Chandos, *The Memoirs of Lord Chandos* (London, 1962), p. 180.

5. Oliver Lyttelton, Minister of Production, HC Deb. (series 5), vol. 381, col. 1108 (14 July 1942).

6. Former naval officer and Labour peer Lord Strabolgi in a debate calling for a scientific general staff: HL Deb. (series 5), vol. 124, col. 104 (29 July 1942).

7. There is little sense of Churchill's extensive interest in inventions, machines and science in prominent biographies – for example, Piers Brendon, *Winston Churchill: A Brief Life* (London, 1984); Martin Gilbert, *Road to Victory: Winston S. Churchill 1941–1945* (London, 1986); Roy Jenkins, *Churchill* (London, 2001); John Keegan, *Churchill* (London, 2002); and Max Hastings, *Finest Years: Churchill as Warlord 1940–1945* (London, 2009). A partial exception is Geoffrey Best, who, however, sees only Lindemann's Statistical Section as a manifestation of Churchill's 'support of invention and ingenuity': Geoffrey Best, *Churchill: A Study in Greatness* (London, 2001), p. 201. J. M. Lee, *The Churchill Coalition, 1940–1945* (London, 1980), p. 29, notes 'The appointment of Winston Churchill as Prime Minister provided both an appropriate focus for popular consent and a source of patronage for technical invention'. Ronald Lewin, *Churchill as Warlord* (London, 1973), sees Churchill as backward in his appreciation of science, though fascinated by its 'by-products' (p. 19). Carlo D'Este, *Warlord: A Life of Churchill at War, 1874–1945* (London, 2009), pp. 523–30, recognizes the enthusiasm for science and is the only study to pick up on Nellie, discussed below. Greater recognition of Churchill's role is beginning to appear: see Nicholas Rankin, *Churchill's Wizards: The British Genius for Deception 1914–1945* (London, 2008), and Taylor Downing, *Churchill's War Lab: Code Breakers, Boffins and Innovators: The Mavericks Churchill Led to Victory* (London, 2010).

8. Churchill's memoirs are deeply misleading on many crucial points, as David Reynolds has brilliantly demonstrated, and deliberately so, in ways reflecting very particular concerns when he was writing them. At the same time they told a very personal story, full of the most extraordinary omissions and lack of generosity to some key figures, for example the hugely important Chief of the Imperial General Staff, Sir Alan Brooke: David Reynolds, *In Command of History: Churchill Fighting and Writing the Second World War* (London, 2004).

9. Winston S. Churchill, *The Second World War*, Vol. II: *Their Finest Hour* (London, 1949), p. 338.

10. Jones, it should be noted, helped Churchill with the sections on 'The Wizard War', and also intervened to change the account of the V-1 and V-2 story as told by Churchill's son-in-law, Duncan Sandys: Reynolds, *In Command of History*, pp. 398–9, 452.

11. Winston S. Churchill, *The Second World War*, Vol. IV: *The Hinge of Fate* (London, 1951), p. 257.

12. The names are not always correct, for example 'a Mr Pryke on Mount-

batten's staff', is Pyke; Harry L. Hopkins, *the* Harry Hopkins, is confused with Charles Hopkins. Watson-Watt is made a Professor.

13. He argued, as Minister of Munitions in the Great War, for the greater use of mechanical means to economize on manpower. See Tim Travers, 'The Evolution of British Strategy and Tactics on the Western Front in 1918: GHQ, Manpower, and Technology', *The Journal of Military History* 54 (1990), pp. 173–200.

14. Winston S. Churchill, *The Second World War*, Vol. III: *The Grand Alliance* (London, 1950), Appendix M.

15. Sir Charles Goodeve, 'The Ice Island Fiasco', unknown newspaper (19 April 1951), The Papers of Sir Charles Goodeve, GOEV 3/1, Churchill Archives Centre.

16. Winston S. Churchill, *The Second World War*, Vol. I: *The Gathering Storm* (London, 1948), p. 459.

17. 'Cultivator No. 6: Notes of First Progress Meeting', 13 March 1940, PREM 3/320/4, The National Archives.

18. Ibid.

19. C. J. W. Hopkins, 'Naval Land Equipment', 13 March 1940, PREM 3/320/6, The National Archives; 'Naval Land Equipment: Report on Progress No. 2', 18 March 1940; 'Report on Progress No. 4', 2 April 1940, PREM 3/320/5, The National Archives.

20. Ibid.

21. 'Naval Land Equipment: Report on Progress No. 16', 17 February 1941, PREM 3/320/5, The National Archives.

22. C. J. W. Hopkins to Minister of Supply, 16 March 1940, PREM 3/320/3, The National Archives.

23. Winston Churchill to Sir Kingsley Wood, 19 March 1940; Sir Kingsley Wood to Winston Churchill, 27 March 1940; Winston Churchill to Sir Kingsley Wood, 27 March 1940, PREM 3/320/3, The National Archives.

24. C. J. W. Hopkins to Churchill, 15 April 1940, PREM 3/320/5, The National Archives.

25. Leslie Burgin to Winston Churchill, 18 April 1940, PREM 3/320/2, The National Archives; 'Naval Land Equipment: Report by the Minister of Supply of a Visit to Lincoln', 17 April 1940, PREM 3/320/5, The National Archives.

26. 'Naval Land Equipment, Report of Progress No. 11' (by C. J. W. Hopkins), Herbert Morrison to Winston Churchill, 30 July 1940, on which Churchill indicated agreement, PREM 3/320/5, The National Archives.

27. 'Naval Land Equipment, Report of Progress No. 15 and No. 16' (by C. J. W. Hopkins); PREM 3/320/5, The National Archives.

28. C. J. W. Hopkins to Commander Thompson, Downing Street, 18 November 1941, PREM 3/320/7, The National Archives.

29. Churchill, *Second World War*, Vol. I: *The Gathering Storm*, p. 567.

30. P. J. Grigg, Secretary of State for War, to Prime Minister, 30 April 1943, with handwritten minute by Churchill, PREM 3/320/7, The National Archives.

31. Prime Minister's Personal Minute to Secretary of State for War, 21 May 1945, PREM 3/320/7, The National Archives.

32. 'Cultivator No. 6', note by the author, in Churchill, *Second World War*, Vol. I: *The Gathering Storm*, Appendix O, pp. 566–8. There is an interesting history, with photographs of this extraordinary project: John T. Turner, *'Nellie': The History of Churchill's Lincoln-Built Trenching Machine*, Occasional Papers in Lincolnshire History and Archaeology, No. 7 (Society for Lincolnshire History and Archaeology, 1988).

33. L. S. Amery to C. S. Wright, Admiralty DSR, 23 July 1940, The Papers of Leopold Amery, AMEL 1/6/5, Churchill Archives Centre.

34. Memorandum for First Lord of the Admiralty, C. S. Wright, DSR, 19 July 1940, Amery Papers, AMEL 1/6/5, Churchill Archives Centre.

35. Wright to Amery, 12 August 1940, and Amery to Wright, 16 August 1940, Amery Papers, AMEL 1/6/5, Churchill Archives Centre.

36. Gough to Amery, 28 August 1940, Amery Papers, AMEL 1/6/5, Churchill Archives Centre.

37. Amery to Lindemann, 26 June 1941, Amery Papers, AMEL 1/6/5, Churchill Archives Centre.

38. The scramblers would have to be changed for each Cabinet meeting: Lindemann to Amery, 30 June 1941, Amery Papers, AMEL 1/6/5, Churchill Archives Centre.

39. Ministry of Works, Division of the Chief Scientific Adviser, 'History of the Research and Experiments Department, Ministry of Home Security, 1939–1945', by A. R. Astbury, p. 112, The Papers of Baron Baker of Windrush, LDBA 1/21, Churchill Archives Centre.

40. J. F. Baker, MS of story of the Morrison shelter, probably the basis of his presentation to the Royal Commission on Awards to Inventors, internal evidence dates it 1950, Baker Papers, LDBA 1/63, Churchill Archives Centre. See also John Fleetwood Baker, *Enterprise versus Bureaucracy: The Development of Structural Air-Raid Precautions during the 2nd World War* (Oxford, 1978).

41. Inaugural address by Lord Privy Seal, Home Office (ARPD), Civil Defence Research Committee, 1st Meeting, 12 May 1939, Baker Papers, LDBA 1/12, Churchill Archives Centre.

42. Calder discusses him at some length after dealing with Bevin, stressing his

role in repression, that he was a human computer and made terrible speeches. He does not get the full measure of his wartime importance: Angus Calder, *The People's War: Britain 1939–1945* (London, 1969), pp. 103–4.

43. The term 'cronies' is extensively used in the secondary literature, though only to refer to a small handful of Churchill's handpicked associates, usually Lindemann, Brenden Bracken, Beaverbrook, and Desmond Morton. See, for example, Best, *Churchill and War*, p. 17. By contrast, my usage recognizes a greater number of them, particularly experts.

44. Harold Macmillan, *The Blast of War: 1939–1945* (London, 1967), p. 85.

45. Simon Ball, 'The German Octopus: The British Metal Corporation and the Next War, 1914–1939', *Enterprise & Society* 5 (2004), pp. 451–89.

46. He had become President of the Board of Trade in October 1940; Churchill had offered him the War Office but could not effect the necessary moves: Lyttelton, *The Memoirs of Lord Chandos*, pp. 191–4.

47. Churchill to Admiral Pound and others, 9 November, in Martin Gilbert (ed.), *The Churchill War Papers*, Vol. 1: *At the Admiralty, September 1939– May 1940* (London, 1993), p. 349.

48. Max Nicholson, unpublished autobiography, Chapter 9, at http://www.maxnicholson.com/chapter9.htm; accessed 28 April 2010.

49. Churchill, *Second World War*, Vol. III: *The Grand Alliance*, p. 131.

50. Francis Keenlyside, 'Leathers, Frederick James, first Viscount Leathers (1883–1965)', rev. Marc Brodie, *Oxford Dictionary of National Biography* (Oxford, 2004). Online edn (January 2008) at http://www.oxford dnb.com/view/article/34457; accessed 18 February 2008.

51. Viscount Samuel, HL Deb. (series 5), vol. 128, col. 555 (15 July 1943).

52. The government also included Ernest Bevin from outside Parliament, and the economist and Labour MP Hugh Dalton.

53. G. D. A. MacDougall, 'The Prime Minister's Statistical Section', in D. N. Chester (ed.), *Lessons of the British War Economy* (Cambridge, 1951), pp. 58–68.

54. He went on: 'He is not the only Minister who has a scientific adviser. Even commanders-in-chief have them, as General Pile, who is here, would blushingly admit': Sir Henry Tizard, Speech at the Parliamentary and Scientific Committee's Luncheon, Tuesday, 3 February 1942, The Papers of A. V. Hill, AVHL II 4/79–80, Churchill Archives Centre.

55. Churchill, *Second World War*, Vol. I: *The Gathering Storm*, p. 368.

56. HC Deb. (series 5), vol. 373, cols. 1395–6 (30 July 1941). In response to a question by R. R. Stokes.

57. Roy Harrod, *The Prof: A Personal Memoir of Lord Cherwell* (London, 1959), p. 255, reporting the story as told by Cherwell. In the general

election campaign Churchill had notoriously warned that if elected Attlee would introduce a Gestapo, a comment regarded as being not only in bad taste but politically inept.

58. See, for example, Andrew Brown, *J. D. Bernal: The Sage of Science* (Oxford, 2005), pp. 196–204.

59. Joubert, C/o Coastal Command to Hill, 20 February 1942, Hill Papers, AVHL I 2/1, Churchill Archives Centre.

60. Harold Hartley and D. Gabor, 'Thomas Ralph Merton. 1888–1969', *Biographical Memoirs of Fellows of the Royal Society* 16 (1 November 1970), pp. 421–40, particularly pp. 424, 425.

61. Simon Courtauld, *As I was Going to St Ives: A Life of Derek Jackson* (Norwich, 2007).

62. Macmillan, *The Blast of War*, pp. 100–102.

63. Thanks to Michael Weatherburn.

64. O. J. M. Lindsay, 'John Anthony Hardinge Giffard, 3rd Earl of Halsbury. 4 June 1908–14 January 2000', *Biographical Memoirs of Fellows of the Royal Society* 47 (November 2001), pp. 240–53.

65. Ibid. He was director of research at Decca 1947–9, when he left to become the first managing director of the new National Research Development Corporation. He was known to Harold Wilson's wife Mary: they had met when she was a typist and he a chemist at the Port Sunlight works.

66. Kenneth Rose, *Elusive Rothschild: The Life of Victor, Third Baron* (London, 2003), Chapter 3.

67. Ibid., pp. 68–71. There are more aristocratic scientists that could be mentioned: Robert Cecil, 2nd Baron Rockley (1901–76), Christ Church, Yale and Royal Artillery, described himself a 'engineer' in *Who's Who* and worked at the Ministry of Supply, 1939–43. Robin Strutt, 4th Baron Rayliegh, who succeeded to the title in 1919, resigned his chair in Physics at Imperial College and continued his spectroscopic researches at his home, Terling Place. Randal Berkeley, 8th Earl of Berkeley (1865–1942), was a physicist.

68. Churchill, *Second World War*, Vol. I: *The Gathering Storm*, pp. 116–20, 183.

69. In June 1940. See Ronald W. Clark, *Tizard* (London, 1965) pp. 232–7.

70. W. K. Hancock and M. M. Gowing, *British War Economy* (London, 1949), pp. 317–18.

71. Harrod, *The Prof*, p. 203.

72. Ibid., p. 204.

73. Ibid.

74. Ibid., p. 205.

75. Ibid., p. 319.
76. Donald MacDougall, *Don and Mandarin: Memoirs of an Economist* (London, 1987), p. 26.
77. G. D. A. MacDougall, 'The Prime Minister's Statistical Section', in Chester (ed.), *Lessons of the British War Economy*, pp. 58–68, particularly p. 59.
78. Harrod, *The Prof*, p. 205.
79. G. D. A. MacDougall, 'The Prime Minister's Statistical Section', in Chester (ed.), *Lessons of the British War Economy*, p. 63.
80. 'I believe,' said Roy Harrod, 'that S Branch was the governing influence in the disposal of our available shipping during the war; anyhow in my time it seemed to be so': Harrod, *The Prof*, p. 199.
81. Ibid.
82. MacDougall, *Don and Mandarin*, p. 32.
83. Hancock and Gowing, *British War Economy*, p. 418.
84. Cherwell to Prime Minister, 30 June 1942, The Papers of Sir Donald MacDougall, MACD 25, Churchill Archives Centre. Hancock and Gowing, referring to similar figures offered in July 1942, suggest the calculations were 'shaky' but gave some 'idea about the magnitudes involved': Hancock and Gowing, *British War Economy*, p. 418.
85. Harrod, *The Prof*, pp. 182–3.
86. Committee of Imperial Defence, Sub-Committee on Air Defence Research, 'High Altitude Balloon Barrages – Note by the Air Ministry', 16 January 1939. Copy in The Papers of Sir Winston Churchill, CHAR 25/18, Churchill Archives Centre.
87. Lindemann to Dr Crow, 24 January 1939, Churchill Papers, CHAR 25/17, Churchill Archives Centre; Committee of Imperial Defence, Sub-Committee on Air Defence Research, 'The use of UP to create an Aerial Minefield – note by the War Office', 31 January 1939, Churchill Papers, CHAR 25/18, Churchill Archives Centre.
88. Churchill to Admiral Pound and others, First Lord's Personal Minute No. 1, 14 November 1939, in Gilbert (ed.), *The Churchill War Papers*, Vol. 1: *At the Admiralty, September 1939–May 1940*, p. 367.
89. Churchill to Admiral Pound, 23 March 1940, in ibid., p. 911.
90. Churchill to Sir Kingsley Wood, Secretary of State for Air, 11 November 1939, in ibid., p. 356.
91. Churchill to Admiral Fraser and others, First Lord's Personal Minute No. 262, 24 March 1940, in ibid., p. 913.
92. Alwyn Crow, 'The Rocket as a Weapon of War in the British Forces', 34th Thomas Hawksley Lecture, November 1947, *Proceedings of the*

Institution of Mechanical Engineers 158 (1948), pp. 15–21, on p. 19. Thanks to Alex Oikonomou.

93. Bruce Taylor, *The Battlecruiser HMS Hood: An Illustrated Biography 1916–1941* (London, 2005), p. 209.

94. Prime Minister to Professor Lindemann, 29 June 1940, in Winston S. Churchill, *The Second World War*, Vol. II: *Their Finest Hour* (London, 1949), p. 565. The use of the term 'Radar' suggests post-war editing.

95. Macmillan, *The Blast of War*, p. 129.

96. Lord Weir to Secretary, Ministry of Supply, 29 November 1940, The Papers of 1st Viscount Weir, WEIR, 20/11, Churchill Archives Centre. Weir complained that the Sticky Bomb was ordered by the War Office without consultation with the Ministry of Supply, despite the Anti-Tank committee considering it ineffective: Lord Weir, Memorandum on S.T. Bombs for Minister of Supply, 20 November 1940, Weir Papers, WEIR 20/11, Churchill Archives Centre.

97. Lord Weir, DGX, 'SC Cordite for UP', 18 June 1941, Weir Papers, WEIR 20/12, Churchill Archives Centre.

98. See Christopher Hinton, 'Memoirs', TS 1970, pp. 80–82, The Papers of Lord Hinton of Bankside, HINT 4/2, Churchill Archives Centre.

99. Viscount Weir to Minister of Supply, 14 November 1940 (marked 'Proximity Fuze and UP Weapon. Weekly report for Prime Minister'), Weir Papers, WEIR 20/11, Churchill Archives Centre.

100. Central Statistical Office, *Statistical Digest of the War* (London, 1951), Table 125, p. 147.

101. David Kendall and Kenneth Post, 'The British 3-Inch Anti-Aircraft Rocket. Part Two: High-Flying Bombers', *Notes and Records of the Royal Society of London* 51 (January 1997), pp. 133–40.

102. E. M. C. Clarke to A. V. Hill, 29 October 1940, Hill Papers, AVHL I 2/3 Pt 2, Churchill Archives Centre.

103. Ibid.

104. Note added to Hill Papers, AVHL I 2/3 Pt 2, Churchill Archives Centre.

105. Note by Hill, n.d., Hill Papers, AVHL I 2/3 Pt 2, Churchill Archives Centre.

106. While the requirement for PE fuzes for bombs to attack aeroplanes was dropped in summer 1941: R. W. Burns, 'Early History of the Proximity Fuze (1937–1940)', *IEE Proceedings-A* 140 (May 1993), pp. 224–36, particularly p. 235.

107. David Kendall and Kenneth Post, 'The British 3-Inch Anti-Aircraft Rocket. Part One: Dive-Bombers', *Notes and Records of the Royal Society of London* 50 (July 1996), pp. 229–39, on p. 236. Kendall was a

mathematician on the rocket programme and Captain Kenneth Post was Sandys's university friend and second-in-command, who attended and spoke at the meeting.

108. Burns, 'Early History of the Proximity Fuze', p. 224.

109. 'Final Report, Directory of Boom Defence, 19th December 1945' ADM 199/848 attached to Sheila Bywater, 'My War', TS, Liddle Collection 1939–1945, Brotherton Library, University of Leeds.

110. D. McKenna, 'Airborne Rockets and Projectors', CAB 102/109, The National Archives.

111. Macmillan, *The Blast of War*, p. 130.

112. Alfred Price, 'The Rocket-Firing Typhoons in Normandy', *Royal Air Force Historical Society Journal* 45 (2009), pp. 109–20, on p. 112.

113. Winston S. Churchill, *The Second World War*, Vol. V: *Closing the Ring* (London, 1952), p. 206.

114. R. V. Jones, *Reflections on Intelligence* (London, 1989).

115. Quoted in Philip Ziegler, *Mountbatten: The Official Biography* (London, 1985), p. 156.

116. Ibid., pp. 198, 206, 209.

117. He was married to the birth control advocate Margaret Pyke and was first cousin to the wartime nutritionist and post-war television eccentric scientist Magnus Pyke.

118. These details are from the accounts of a frequent visitor to the house, Elias Canetti in his *Party in the Blitz: The English Years*, trans. Michael Hofmann (London, 2005), pp. 180–83.

119. Details from Brown, *J. D. Bernal*.

120. Brown has a lot on Bernal and Habbakuk – but does not see it as the disaster it was and never takes Bernal – or the methods of operational research – to task for it: Brown, *J. D. Bernal*.

121. Solly Zuckerman, *From Apes to Warlords 1904–46* (London, 1978), p. 159.

122. Sir Charles Goodeve, 'The ice island fiasco', unknown newspaper, (19 April 1951), Goodeve Papers, GOEV 3/1, Churchill Archives Centre.

123. Ibid.

124. Zuckerman, *From Apes to Warlords*, pp. 164–5.

125. Ritchie Calder hints at an explanation of this kind too, but notably makes no reference that his endorsement of Habbakuk represented a gross failure of scientific advice and/or Bernal's judgement: Ritchie Calder, 'Bernal at War', in Brenda Swann and Francis Aprahamiam (eds.), *J. D. Bernal: A Life in Science and Politics* (London, 1999), pp. 160–90. Calder's piece was unfinished at his death in 1982.

126. Churchill, *Second World War*, Vol. V: *Closing the Ring*, p. 516.

127. M. M. Postan, *British War Production* (London, 1952), p. 282.

128. Ibid., pp. 283–4.

129. Ibid., p. 282.

130. For 'Port Winston', see *D-Day Secret Revealed. Factory-Made Invasion Port*, British Pathé (October 1944), http://www.britishpathe.com/record.php?id=23489; accessed 2 April 2010.

131. R. W. Crawford, 'Mulberry "B" D+4–D+147 1944, 10 June to 31 October', 5 November 1944, available at http://www.ibiblio.org/hyperwar/ETO/Overlord/MulberryB/index.html; accessed 18 September 2009.

132. Guy Hartcup, *Code Name Mulberry: The Planning, Building and Operation of the Normandy Harbours* (Newton Abbot, 1977).

133. Roland G. Ruppenthal, *United States Army in World War II. European Theater of Operations: Logistical Support of the Armies*, Vol. I: *May 1941–September 1944* (Washington DC, 1953), p. 320.

134. D. J. Payton-Smith, *Oil: A Study of War-Time Policy and Administration* (London, 1971), p. 411.

135. Ibid., pp. 445–9.

136. An instance is Andrew Roberts, *The Storm of War: A New History of the Second World War* (London, 2009), p. 606.

137. Margaret Gowing, *Britain and Atomic Energy 1939–1945* (London, 1964), p. 37.

138. Ibid., p. 77.

139. Appendix I of MAUD report of 1943, in Gowing, *Britain and Atomic Energy*, pp. 414–15.

140. Ibid., p. 92.

141. Ibid., Chapter 3.

142. Ibid., p. 109.

143. Ibid., pp. 150–51.

144. Ibid., pp. 155–7.

145. Ibid., pp. 158–61.

146. Ibid., p. 162.

147. There were some fifty British scientists in the USA, and a maximum of 120 in Canada working on the bomb: DSIR, *Report for the Year 1947–48, with a Review of the Years 1938–48*, Cmd 7761 (1949), p. 18.

148. Quoting a post-war document by Groves and Jannerone in Robert S. Norris, *Racing for the Bomb: General Leslie R. Groves, the Manhattan Project's Indispensable Man* (South Royalton, Vt, 2002), p. 343.

149. See, for example, Gowing, *Britain and Atomic Energy*, p. 353.

150. Interrogation of Albert Speer by Lieutenant-General Evetts (Senior

Military Adviser, Ministry of Supply) and Major-General Jefferis, 25 June 1945, The Papers of Colonel R. S. Macrae, MCRA 1/3, Churchill Archives Centre. For a drastically revised account of Speer's production record, see Adam Tooze, *The Wages of Destruction: The Making and Breaking of the Nazi Economy* (London, 2006).

5. POLITICS AND PRODUCTION

1. Historians have tended to see 1941 as the year of concern over production and early 1942 as a period of concern over military performance: Angus Calder, *The People's War: Britain 1939–1945* (London, 1969), Chapter 5; Kevin Jefferys, *The Churchill Coalition and Wartime Politics, 1940–1945* (Manchester, 1991), Chapters 3 and 4. Richard Croucher, *Engineers at War 1939–1945* (London, 1982), correctly notes the centrality of production in 1942 also.

2. Argonaut, *Give Us the Tools: A Study of the Hindrances to Full War Production and How to End Them* (London, 1942), p. 1. Other relevant 1942 titles are Nicholas Davenport, *Vested Interests or Common Pool* (London, 1942); Mass-Observation, *People in Production* (London, 1942); J. T. Murphy, *Victory Production!* (London, 1942).

3. Sir Stafford Cripps, address to the BAAS conference on 'Mineral Resources and the Atlantic Charter', 25 July 1942, *The Advancement of Science* 2 (October 1942), p. 240.

4. Mass-Observation, *People in Production*, p. 14. Emphasis in the original.

5. Ibid., p. 56.

6. Cripps, address to the BAAS conference on 'Mineral Resources and the Atlantic Charter', 25 July 1942, *The Advancement of Science* 2 (October 1942), p. 240.

7. Quoted in Calder, *The People's War*, p. 272. Eric Estorick, *Stafford Cripps: A Biography* (London, 1949), p. 293.

8. His crucial anti-Churchill speech, noted in Paul Addison, *The Road to 1945: British Politics and the Second World War* (London, 1975), p. 200, was on 8 February 1942 in Bristol – after he learned that Beaverbrook had been offered the Ministry of Production (announced 4 February 1942). The speech is partially produced in: 'Total Effort Now Clear Enunciation of Peace Aims, Sir S. Cripps's Appeals', *The Times* (9 February 1942). There is no mention of production and science, just a call for more 'effort' and help to the Russians. 'There seems to be a lack of urgency in the atmosphere. It is almost as if we are spectators rather

than participants.' In a broadcast the same night he called for more material help for the Russians: 'arms', 'raw materials' and 'foodstuffs'.

9. For example, in A. J. P. Taylor, *English History, 1914–1945* (Oxford, 1965; Harmondsworth, 1975), pp. 672–3; Calder, *The People's War*, pp. 299–302; John Charmley, *Churchill: The End of Glory, a Political Biography* (London, 1993), pp. 500–503; Geoffrey Fry, *The Politics of Crisis* (London, 2001); Martin Gilbert, *Road to Victory: Winston S. Churchill 1941–1945* (London, 1986), pp. 137–40. Nor does Jefferys, *The Churchill Coalition and Wartime Politics*, note the centrality of production to the censure debate of July 1942 (pp. 99–100), though he most certainly recognizes the importance of discussion of production between 1940 and the autumn of 1942 more generally. The older Henry Pelling, *Britain and the Second World War* (London, 1970), pp. 156–7, is better. George Malcolm Thomson, *Vote of Censure* (London, 1968), gives some attention to the production issue, but not as much as I think it merits.

10. Jason Tomes, 'Milne, Sir John Sydney Wardlaw- (1879–1967)', *Oxford Dictionary of National Biography* (Oxford, 2004). Online edn (Oxford, October 2005) at http://www.oxforddnb.com/view/article/76640; accessed 6 August 2008.

11. Cripps to Churchill, 2 July 1942, in Winston S. Churchill, *The Second World War*, Vol. IV: *The Hinge of Fate* (London, 1951), pp. 354–6.

12. HC Deb. (series 5), vol. 381, col. 277 (1 July 1942).

13. Ibid., vol. 381, col. 528 (2 July 1942).

14. Ibid., vol. 381, col. 529 (2 July 1942).

15. Ibid., vol. 381, col. 530 (2 July 1942). Michael Foot's biography of Bevan gushes about Bevan without analysing the content of the charges, and without noting that Bevan was attacking in part at least on grounds similar to those of some very right-wing critics, or the independents who were important in this story. Foot does not report the centrality of production to the debate either: Michael Foot, *Aneurin Bevan: A Biography*, Vol. 1: *1897–1945* (London, 1967).

16. Murphy, *Victory Production!*, p. 143.

17. Churchill quoted Bevan's speech at length in his memoirs as an example of the outrageousness of the criticism he faced, though without naming Bevan: Churchill, *Second World War*, Vol. IV: *The Hinge of Fate*, p. 359.

18. HC Deb. (series 5), vol. 381, col. 257 (1 July 1942).

19. Churchill, HC Deb. (series 5), vol. 381, col. 596 (2 July 1942).

20. Ibid., vol. 381, col. 599 (2 July 1942).

21. Jaime Reynolds and Ian Hunter, 'Tom Horabin: Liberal Class Warrior', *Journal of Liberal Democrat History* 28 (Autumn 2000), pp. 17–21.

22. Robert Crowcroft, '"What is Happening in Europe?" Richard Stokes, Fascism, and the Anti-War Movement in the British Labour Party during the Second World War and After', *History* 93 (October 2008), pp. 514–30.

23. 'Obituary. Sir Herbert Geraint Williams, 1884–1954', *ICE Proceedings* 4 (1955), p. 112.

24. Large contracts were placed before and at the beginning of the war in Switzerland, notably for the 20mm Oerlikon anti-aircraft cannon and the rights to manufacture it: Neville Wylie, 'British Smuggling Operations from Switzerland, 1940–1944', *The Historical Journal* 48 (December 2005), pp. 1077–1102. They were for the navy, and the naval officer Steuart Mitchell, later a senior procurement officer, escaped from Switzerland with the plans for the Oerlikon gun in 1940. See Edward Terrell, *Admiralty Brief* (London, 1958), p. 41.

25. 'MARC' alluding to the owner of Hispano-Suiza, Marc Birkigt. Note of a visit 25.7.42 to Captain Dawson, Ministry Of Aircraft Production, Kendall Security Service, MI5, file KV/2/2780 image 59, The National Archives. The firm was under surveillance precisely because of its foreign ownership. Following the fall of France the firm was in a rather odd position, and the government nominated its own directors and was seriously considering buying it up: HC Deb. (series 5), vol. 383, cols. 1288–9 (7 October 1942). Some readers may recognize the name from the recent past. BMARC was owned by Hispano-Suiza into the late 1960s, when the Hispano business merged with Oerlikon (also Swiss). In 1988 it was sold to Astra Holdings, when it was subject to a major scandal: it illegally sold arms to Iran while a Conservative MP, Jonathan Aitken, was on its board.

26. David Lee, 'Longmore, Sir Arthur Murray (1885–1970)', rev. Christina J. M. Goulter, *Oxford Dictionary of National Biography* (Oxford, 2004). Online edn (2008) at http://www.oxforddnb.com/view/article/34593; accessed 1 August 2008. Paul Addison, 'By-Elections of the Second World War', in Chris Cook and John Ramsden (eds.), *By-Elections in British Politics* (London, 1973), pp. 165–90, sees Kendall only as managing director of a 'local engineering firm'. Calder overplays the welfare angle of Kendall's position and downplays the production and engineering angles: Calder, *The People's War*, pp. 289–90. Jefferys, *The Churchill Coalition and Wartime Politics*, notes only that Kendall was the manager of an armaments factory and not the importance of production in the election (pp. 146–7).

27. Calder, *The People's War*, p. 289.

28. *Reveille* (16 March 1942), cutting in Kendall Security Service, MI5, file KV/2/2780 image 75, The National Archives.

29. *Picture Post* (18 April 1942), cutting in Kendall Security Service, MI5, file KV/2/2780 image 66, The National Archives.

30. Report of speech by Sergeant 339 to Superintendent Good, 12 March 1942, Kendall Security Service, MI5, file KV/2/2780 image 75, The National Archives.

31. See Barbara Stoney, *Twentieth Century Maverick: The Life of Noel Pemberton Billing* (East Grinstead, 2004), and Addison, 'By-Elections of the Second World War', in Cook and Ramsden (eds.), *By-Elections in British Politics*, pp. 165–90.

32. They were Gordon Reakes, William Brown (ex-Labour MP) and Tom Driberg, in 1942; three Common Wealth candidates were elected in 1943, 1944 and 1945, one independent in 1944, and another in 1945.

33. In February 1942 four Liberal Nationals deserted their party (E. Granville, Sir Murdoch MacDonald, Leslie Hore-Belisha, and Sir H. Morris-Jones) and two National Labour members (Stephen King-Hall and K. Lindsay); in May 1942 Captain Cunningham-Reid defected from the Tories. Of these all, except King-Hall and Lindsay, who voted with the government, and MacDonald, who abstained, voted to censure the government in July 1942.

34. See Addison, *The Road to 1945*, pp. 40–42. The usual story is that these by-elections were indicative of a move to the left: for example, Jefferys, *The Churchill Coalition and Wartime Politics*, pp. 146–7, and Stephen Brooke, *Labour's War: The Labour Party during the Second World War* (Oxford, 1992), pp. 62–3.

35. Richard Temple, 'Brown, William John (1894–1960)', *Oxford Dictionary of National Biography* (Oxford, 2004). Online edn (January 2008) at http://www.oxforddnb.com/view/article/32118; accessed 1 August 2008.

36. *The Times* (15 May 1942). It should be noted that production is not mentioned in a defence of Reid published in 1945: R. J. Ellis, *He Walks Alone: The Public and Private Life of Captain Cunningham-Reid, DFC, Member of Parliament, 1922–1945* (London, 1945).

37. Calder, *The People's War*, p. 291.

38. While the evidence in the MI5 file is not completely convincing, the evidence of close connections to Mosleyites after the war, when Kendall appeared to be involved in gun-running, is compelling. It turned out that he was, or became, a member of, or associated with, the British National Party, made up of former members of the BUF, and after the war

associated in business with some very dubious known Mosleyites: Kendall Security Service, MI5, file KV/2/2780, The National Archives.

39. HC Deb. (series 5), vol. 383, cols. 1224–320 (7 October 1942). After the war Kendall was allocated, in controversial circumstances, some of the wartime factories, in which he intended to build a people's car and light tractors. He was financed by an Indian majarajah, who appears to have lost a great deal of money: HC Deb. (series 5), vol. 441, cols. 525–83 (2 July 1942). On the Kendall–Beaumont tractor and cheap car, see *Personality meets Dennis Kendall MP*, British Pathé (August 1945), http://www.britishpathe.com/record.php?id=48195, where Kendall and his machines are presented as futuristic. It has been suggested that the fiasco influenced the thinking of Margaret Thatcher, whose father was a local politician in the town. See Karl Ludvigsen, *Battle for the Beetle* (London, 2000).

40. Roger Cooter, 'The Rise and Decline of the Medical Member: Doctors and Parliament in Edwardian and Interwar Britain', *Bulletin of the History of Medicine*, 78 (2004), pp. 59–107, calculated from Appendix A.

41. Robert Self, *Neville Chamberlain: A Biography* (Aldershot, 2006), pp. 21ff.

42. Ibid., p. 27.

43. Cripps, address to the BAAS conference on 'Mineral Resources and the Atlantic Charter', 25 July 1942, *The Advancement of Science* 2 (October 1942), p. 240.

44. Leonard Frank Plugge (1889–1981), Conservative MP for Chatham, 1935–45. His International Broadcasting Company had studios in Portland Place and a staff of nearly 180 in 1939. He was also an inventor. His daughter, Gale Benson, a friend of Michael X, was murdered in Trinidad in 1972.

45. It had little influence on warlike research policy, but had more success when it turned in 1943 to considering post-war civil research. S. A. Walkland, 'Science and Parliament: The Origins and Influence of the Parliamentary and Scientific Committee', *Parliamentary Affairs* XVII (1963), pp. 317–18.

46. Group Captain W. Helmore, CBE, Ph.D., M.Sc., FCS, FRAeS, *Air Commentary* (London, 1942), p. 76. The book is illustrated with RAF portraits by the official RAF artist, Eric Kennington. Helmore covered the Schneider races, RAF displays and official air events from 1926. He was an 'RAF War Commentator', 1941–3, and Conservative MP for the Watford Division of Hertfordshire, 1943–5. As well as holding various technical advisory positions he was a member of the Brabazon Committee

on Civil Aviation, 1943–5. He probably also wrote Flight Commander, *Cavalry of the Air* (London, 1918).

47. Richard Lawson and Doug Lawson, 'Obituary: Hugh Lawson', *Independent* (21 May 1997). The other two elected as Common Wealth candidates were both aviators, John Loverseed (fighters) and Ernest Millington (bombers).

48. Tom Driberg, 'The Man Who Loved Tanks', *Leader Magazine* (25 March 1950).

49. The Ransome founders were of the same family as the writer Arthur Ransome. The firm had been separated from the famous agricultural machinery makers of Ipswich, Ransome's, in the 1860s.

50. R. R. Stokes, report of a speech, 8 October 1939, Stokes Papers, Box 18, Bodleian Library, University of Oxford.

51. Note for the press of a speech at Caernarvon, 13 June 1942, Stokes Papers, Box 22, Bodleian Library, University of Oxford.

52. HC Deb. (series 5), vol. 387, col. 929 (11 March 1943).

53. Oliver Lyttelton, Viscount Chandos, *The Memoirs of Lord Chandos* (London, 1962), p. 317.

54. Richard Stokes, HC Deb. (series 5), vol. 381, col. 1180 (14 July 1942).

55. Lyttelton, *The Memoirs of Lord Chandos*, p. 317.

56. *Wartime Tank Production*, Cmd 6865 (London, 1946).

57. Although A. V. Hill and his role as MP are well-known, the story told here about his election and his position with respect to Churchill is not.

58. He would leave Cambridge in 1943 to become the first Professor of Social Medicine, at Oxford.

59. See John Ryle, 'Peace Aims', *Federal Union News* (10 February 1940), available online at http://www.federalunion.org.uk/archives/peaceaims.shtml.

60. *The Times* (26 February 1940).

61. Each university MP had a notional average constituency of around 10,000, smaller than most ordinary constituencies. It would be wrong to assume that the politics of the learned were of the left. We can surmise this from the electoral preferences of British graduates. For up to and including the 1945 general election graduates had a second vote in university seats, electing twelve MPs. In our period the universities of Oxford and Cambridge were represented by two MPs each, London by one, Combined English Universities by two, Combined Scottish Universities by three, Queen's Belfast by one (one in thirteen Ulster seats) and the University of Wales by one. In 1935 the total electorate consisted of 134,000 graduates, though only around 80,000 voted: Millicent B. Rex,

'The University Constituencies in the Recent British Election', *The Journal of Politics* 8 (May 1946), pp. 201–11.

62. In our period not a single Labour man or woman sat for a university seat, and only Wales was represented by a Liberal. To give some examples, one Scottish university seat went in a by-election in 1938 to Sir John Anderson, who replaced Ramsay MacDonald on his death after his brief tenure of the seat. The notable Combined English Universities MP of our period was the independent Eleanor Rathbone; the most significant University of London MP was the national independent doctor Sir Ernest Graham-Little. Oxford was represented by two independents, the humorist A. P. Herbert and Sir Arthur Salter (who had defeated, among others, Frederick Lindemann). Cambridge was represented by two Conservatives, including the Conservative historian Sir Kenneth Pickthorn and from 1940 Hill.

63. The only Nobel laureates who were MPs when the prize was awarded were Sir Austen Chamberlain, Peace Prize 1925; Arthur Henderson, Peace Prize 1934; and Sir Winston Churchill, who was awarded the Literature Prize in 1953 while Prime Minister. Sir Norman Angell got his Peace Prize after his brief career as an MP, as did Sir John Boyd-Orr.

64. D. K. Hill, 'A. V. Hill, Personal Memoirs by a son and comments on landmarks and events in his life from boyhood onwards', TS 2002, The Papers of A. V. Hill, AVHL II 8/7, IV, p. 1, Churchill Archives Centre.

65. Such a violation of the rules which prevented MPs from being even advisory officials was possible as a result of a wartime dispensation that allowed MPs, with the permission of the PM, to hold government positions.

66. Debate on Science and the War, HL Deb. (series 5), vol. 124, col. 76 (2 July 1942).

67. Hill supported Churchill in the vote of confidence in January 1942, when only one MP opposed him, suggesting that he did not abstain on principle. The *Manchester Guardian* reported that his name was on the motion of no-confidence of July 1942 but was withdrawn as it had been put there without his permission: *Manchester Guardian* (Friday, 26 June 1942).

68. 'Combined Operations and a Great General Staff. Private, not for circulation', n.d., but probably 1942, Hill Papers, AVHL I 2/1, Churchill Archives Centre.

69. HC Deb. (series 5), vol. 378, cols. 131–2 (24 February 1942).

70. Ibid.

71. But the theme was taken up in the censure debate by the Labour doctor Leslie Haden-Guest: 'I learned the day before yesterday to my astonishment that the Prime Minister, either in his capacity of Prime Minister or as

Minister of Defence, assisted by his personal assistant, Lord Cherwell, has been conducting a series of experiments of a scientific character, of a most amazing character, investigating some new weapon, and that a large sum has been spent. I do not know whether it is £10,000,000, £20,000,000 or £30,000,000. I hope the Leader of the House will inform the Prime Minister that I have made this statement and ask him to let the House know what authority there is for his conducting experiments of that kind and spending that amount of money without Parliament knowing anything about it. If I am wrong, I will withdraw, but I have every reason to believe, because I have the information from the highest authority, that I am right, and it is a thing about which the country ought to be informed. This strikes me as an abuse of the Prime Minister's powers and authority to do that kind of thing which his training does not equip him to do': Dr. Haden Guest (Islington, North), HC Deb. (series 5), vol. 381, cols. 457–61 (1 July 1942).

72. *The Times* (1 July 1942).

73. *The Times* (21 April 1942).

74. Hill, HC Deb. (series 5), vol. 378, cols. 127–9 (24 February 1942).

75. A. V. Hill, letter to *The Times,* 21 April 1942.

76. 'Scientific Advisory Councils in the Service and Supply Ministries', 1st draft, 22 August 1940, Hill Papers, AVHL I 2/1, Churchill Archives Centre. A later version was sent to the Lord President, Neville Chamberlain.

77. *Endeavour: A Quarterly Designed to Record the Progress of the Sciences in the Service of Mankind*, Vol. 1, No. 3 (July 1942).

78. They were Prof. (Sir) Ian Heilbron FRS (1886–1959), Prof. (Sir) Thomas Merton FRS (1888–1969) and (Sir) William Stanier (1876–1965): William McGucken, 'The Central Organisation of Scientific and Technical Advice in the United Kingdom during the Second World War', *Minerva* 17 (1979), pp. 33–69. See also Lyttelton, *The Memoirs of Lord Chandos*, pp. 169–70.

79. 'Appointment of Scientific Advisers to the Minister of Production', memorandum by the Lord President and the Minister of Production, 3 September 1942, War Cabinet, WP (42) 389, CAB 123/180, The National Archives.

80. The scientists were given a sop in the form of the War Cabinet Scientific Advisory Committee (SAC) formed in 1940, chaired at first by Lord Hankey: McGucken, 'The Central Organisation of Scientific and Technical Advice'. It consisted of Sir William Bragg (replaced by Henry Dale, when he became PRS), A. C. G. Egerton, A. V. Hill, and the secretaries of the DSIR, ARC and MRC. An Engineering Advisory Committee was formed in 1941, also under Lord Hankey's chairmanship. The

members were Lord Falmouth, Sir Henry Tizard, A. P. M. Fleming, W. T. Halcrow, C. C. Paterson (GEC), H. R. Ricardo and Dr A. Robertson. It soon died. The SAC was called a 'Brains Trust' by the *Daily Telegraph* and *Daily Herald* (3 October 1940). It was advisory and was not concerned with the running of the services' R&D programmes. After the war it was turned into the Advisory Council on Scientific Policy, which dealt with civil matters.

81. Lyttelton, *The Memoirs of Lord Chandos*, pp. 169–70. For Chandos's awful comments on the engineers and senior managers of AEI, which he headed for a long period after the war, see R. Jones and Oliver Marriott, *Anatomy of a Merger* (London, 1970), p. 234.

82. See David Edgerton, 'British Scientific Intellectuals and the Relations of Science, Technology, and War', in Paul Forman and Jose Manuel Sanchez-Ron (eds.), *National Military Establishments and the Advancement of Science and Technology* (Dordrecht, 1996), pp. 1–35.

83. J. G. Crowther, O. J. R. Howarth and D. P. Riley, *Science and World Order* (Harmondsworth, 1942). It was based on a conference which took place the previous year but is best thought of as a fresh book.

84. Among his students at this Bristol public school were Neville Mott, Conrad Waddington, Charles Coulson, John Kendrew, Brian Pippard and Trevor Williams, who succeeded him as editor in 1955.

85. *Endeavour: A Quarterly Designed to Record the Progress of the Sciences in the Service of Mankind*, Vol. 1, No. 1 (January 1942), p. 3.

86. *Endeavour*, Vol. 1, No. 3 (July 1942) has responses from Bernal and Haldane. Just after the war *Endeavour* gave a lot of attention to E. F. Caldin, a Catholic chemist and philosopher of science.

87. Max Stadler, 'Assembling Life: Models, the Cell, and the Reformations of Biological Science, 1920–1960', unpubl. Ph.D. thesis, University of London (2010).

88. Sir Henry Tizard to Prof. J. Lennard-Jones, 11 October 1938, Tizard Papers, HTT 65, Imperial War Museum.

89. Sir Henry Tizard to A. V. Hill, 15 February 1939, Hill Papers, AVHL I 2/1, Churchill Archives Centre.

90. Margaret Gowing, *Britain and Atomic Energy 1939–1945* (London, 1964), pp. 78, 80, 92–3. For Tizard's opposition, see further note dated 8/7/43, Tizard Papers, HTT 20/21, Imperial War Museum.

91. P. M. S. Blackett, 'Scientists at the Operational Level' (December 1941), reprinted in *The Advancement of Science* 5, No. 17 (1948), pp. 27–9, particularly p. 28. From the context it seems unlikely that the accusation is directed only at service personnel.

92. Ibid.

93. P. M. S. Blackett, 'A Note on Certain Aspects of the Methodology of Operational Research' (1943), reprinted in *The Advancement of Science* 5 (1948), p. 31. See Erik P. Rau, 'Technological Systems, Expertise and Policy Making: The British Origins of Operational Research', in Michael Allen and Gabrielle Hecht (eds.), *Technologies of Power: Essays in Honor of Thomas Parke Hughes and Agatha Chipley Hughes* (Cambridge, Mass., 2001), pp. 215–52, and Paul Crook, 'Science and War: Radical Scientists and the Tizard–Cherwell Area Bombing Debate in Britain', *War & Society* 12 (1994), pp. 69–101.

94. 'Notes of First Informal Meeting of independent scientific advisers held in Sir Henry Tizard's office, MAP, on Monday 8th June 1942', Tizard Papers, HTT 298, Imperial War Museum.

95. Basil Liddell Hart, 'Early Efforts towards Scientific Military (Operational) Research', attached to letter to Richard Whiddington, 29 March 1949, Whiddington Papers, Brotherton Library, University of Leeds.

96. Hart to Whiddington, 29 March 1949, Whiddington Papers, Brotherton Library, University of Leeds.

97. See Alex Danchev, *Alchemist of War: The Life of Sir Basil Liddell Hart* (London 1998), Chapters 7 and 8.

98. For a stimulating new approach, drawn on here, see William Thomas, 'The Heuristics of War: Scientific Method and the Founders of Operations Research', *The British Journal for the History of Science* 40 (2007), pp. 251–74.

99. I owe this important point to Dominique Pestre.

100. M. Postan, D. Hay and J. D. Scott, *Design and Development of Weapons* (London, 1964), p. 384; Maurice W. Kirby, *Operational Research in War and Peace: The British Experience from the 1930s to 1970* (London, 2003), p. 91–4.

101. Ibid., pp. 95–107.

102. Terrell, *Admiralty Brief*, p. 155.

103. Malcolm Llewellyn-Jones, 'A Clash of Cultures: The Case for Large Convoys', in Peter Hore (ed.), *Patrick Blackett: Sailor, Scientist and Socialist* (London, 2003), pp. 138–66.

104. Butt Report in Sir Charles Kingsley Webster and Noble Frankland, *The Strategic Air Offensive against Germany 1939–1945*, Vol. IV: *Annexes and Appendices* (London, 1961), Appendix 13, pp. 205–13.

105. Randall T. Wakelam, *The Science of Bombing: Operational Research in RAF Bomber Command* (Toronto, 2009), p. 45.

106. Ministry of Works, Division of the Chief Scientific Adviser, 'History of

the Research and Experiments Department, Ministry of Home Security, 1939–1945. Compiled from Official Sources by A. R. Astbury, Technical Adviser, Ministry of Works', p. 2. Copy in The Papers of Baron Baker of Windrush , LDBA 1/21, Churchill Archives Centre.

107. Ibid., pp. 5–57.

108. R. J. Desmarais, 'Science, Scientific Intellectuals and British Culture in the Early Atomic Age, 1945–1956: A Case Study of George Orwell, Jacob Bronowski, J. G. Crowther and P. M. S. Blackett', unpublished Ph.D. thesis, Imperial College London (2009).

109. Rt Hon. Lord Hankey, 'British Air Power', *Sunday Times* (12 July 1942).

110. Sir Charles Kingsley Webster and Noble Frankland, *The Strategic Air Offensive against Germany 1939–1945*, Vol. 1: *Preparation* (London, 1961), pp. 331–6, is a very clear summary of the correspondence between Tizard and Cherwell, in Tizard Papers, HTT 353, Imperial War Museum.

111. C. P. Snow, *Science in Government* (London, 1961). For a clear defence of Lindemann against Snow's quite bizarre arguments and insinuations, see Thomas Wilson, *Churchill and the Prof* (London, 1995).

112. See Tizard, 'Estimates of Bombing Effect', 20 April 1942, and Cherwell to Tizard, 22 April 1942, in Blackett Papers, D66, Royal Society. Nowhere is there any reference to moral questions. See also Tizard Papers, HTT 353, Imperial War Museum. Tizard and Blackett were arguing, in different ways, for the use of bombers against both submarines and enemy merchant ships (Tizard particularly emphasizing the latter), rather than the bombing of land targets, but in the very particular context of early 1942.

113. Adrian Fort, *Prof: The Life of Frederick Lindemann* (London, 2004), pp. 275–6.

114. Kenneth Hutchison, J. A. Gray and Harrie Massey, 'Charles Drummond Ellis. 11 August 1895–10 January 1980', *Biographical Memoirs of Fellows of the Royal Society* 27 (November 1981), pp. 199–233, on p. 216.

115. P. B. Moon, 'George Paget Thomson. 3 May 1892–10 September 1975', *Biographical Memoirs of Fellows of the Royal Society* 23 (November 1977), pp. 529–56.

116. British Gaumont Newsreel, 9 June 1942, www.Itnsource.com Ref. BGX408230132 0.

117. See S. W. H. Zaidi, 'Technology and the Reconstruction of International Relations: Liberal Internationalist Proposals for the Internationalisation of Aviation and the International Control of Atomic Energy in Britain, USA and France, 1920–1950', unpublished Ph.D. thesis, University of London (2009). The following blistering critique diagnoses the

condition rather well: 'Their pacifism acquired a messianic character – they were less concerned with saving their fellow countrymen and more with saving all mankind from war. Their own security made them more accessible than any other nations to utopian dreams of universal peace – and blinder to the danger inherent in such utopian dreams ... Monstrous proposals, like the proposal to create an International Air Force that would emerge – from some Alpine stronghold, presumably – and bomb the cities of the alleged aggressor, found a considerable following ... Such inhuman phantasmagoria had an affinity with the secular religions of the European continent. Indeed, English militant pacifism had something in common with the Marxian dreams of a universal realm of peace, justice, and well-being ... The threat of universal war as the means of establishing universal peace is a peculiarly English conception that has crystallised in the doctrine of "sanctions". This doctrine is analogous to the doctrine of the proletarian dictatorship which would establish social peace by making class war permanent and universal. Sanctions are the counterpart of the revolutionary terror – the purpose of either is peace, but the effect of both is the consolidation, through war or the threat of war (whether between classes or nations), of power in the hands of those who hold it': F. A. Voigt, *Unto Caesar* (London, 1938), pp. 211–12.

118. J. M. Spaight, *Bombing Vindicated* (London, 1944), p. 152.

119. Stephen King-Hall, *Total Victory* (London, 1941), p. 217. See also Sir William Beveridge, *The Price of Peace* (London, 1945), p. 54.

120. King-Hall, *Total Victory*, p. 219.

121. M. J. B. Davy, *Air Power and Civilization* (London, 1941), pp. 196, 161

122. Zaidi, 'Technology and the Reconstruction of International Relations' p. 83.

123. Prime Minister's Personal Minute, Foreign Secretary General Ismay for COS Committee (29 August 1944), CAB 120/837, The National Archives, quoted in ibid., p. 6.

124. Winston Churchill, 'The Soviet Danger: "The Iron Curtain"', in David Cannadine (ed.), *The Speeches of Winston Churchill* (London, 1989), pp. 295–308, quoted in Zaidi, 'Technology and the Reconstruction of International Relations', pp. 6–7.

125. Polyglot, 'Letter to the Editor: International Policing', *Manchester Guardian* (8 May 1942), quoted in Zaidi, 'Technology and the Reconstruction of International Relations', pp. 6, 85.

126. Peter Clarke, *The Cripps Version: The Life of Sir Stafford Cripps 1889–1952* (London, 2002), pp. 361–2, 373–4.

127. Squadron Leader John Strachey, 'The Story of a Target', Home Service broadcast, printed in *The Listener* (1 June 1944).

128. Vera Brittain, *England's Hour* (London, 2005; first publ. 1941), pp. 128, 225.

129. Ibid., pp. 225–6.

130. Ibid.

131. *No Part of the Reich is Safe (The Lancaster Bomber)*, British Movietone News (August 1942), available at: http://www.youtube.com/watch?v=cLTP1U4Lz9k; accessed 10 April 2010.

132. Crowther, Howarth and Riley, *Science and World Order*, p. 12.

133. Cripps, address to the BAAS conference on 'Mineral Resources and the Atlantic Charter', p. 242.

134. Ibid., p. 243.

135. See R. G. D. Allen, 'Mutual Aid between the US and the British Empire, 1941–45', *Journal of the Royal Statistical Society* 109 (1946), pp. 243–77.

136. Sir William Beveridge, BBC Home Service, 2 December 1942, at http://www.bbc.co.uk/archive/nhs/5139.shtml; accessed 23 April 2010. The Atlantic Charter was also mentioned by Attlee in an informative broadcast on the eve of the inauguration of the National Health Service and the new National Insurance, National Assistance and Industrial Injuries provisions: BBC Home Service, 4 July 1948, at http://www.bbc.co.uk/archive/nhs/5147.shtml; accessed 23 April 2010.

137. Calder, *The People's War*, p. 138. Peter Mandler makes a similar observation about Calder's *Myth of the Blitz* in *The English National Character: the history of an idea from Edmund Burke to Tony Blair* (London, 2006), p. 287.

138. I draw these comments from examining the use of the term in the *Manchester Guardian*.

139. Advertisement in the *Manchester Guardian*, 29 August 1942.

140. E. H. Carr, *Nationalism and After* (London, 1945), p. 48. See also Peter Wilson, 'The New Europe Debate in Wartime Britain', in Philomena Murray and Paul Rich (eds.), *Visions of European Unity* (Boulder, Colo., 1996), pp. 39–62.

141. Ivor Montagu, *The Traitor Class* (London, 1940).

142. James Hinton, 'Coventry Communism: A Study of Factory Politics in the Second World War', *History Workshop* 10 (Autumn 1980), pp. 90–118. See also the outstanding Richard Croucher, *Engineers at War 1939–1945* (London, 1982), on both the Communist Party and on the centrality of production in the politics of 1942.

143. Argonaut, *Give Us the Tools*, pp. 25–7. See the very positive review in the Communist *Labour Monthly* 24 (1942), p. 160.

144. Argonaut, *Give Us the Tools*, Chapter V.

145. Celticus (Aneurin Bevan MP), *Why Not Trust the Tories?* (London, 1944), pp. 60–64.

146. Murphy, *Victory Production!*, p. 11.

147. Ibid., pp. 147–64.

148. The following mention Wilson, but not his politics: Walkland, 'Science and Parliament', and Roy MacLeod, 'Science for Imperial Efficiency and Social Change: Reflections on the British Science Guild, 1905–1936', *Public Understanding of Science* (1994), pp. 155–93. Wilson's biography does not mention his involvement with science: John Marlowe, *Late Victorian: The Life of Sir Arnold Talbot Wilson* (London, 1967). Some studies of the film identify the man, but none the politics: Michael Weatherburn, 'Arnold T. Wilson, the New Victorians and the Forgotten Technocrats of Inter-War Britain', unpublished M.Sc. dissertation, Imperial College London (2009). See David Edgerton, *England and the Aeroplane: An Essay on a Militant and Technological Nation* (London, 1991), p. 60.

149. In Edgerton, *England and the Aeroplane*, p. 60, I incorrectly state the date of release as 1941, and mislead slightly on the treatment of Lady Houston – it is clear she is attacking the British government of 1929. On the connections between the far right and aviation in Britain, see Edgerton, *England and the Aeroplane*; Waqar H. Zaidi, 'The Janus-Face of Techno-Nationalism: Barnes Wallis and the "Strength of England"', *Technology and Culture* 49 (2008), pp. 62–88; and Patrick Zander, 'Right Modern: Technology, Nation, and Britain's Extreme Right in the Interwar Period (1919–1940)', Ph.D. thesis, Georgia Institute of Technology (2009).

150. Anthony Aldgate and Jeffrey Richards, *Britain Can Take It: British Cinema in the Second World War*, 2nd edn (London, 2007), pp. 327–8, and see also pp. 61–6. Apart from the gentlemanly ruralism the politics of the film is not commented on in Aldgate and Richards or in the chapter on the film in S. P. Mackenzie, *The Battle of Britain on Screen: 'The Few' in British Film and Television Drama* (Edinburgh, 2007), pp. 23–46.

151. Yet the crucial war period is curiously missing from our accounts of state secrecy, including David Vincent, *The Culture of Secrecy: Britain 1832–1998* (Oxford, 1998).

152. Ian McLaine, *Ministry of Morale: Home Front Morale and the Ministry of Information in World War II* (London, 1979).

153. Austin Robinson, 'The National Institute: The Early Years', *National Institute Economic Review* 124 (1988), pp. 63–6.

154. Richard Overy, *Why the Allies Won* (London, 1995), p. 198.

155. Calder, *The People's War*; Addison, *The Road to 1945*, Chapter V, 'Two Cheers for Socialism, 1940–1942'; Malcolm Smith, *Britain and 1940: History, Myth and Popular Memory* (London, 2000), p. 107. This picture has been challenged, correctly, by Steven Fielding, who stresses the anti-party feeling in 1942: Steven Fielding, 'The Second World War and Popular Radicalism: The Significance of the "Movement Away from Party"', *History* 80 (1995), pp. 38–58. Keith Middlemas, *Politics in Industrial Society: The Experience of the British System since 1911* (London, 1979), makes the important point that the 1940 settlement came from the elite, not from public opinion, and that political programmes for reform followed rather than preceded this. Programmes of reform after 1943 benefited from a state already acting in particular wartime ways; the key change was in the image of the state. The state, not the party, was the agent guarantor of reconstruction.

156. Addison, *The Road to 1945*, Chapter V, 'Two Cheers for Socialism, 1940–1942', p. 132.

157. Jeffrey Richards and Anthony Aldgate, *Best of British: Cinema and Society from 1930 to the Present* (London, 1999), Chapter 5, is an excellent treatment of the film which correctly points to the significance of the situation in 1942 to the government response to the proposals to make it. Other treatments of the film, and there are many, attempt to see it in class terms, in terms of consensus and as a critique of Britain, sometimes one deliberately engineered by the state.

6. SONS OF THE SEA

1. Churchill, referring to the war in general, but alluding directly to a comparison with the threat of invasion as well as the Battle of Britain: Winston S. Churchill, *The Second World War*, Vol. II: *Their Finest Hour* (London, 1949), pp. 529, 530.

2. Duncan Redford, 'The March 1943 Crisis in the Battle of the Atlantic: Myth and Reality', *History* 92 (2007), pp. 64–83.

3. W. K. Hancock and M. M. Gowing, *British War Economy* (London, 1949), pp. 432, 434.

4. Erin Weir, 'German Submarine Blockade, Overseas Imports, and British Military Production in World War II', *Journal of Military and Strategic Studies* (Spring–Summer 2003), article 1, makes the point that reduced imports were accommodated.

5. Combined Production and Resources Board, *The Impact of the War on*

Civilian Consumption in the United Kingdom, the United States and Canada: A Report to the Combined Production and Resources Board from a Special Combined Committee on Nonfood Consumption Levels (London, 1945), Table 5 and p. 14.

6. Central Statistical Office, *Annual Abstract of Statistics*, No. 85, 1937–1947 (London, 1948), Table 195, p. 166.

7. The post-war *Statistical Digest of the War* also excludes munitions from its tables of imports. Shipping statistics were based on cargo tonnages, which exclude government stores. Ships loading or unloading only government stores (which include armaments and munitions) were classified as 'in ballast': Central Statistical Office, *Statistical Digest of the War* (London, 1951), p. 230.

8. Of a total import value of £1,205.6m in 1942 and £1,8774.7m in 1943, the non-aircraft, non-munitions imports were £1,003m and £1,216.8m respectively, suggesting munitions imports were around £200m in 1942, and £600m in 1943, fully one third of British imports in the latter year: *Statistics Relating to the War Effort of the United Kingdom 1944*, Cmd 6564 (London, 1944), Appendix E, which shows an 18 per cent increase in total imports by 1943, and a fall of 21 per cent excluding munitions. Another way of making the calculation is to assume a $7 to £1 exchange rate for munitions (since US armaments were relatively expensive), giving £343m imports of US munitions in 1944, 20 per cent of overall imports. R. G. D. Allen, 'Mutual Aid between the US and the British Empire, 1941–45', *Journal of the Royal Statistical Society* 109 (1946), pp. 243–77, on p. 258. Using the same exchange rate as for civil goods increases the proportion considerably, reaching perhaps 40 per cent for 1944. The value of non-munitions supplied under Lend-Lease to the UK in 1944 was $2.4bn, while the munitions were worth $3.8bn. We know that Britain was importing (for British rather than Empire use) munitions from the USA in 1944 worth £842m, implying, given the same exchange rate for munitions and non-munitions, a value of non-munitions imports of £532m. The data are from Table 4 of Allen, 'Mutual Aid between the US and the British Empire', p. 250.

9. Joel Hurstfield, *The Control of Raw Materials* (London, 1953), p. 98.

10. The figures are of ships with cargo and ballast, the last including, perversely, government stores: CSO, *Annual Abstract of Statistics*, 1937–1947, Table 229.

11. UK entrances, non-tankers, December 1944, MT 65/136, The National Archives. With thanks to Malcolm Llewellyn-Jones.

12. C. B. A. Behrens, *Merchant Shipping and the Demands of War* (London, 1958), Appendix VIII, p. 69.

13. CSO, *Statistical Digest of the War*, Table 153.

14. Frank H. House, *Timber at War: An Account of the Organisation and Activities of the Timber Control 1939–1945* (London, 1965), p. 138.

15. Tony Lane, *The Merchant Seamen's War* (Liverpool, 1990), p. 22.

16. http://www.red-duster.co.uk; accessed 21 September 2009.

17. From the complete list of Empire ships at http://www.mariners-l.co.uk/EmpireU.html#empw; accessed 15 May 2008. The complete listings available of Blue Star ships do not list managed ships, except incidentally when they were later owned by Blue Star.

18. Frank C. Bowen, *The Flag of the Southern Cross 1939–1945* (London, n.d. but 1947).

19. W. J. Harvey and R. J. Solly, *BP Tankers: A Group Fleet History* (London, 2005), pp. 25, 29.

20. Edward Terrell, *Admiralty Brief* (London, 1958), p. 169.

21. Ibid., Chapters 13 and 14.

22. CSO, *Statistical Digest of the War*, Table 72 for losses; 1941 food imports of 14 million tons.

23. Redford, 'The March 1943 Crisis in the Battle of the Atlantic', p. 69. The myth lives on, as is evident in Andrew Roberts, *The Storm of War: A New History of the Second World War* (London, 2009), p. 606, where it is argued that it was as well that Overlord was postponed past 1943 given conditions at sea.

24. Ministry of Information, *What Britain Has Done 1939–1945: A Selection of Outstanding Facts and Figures* (London, 2007; first publ. 1945), p. 341.

25. Redford, 'The March 1943 Crisis in the Battle of the Atlantic'.

26. David Syrett (ed.), *The Battle of the Atlantic and Signals Intelligence: U-Boat Tracking Papers, 1941–1945* (Aldershot, 2007), p. 11.

27. Ministry of Information, *What Britain Has Done*, p. 33.

28. See the extraordinary website http://www.uboat.net/, which details every U-boat and every merchant ship attacked!

29. Lord Woolton, HL Deb. (series 5), vol. 118, col. 380 (18 February 1941).

30. J. G. Ballard, *Miracles of Life: Shanghai to Shepperton: An Autobiography* (London, 2008), p. 116. On *Arawa* see Kenneth Chapman, *Peace, War and Friendships* (East Hoathly, 2006), p. 77.

31. http://www.oceanlinermuseum.co.uk/DM%20History.htm; accessed 23 January 2008.

32. Bowen, *The Flag of the Southern Cross*, p. 80.

33. Hancock and Gowing, *British War Economy*, p. 419.
34. Harold Macmillan, *The Blast of War: 1939–1945* (London, 1967), p. 173.
35. S. J. Butlin and C. B. Schedvin, *Australia in the War of 1939–1945. Series 4 – Civil – Vol. IV: War Economy, 1942–1945* (Canberra, 1977), p. 516.
36. For a recent popular account which reads history in this way, see James Fergusson, *The Vitamin Murders: Who Killed Healthy Eating in Britain?* (London, 2007), particularly Chapter 1, 'The Golden Age of Nutrition'. For a cultural historical version of the story, with a colonial twist, see James Vernon, *Hunger: A Modern History* (London, 2007).
37. Ina Zweiniger-Bargielowska, *Austerity in Britain: Rationing, Controls and Consumption, 1939–1955* (Oxford, 2000), p. 31.
38. Ibid., p. 32.
39. Angus Calder, *The People's War: Britain 1939–1945* (London, 1969), p. 277.
40. Karl Brandt, *The Management of Agriculture and Food in the German-Occupied and Other Areas of Fortress Europe* (Stanford, Calif., 1954), p. 559.
41. 'A World Broadcast on the German Invasion of Russia. June 22, 1941', in Winston Churchill and Charles Eade (ed.), *The War Speeches of the Rt. Hon. Winston S. Churchill*, Vol. 1 (London, 1952), pp. 450–54.
42. Zweiniger-Bargielowska, *Austerity in Britain*, p. 32.
43. Arthur Marwick, *The Explosion of British Society 1914–1970* (London, 1971; first published 1963), p. 100.
44. Charles Webster, 'Government Policy on School Meals and Welfare Foods, 1939–70', in David F. Smith (ed.), *Nutrition in Britain: Science, Scientists and Politics in the Twentieth Century* (London, 1997), pp. 190–213.
45. Royal Army Service Corps, *The Story of the Royal Army Service Corps, 1939–1945* (London, 1955).
46. Mass-Observation, *War Factory* (London, 1987; first publ. 1943), pp. 73–6.
47. It is not mentioned in Calder, *People's War*, or in Zweiniger-Bargielowska, *Austerity in Britain*, nor in the official history on food (R. J. Hammond, *Food*, 3 vols. (London, 1951–62)).
48. Royal Army Service Corps, *The Story of the Royal Army Service Corps*, pp. 463–8.
49. Figures from the National Food Survey give around 8 ounces per week of beef, 5–6 ounces of lamb, and 4–5 ounces of pork, bacon or ham. It should be noted that wartime beef consumption was at the level of beef consumption in 1990, that lamb consumption was much higher during the war, and that only pork, bacon and ham consumption was much

lower. The really big change in meat consumption compared with the war was the increase in poultry to second position from a negligible position in the 1940s: David Buss, 'The Changing Household Diet', in J. M. Slater, *Fifty Years of the National Food Survey 1940–1990* (London, 1991), pp. 47–54, particularly Table 5.4, p. 51.

50. Royal Army Service Corps, *The Story of the Royal Army Service Corps*, p. 106.

51. War Office, *Handbook of Mobile Refrigerating Equipment 1945* (London, 1945).

52. Royal Army Service Corps, *The Story of the Royal Army Service Corps*, pp. 106–8.

53. Ibid., pp. 469–74.

54. Hans Krebs, 'Feeding the People', lecture, 5 September 1942, p. 8, Krebs Papers, H 114, University of Sheffield.

55. This is major theme of Mancur Olson Jr., *The Economics of the Wartime Shortage: A History of British Food Supplies in the Napoleonic Wars and in World Wars I and II* (Durham, NC, 1963). This excellent study does not, however, sufficiently note the significance of continuing imports and overplays the dangers of sea transport. Alan S. Milward, *War, Economy and Society 1939–1945* (London, 1977), while exceedingly perceptive, also underplays the continued significance of imports.

56. See Ministry of Food cinema, *Food Flash*, Imperial War Museum video.

57. World Wide Films for Ministry of Information, *Rationing in Britain* (1944), written and spoken by Mary Welsh. Historians report from official figures that food imports fell, according to Calder to 'less than half their pre-war level' (Calder, *The People's War*, p. 276); another that 'food imports were halved from an average of 22 million tons before the war to between 10.6 million and 11.5 million tons between 1942 and 1944. A combination of increased home production, conservation, regulation of distribution, and transformation of dietary patterns prevented starvation': Zweiniger-Bargielowska, *Austerity in Britain*, p. 36. See also Juliet Gardiner, *Wartime: Britain 1939–1945* (London, 2004), p. 144, and Alan Milward, who stresses the great increases in domestic production and provides an even more dramatic picture of cuts in his superb global account of food in the Second World War: Milward, *War, Economy and Society 1939–1945*, Chapter 8, 'War, Agriculture and Food', p. 253. Wilt notes the halving of food but says that one reason was the fall in animal feed, though he does not state that this was by far the most important reason: Alan F. Wilt, *Food for War: Agriculture and Rearmament in Britain before the Second World War* (Oxford, 2001), p. 224.

58. CSO, *Statistical Digest of the War*, Table 147.

59. Scientific Adviser's Division and Supply Plans Division, Ministry of Food, 'Survey of War-time Nutritional Administration in the United Kingdom', June 1943, p. 24, MAF 98/46hg, The National Archives.

60. Calculated from CSO, *Statistical Digest of the War*, Table 147.

61. Ministry of Food, lectures on the administration of food control, rationing and distribution, May/June 1944 (arranged by the British Council for Allied governments), lecture by Mr Harold Jones, 1 June 1944, British Library of Political Science (London School of Economics), Library 42 (f27).

62. Bowen, *The Flag of the Southern Cross*, p. 9.

63. Ibid., p. 10.

64. Ibid., p. 9.

65. CSO, *Annual Abstract of Statistics*, 1937–1947, Table 179.

66. Imperial War Museum, MGH 4137.

67. Anon., *Science in War* (Harmondsworth, 1940), p. 80.

68. Ministry of Information, *Land at War: The Official Story of British Farming, 1939–1944* (London, 1945; facsimile publ. by HMSO, 2001), pp. 7–8.

69. This is a major theme of Olson, *The Economics of the Wartime Shortage*, p. 125, citing Keith A. Murray, *Agriculture* (London, 1955), p. 242.

70. Recent calculations show a highly negative total factor of productivity growth in British agriculture during the war; that is, output increased much less than the inputs of capital, labour and land. However, the measure of output used was the financial value of output, which rose only 8 per cent; the nutritional value, which was what was being aimed at, subject to constraints, rose much more than this. There were hugely important changes in the nature of crops, and of total output, but this should not be taken to imply overall change in the technique of agriculture. See Paul Brassley, 'Wartime Productivity and Innovation, 1939–1945', in Brian Short, Charles Watkins and John Martin (eds.), *The Front Line of Freedom: British Farming in the Second World War* (The Agricultural History Review Supplement Series 4) (London, 2006), pp. 16–54. For a parallel discussion of an earlier correction of over-optimistic assessments of British agriculture which spells out the large increase in productivity if measured in calories produced per agricultural worker, see Stephen Broadberry and Peter Howlett, 'The United Kingdom: "Victory at all costs"', in Mark Harrison (ed.), *The Economics of World War II: Six Powers in International Comparison* (Cambridge, 1998), pp. 43–80, particularly Table 2.16, p. 63.

71. CSO, *Statistical Digest of the War*, Table 4.6.

72. Ibid., Table 5.1.

73. Estimated from Peter Dewey, 'The Supply of Tractors in Wartime', in Short, Watkins and Martin (eds.), *The Front Line of Freedom*, pp. 89–100, particularly p. 95.

74. Ibid. The dominance of the Fordson is also clear from wartime surveys: see Brian Short et al., *The National Farm Survey* (Oxford, 2000), p. 186.

75. Ibid., p. 181.

76. Abigail Woods, 'The Farm as Clinic: Veterinary Expertise and the Transformation of Dairy Farming, 1930–1950', *Studies in the History and Philosophy of the Biological & Biomedical Sciences*, 38 (2007), pp. 462–87. See also, on artificial insemination, essentially a post-war phenomenon: Sarah Wilmot, 'From "Public Service" to Artificial Insemination: Animal Breeding Science and Reproductive Research in Early Twentieth-Century Britain', ibid., pp. 411–41.

77. See Hammond, *Food*, Vol. III: *Studies in Administration and Control* (1962).

78. Ministry of Food, lectures on the administration of food control, rationing and distribution, May/June 1944 (arranged by the British Council for Allied governments), lecture by Mr Harold Jones, 1 June 1944, LSE Library 42 (f27).

79. 'Before 1939 there were in use in England and Wales some 11,500 slaughterhouses. By 1953, under the control scheme, these had been reduced to 482, of which 119 are public and 358 are privately owned slaughterhouses and 5 were built by the Government after the 1950s': *Slaughterhouses (England and Wales): Interim Report 1953/4*, Cmd 9060 (London, 1954).

80. Richard Blundel and Angela Tregear, 'From Artisans to "Factories": The Interpenetration of Craft and Industry in English Cheese-Making, 1650–1950', *Enterprise and Society* 7 (2006), pp. 728–31.

81. Hammond, *Food*, Vol. III: *Studies in Administration and Control*, pp. 588–9.

82. Royal Army Service Corps, *The Story of the Royal Army Service Corps*, p. 106.

83. E. M. H. Lloyd, *Food and Inflation in the Middle East 1940–1945* (Stanford, Calif., 1956).

84. E. H. Carr, *Nationalism and After* (London, 1945), p. 48.

85. Before the war, Middle East meant Arabia, Iraq, Iran and Afghanistan; further west came the Near East, which often included Greece and Bulgaria as well as the Levant and Egypt.

86. Benedict Anderson, *The Spectre of Comparisons: Nationalism, Southeast Asia and the World* (London, 1998), p. 3. It was first based in New

Delhi but from April 1944 was in Kandy, Ceylon. The main area of operation was Burma and points south and later east.

87. Krebs, 'Feeding the People'.

88. However, the Ministry of Health was advised on nutritional matters by Edward Mellanby of the Medical Research Council: see David F. Smith, 'Nutrition Science and the World Wars', in Smith (ed.), *Nutrition in Britain*, pp. 142–65. This article has no information on food policy and its relation to nutrition science. F. G. Young, 'Jack Cecil Drummond. 1891–1952', *Obituary Notices of Fellows of the Royal Society* 9 (November 1954), pp. 98–129.

89. I. Dennehy, J. C. Drummond and A. N. Duckham, 'A Survey of Wartime Nutrition with Special Reference to Home Production of Foods and Import Policy', 12 May 1940, MAF 98/46, The National Archives.

90. Scientific Adviser's Division and Supply Plans Division, Ministry of Food, 'Survey of War-time Nutritional Administration in the United Kingdom', June 1943, p. 3, MAF 98/46, The National Archives.

91. Hammond, *Food*, Vol. III: *Studies in Administration and Control*, p. 591.

92. Hancock and Gowing, *British War Economy*, p. 423.

93. Hammond, *Food*, Vol. III: *Studies in Administration and Control*, pp. 591–624.

94. Anon., *Science in War*, p. 68.

95. Krebs, 'Feeding the People'.

96. Hancock and Gowing, *British War Economy*, p. 423.

97. For example, in Scientific Adviser's Division and Supply Plans Division, Ministry of Food, 'Survey of War-time Nutritional Administration in the United Kingdom', June 1943, MAF 98/46, The National Archives.

98. Lord Woolton, HL Deb. (series 5), vol. 118, cols. 383–4 (18 February 1941).

99. Hans Krebs to Prof. H. Raistrick, reporting conversation with John Maud, Krebs Papers, B 34, University of Sheffield.

100. Hammond, *Food*, Vol. III: *Studies in Administration and Control*, p. 596.

101. *Report of the Medical Research Council for the Years 1939–45*, Cmd 7335 (London, 1947–8). See also Uwe Spiekermann, 'Brown Bread for Victory: German and British Wholemeal Politics in the Inter-War Period', in Frank Trentmann and Flemming Just (eds.), *Food and Conflict in Europe in the Age of the Two World Wars* (London, 2006), pp. 143–71; Margaret Ashwell, 'Elsie May Widdowson, C.H. 21 October 1906–14 June 2000', *Biographical Memoirs of Fellows of the Royal Society* 48 (December 2002), pp. 483–506, on p. 493. The experiments

were conducted on volunteers by McCance and Widdowson. With their advice similar experiments were conducted in 1941 on volunteer conscientious objectors in Sheffield under the direction of Hans Krebs, who worked on 85 per cent extraction, whereas McCance and Widdowson had done 92 per cent: Hans Krebs, *Reminiscences and Reflections* (Oxford, 1981), pp. 119–20; Kenneth Mellanby, *Human Guinea Pigs* (London, 1945), p. 40.

102. Young, 'Jack Cecil Drummond'.

103. *The Economist* (16 March 1940), p. 485.

104. *The Economist* (1 April 1939), p. 38. In 1938 Mexico, a very major producer, had nationalized its oil industry, dominated by the Mexican Eagle company, owned by Shell, and as a result Britain, the USA and the Netherlands boycotted its oil until 1940.

105. D. J. Payton-Smith, *Oil: A Study of War-Time Policy and Administration* (London, 1971), pp. 260, 392.

106. Sir Charles Kingsley Webster and Noble Frankland, *The Strategic Air Offensive against Germany 1939–1945*, Vol. IV: *Annexes and Appendices* (London, 1961), Appendix 49, Table xxxiv.

107. Lindemann to Churchill, PM 47, 20 May 1942, The Papers of Sir Donald MacDougall, MACD 25, Churchill Archives Centre.

108. Payton-Smith, *Oil*, Tables 52 and 45, p. 453.

109. Ibid., p. 57. See also Gavin Bailey, 'The Narrow Margin of Criticality: The Question of the Supply of 100-Octane Fuel in the Battle of Britain', *English Historical Review* CXXIII (April 2008), pp. 394–411.

110. Payton-Smith, *Oil*, pp. 55, 93–4.

111. On Thornton, see ibid., pp. 273–7.

112. Ibid., p. 279.

113. Allen, 'Mutual Aid between the US and the British Empire', Table 8.

114. Ibid., Table 11. On p. 262 Allen makes clear that a good proportion of the petroleum that came in on Lend-Lease went back in reciprocal aid, pointing out that the pot is much the same and that Lend-Lease figures are gross not net.

115. Payton-Smith, *Oil*, p. 384.

116. John W. Frey and H. Chandler Ide (eds.), *A History of the Petroleum Administration for War 1941–1945* (Washington DC, 2005; reprint of 1946 original), p. 208.

117. Kendall Beaton, *Enterprise in Oil: A History of Shell in the United States* (New York, 1957), pp. 567–8. See also R. W. Ferrier and J. H. Bamberg, *The History of the British Petroleum Company*, Vol. 2: *The Anglo-Iranian Years 1928–1954* (Cambridge, 1994), pp. 198–205.

118. Payton-Smith, *Oil*, p. 271. Ferrier and Bamberg, *The History of the British Petroleum Company*, Vol. 2: *The Anglo-Iranian Years*, p. 246.

119. Beaton, *Enterprise in Oil*, pp. 535, 561–4.

120. Payton-Smith, *Oil*, p. 273.

121. Archibald Sinclair, Secretary of State for Air, '100 Octane Fuel: Completion of Thornton Plant', WP(G)(40)292, 5 November 1940, CAB 67/8/92, The National Archives.

122. Payton-Smith, *Oil*, pp. 55–7.

123. My reading of document on 'Government Fuel Plants at Heysham and Thornton', December 1940, in The Papers of Harold Hartley–Trimpell, HART T 3, Churchill Archives Centre.

124. Note of a meeting, Air Ministry, 18 August 1939 (Tedder in chair), Hartley–Trimpell Papers, HART T 9, Churchill Archives Centre.

125. Payton-Smith, *Oil*, p. 277.

126. J. D. Rose, 'Ronald Holroyd. 1904–1973', *Biographical Memoirs of Fellows of the Royal Society* 20 (December 1974), pp. 240, 241.

127. Notes of a meeting held in Sir Arthur Street's room at the Air Ministry, Thursday, 15 December 1938, Hartley–Trimpell Papers, HART T 4, Churchill Archives Centre.

128. 'Completion of the Thornton Plant', memorandum from the Minister of Aircraft Production, 8 April 1941, War Cabinet, WP (41) 80, CAB 66/16/3, The National Archives.

129. Payton-Smith, *Oil*, pp. 384–5.

130. See F. Godber to Sir Harold Hartley, 2 May 1941, with note to Godber from Oriel, 30 April 1941, and Godber to Hartley, 4 July 1941, Hartley–Trimpell Papers, HART T 1D, Churchill Archives Centre.

131. Notes of a meeting, at the Air Ministry, to discuss the future of Heysham, 6 August 1941, Hartley–Trimpell Papers, HART T 3, Churchill Archives Centre.

132. Trimpell Ltd, *Annual Report 1944*, 26 January 1945, Hartley–Trimpell Papers, HART T 6A, Churchill Archives Centre.

133. Kenneth Gordon, 'Progress in the Hydrogenation of Coal and Tar', *Journal of the Institute of Fuel* 20 (1946), pp. 42–58.

134. There were also other methods of getting petrol from coal, such as the recovery of 'benzole' by 'scrubbing' the output of coke-ovens (coke was made by heating coal), which produced about the same quantity as fuel hydrogenation in Britain: Anthony S. Stranges, 'From Birmingham to Billingham: Synthetic Fuels in Great Britain, 1910–1945', *Technology and Culture* 26 (1985), pp. 726–57.

135. This calculation is a little generous to the tanker – it implies roughly one

round trip to the refineries per month, which was only just possible, every month of the year. Payton-Smith, *Oil*, pp. 22–4. This was the last of the civil official histories to appear.

136. Ibid., pp. 22–4.

137. *The Economist* (12 February 1938), p. 346.

138. Rose, 'Ronald Holroyd. 1904–1973', pp. 235–45.

139. Captain Bernard Acworth, 'Sea Power Surrendered', *English Review* LX (1935), pp. 195–200.

140. *British Union Quarterly* 2 (April–June 1938), p. 14.

141. Payton-Smith, *Oil*, pp. 93–4.

142. CSO, *Annual Abstract of Statistics*, 1937–1947, Table 202.

143. Alan S. Milward and George Brennan, *Britain's Place in the World: A Historical Enquiry into Import Controls, 1945–60* (London, 1996), pp. 190–94.

144. Duncan Burn, 'The Oil Industry', in Duncan Burn (ed.), *The Structure of British Industry*, Vol. 1 (Cambridge, 1958), p. 185.

145. *United Harvest*, Greenpark Production (1947). *The Balance*, directed by Paul Rotha, Central Office of Information (1947). In relation to exports and imports it dealt with New Zealand, South America, Canada, West Africa and Sweden. Canada was as foreign as Sweden. *Family Portrait*, written and directed by Humphrey Jennings (1951). The film was made for the Festival of Britain, 1951. The themes were the unity of old and new, poetry and prose, science and literature.

7. WORLDS OF WAR

1. Frank H. House, *Timber at War: An Account of the Organisation and Activities of the Timber Control 1939–1945* (London, 1965), pp. 48–51. The project was cancelled in June 1944.

2. Ministry of Information, *Land at War: The Official Story of British Farming, 1939–1944* (London, 1945; facsimile published by HMSO, 2001), p. 15.

3. Ibid., pp. 23, 50.

4. This calculation excludes producers of raw materials and food, but includes construction workers: David Edgerton, *Warfare State: Britain, 1920–1970* (Cambridge, 2005), p. 75.

5. The air force was the most feminine of the services, approaching 20 per cent women, the navy next, followed by the army: H. M. D. Parker, *Manpower: A Study of War-Time Policy and Administration* (London, 1957), Table 26.

6. Tony Lane, *The Merchant Seamen's War* (Liverpool, 1990), is an example of the historiography which criticizes the concept of the 'People's War'.

7. Philip Ollerenshaw, 'War, Industrial Mobilisation and Society in Northern Ireland, 1939–1945', *Contemporary European History* 16 (2007), pp. 169–97, particularly p. 181.

8. Ibid.

9. J. T. Murphy, *Victory Production!* (London, 1942), pp. 138–9.

10. *They Fight by Night* (1943), on Imperial War Museum video, about bombers and factories and trains, etc., at night.

11. Christian Wolmar, *Fire and Steam: A New History of the Railways in Britain* (London, 2007), Chapter 13.

12. Her first trip was from the Tyne, on 15 November 1941, in ballast, arriving at St John, Newfoundland. She left St John on 24 December 1941 carrying over 8,000 tons of grain. She arrived in Sydney, Nova Scotia, on the 27th and sailed on to Halifax on 3 January 1942, arriving on the 5th. On leaving Halifax she was forced to put back because of heavy weather, sailing again on 10 February, arriving at Loch Ewe on 26 February. She then sailed to London. There she underwent repairs and sailed for the Tyne on 14 March, arriving two days later. After more repairs she sailed from the Tyne on 29 March to Halifax via Loch Ewe, arriving at Halifax on 18 April, and sailed from Halifax on 30 April (again with grain). She arrived in Liverpool on 16 May. She was repaired again and sailed on the 26 May, crossing to New York, where she arrived on 15 June. She sailed to Tampa, Florida, arriving on 1 July, and sailed on 8 July with 8,943 tons of phosphate concentrate and 116 tons of landing craft. She sailed north, arriving in Halifax on 3 August. More repairs followed, and she sailed on 11 August, arriving in Glasgow on the 28th. Yet more repairs followed and she sailed for West Africa on 29 August with aircraft, with many stops. She arrived in Lagos on 21 November. She left on 28 November, crossing to Guantanamo Bay in Cuba, going on to New York and Loch Ewe. She arrived in Hull on 16 February 1943 with an unknown cargo and was then transferred to the Greek flag as the *Kyclades*: BT 389/11, The National Archives.

13. Central Statistical Office, *Statistical Digest of the War* (London, 1951), Table 34, p. 32.

14. Central Statistical Office, *Annual Abstract of Statistics*, No. 84, 1935–1946 (London, 1948), Table 183, p. 156.

15. Sir Arthur T. Harris, *Despatch on War Operations: 23rd February 1942 to 8th May 1945* (London, 1995), Appendix F, para 54.

16. Ibid., para 52.

17. F. E. Nancarrow, 'The Telecommunications Network for Defence, Part I: The Defence Teleprinter Network', *Post Office Electrical Engineers' Journal* 38 (January 1946; Victory number), p. 122.

18. H. R. Harbottle, 'The Telecommunications Network for Defence, Part II: The Network of Telephone Circuits and the Defence Telecommunications Control', *Post Office Electrical Engineers' Journal* 38 (January 1946; Victory number), p. 125.

19. CSO, *Annual Abstract of Statistics*, No. 84, 1935–1946 (London, 1948), Table 183, p. 156. The figures refer to internal UK building and civil engineering only.

20. Edgerton, *Warfare State*, p. 77.

21. W. Hornby, *Factories and Plant* (London, 1958), pp. 240–42.

22. Ibid., pp. 262–3.

23. Michael Nevell, John Roberts and Jack Smith, *A History of the Royal Ordnance Factory Chorley* (Lancaster, 1999), pp. 24–5.

24. Ibid., p. 60.

25. Ibid., p. 13.

26. Hornby, *Factories and Plant*, p. 101.

27. Ibid., p. 118.

28. http://www.subbrit.org.uk/rsg/sites/r/rhydymwyn/history.html; accessed 28 April 2010.

29. Hornby, *Factories and Plant*, p. 159.

30. Tim Houghton, 'Chemical Offensive: British Chemical Warfare Research, 1914–1945', unpublished M.Sc. dissertation, Imperial College London (2004).

31. War Cabinet, WP (42) 171, report by chief of staff on chemical warfare, CAB 66/24/1.

32. Hornby, *Factories and Plant*, pp. 262–3.

33. Ibid., p. 253.

34. L.X.V., 'Ford Motor Company ... produced 45,000,000 h.p. of Rolls-Royce Merlin Engines ...', *Ford Times* (November–December 1945).

35. J. C. Crellin, 'Ford in Manchester', undergraduate dissertation, University of Manchester (1987). Based on extensive interviews as well as other sources.

36. Ford War Record, Imperial War Museum film collection, FWR 25/01-08.

37. Imperial War Museum film collection, FWR 25/02. Part 2 of an eight-part/eight-reel film record of the production processes in the Manchester Merlin factory.

38. Merlins were produced on an even larger scale by the Packard Company in Detroit, Michigan, for the RAF, and for the USAAF, whose

Mustang fighters were powered by Merlins. Rolls-Royce itself was producing around 60 per cent of the British production of its engines, from its original plant in Derby, but also from large new government-owned factories in Crewe and Glasgow (Hillingdon). Hornby, *Factories and Plant*, pp. 256, 258.

39. At the end of the war two-thirds of Bristol engines came from car firms operating shadow factories. The first Bristol shadow group, in and around Coventry, were making Pegasus and Mercury engines in new factories well before the war. A second group of shadow factories were built to make the sleeve-valved Hercules which came into production during the war.

40. J. E. Embrey, 'Aeroengine Production: Expansion of Capacity 1935–1945', CAB 102/53, The National Archives.

41. W. C. Hornby, 'Aircraft Production: Expansion of Capacity, 1935–1945', p. 99, CAB 102/54, The National Archives.

42. Ibid.

43. John Dummelow, *1899–1949* (history of Metropolitan-Vickers) (Manchester, 1949), pp. 166, 176.

44. Cecil Beaton, introduction to *War Pictures by British Artists*, Second Series: *Production* (London, 1943), p. 6.

45. Ministry of Aircraft Production/MoI, *Speed Up on Stirlings*, UKY 422, Imperial War Museum.

46. William Holt, 'Democracy Marches' (20/21 May 1941), available at http://www.wartimenewport.virtuallyhere.co.uk; accessed 13 April 2010.

47. M. J. Lipman, *Memoirs of a Socialist Business Man* (London, 1980). Lipman remained a socialist and funded the work of socialist intellectuals.

48. Frank Platt, *Report on the Cotton Textile Mission to the United States of America* (London, 1944).

49. Oliver Lyttelton, Minister of Production, HC Deb. (series 5), vol. 381, col. 1108 (14 July 1942).

50. Eric Mensforth, 'Airframe Production', *Proceedings of the Institution of Mechanical Engineers* 156 (1947), pp. 24–38. Hornby, *Factories and Plant*, pp. 34–5.

51. M. M. Postan, *British War Production* (London, 1952), p. 342.

52. Ibid.

53. Ibid., pp. 342–3.

54. Data from I. B. Holley, quoted in Jonathan Zeitlin, 'Flexibility and Mass Production at War: Aircraft Manufacture in Britain, the United States, and Germany, 1939–1945', *Technology and Culture* 36 (January 1995), pp. 46–79, particularly p. 59. It is also clear that British productivity rose substantially during the war: Sebastian Ritchie, 'A New Audit of

War: The Productivity of Britain's Wartime Aircraft Industry Reconsidered', *War & Society* 12 (1994), pp. 125–47.

55. Erik Lund, 'The Industrial History of Strategy: Reevaluating the Wartime Record of the British Aviation Industry in Comparative Perspective, 1919–1945', *Journal of Military History* 66 (January 1998), pp. 75–99, on p. 75; Richard Overy, *The Air War, 1939–1945* (London, 1980); Ritchie, 'A New Audit of War'.

56. H. Duncan Hall, *North American Supply* (London, 1955), p. 179.

57. *The Economist* (3 October 1942), p. 416.

58. Jonathan Zeitlin, 'Rationalization Misconceived: Productive Alternatives and Public Choices in British Shipbuilding, 1940–1990', paper prepared for the annual meeting of the Society for the History of Technology (SHOT), Toronto, 17–20 October 2002. Also British merchant yards used 25 per cent fewer man-hours during the war; contrary to the clichés, US shipbuilding was very labour-intensive: Peter Elphick, *Liberty: The Ships That Won the War* (London, 2001), p. 60. Another estimate was that during the war British merchant ships were built at 60 per cent of the cost of US ships per ton: R. O. Roberts, 'Comparative Shipping and Shipbuilding Costs', *Economica* 14 (November 1947), pp. 296–309, on p. 309.

59. W. H. Mitchell and L. A. Sawyer, *Empire Ships of World War II* (Liverpool, 1965).

60. See the contemporary estimates in Lewis Johnman and Hugh Murphy, *British Shipbuilding and the State since 1918: A Political Economy of Decline* (Exeter, 2002), Table 12, p. 86.

61. http://uboat.net; accessed 13 April 2010.

62. Lane, *The Merchant Seamen's War*, pp. 23–4. Gregory Clark, *A Farewell to Alms: A Brief Economic History of the World* (Princeton, NJ, 2007), Chapters 16 and 17, makes the point that in the case of textiles the same machinery was used with very different quantities of labour in both Britain and India. As in the case of many textile enterprises, the management was the same – British. The perceived and actual inferiority of Eastern workers seems to be the critical variable.

63. *Pathé Gazette Takes Pride in Presenting the First Pictures of New British Battleships*, Pathé Gazette (1942), depicts the commissioning of *Howe* and *Anson*.

64. CSO, *Statistical Digest of the War*, Table 111, p. 133; Table 112, p. 134. The index above somewhat exaggerates the weight of battleships in annual production since this calculation attributes all the work to the year of completion. Battleships took years to make, but that is compensated for at least partially by the fact that work was proceeding on two

battleships completed the following year. Five battleships, all of the same class, were completed during the war, between 1940 and 1942.

65. In the 1930s the Lots Road power station, which supplied London Underground, generated 105MW.

66. Edgerton, *Warfare State*, p. 32.

67. CSO, *Statistical Digest of the War*, Table 112, p. 134.

68. Ibid., Table 131, p. 153.

69. One source gives £42,000 as the purchase cost of a Lancaster in 1943: www.lancaster-archive.com; accessed 13 April 2010.

70. Assuming 5 tons for each Lancaster; and for each battleship heaviest ordnance only on one loading.

71. Norman Franks, *Claims to Fame: The Lancaster* (London, 1995), pp. 8–9.

72. Ibid., p. 87.

73. Sir Charles Kingsley Webster and Noble Frankland, *The Strategic Air Offensive against Germany 1939–1945*, Vol. IV: *Annexes and Appendices* (London, 1961), Appendix 41.

74. Kenneth Philip Werrell, 'At Long Last, a History of the Eighth Air Force', *Air University Review* XXII (January–February 1971), pp. 72–8. Figures are not available in the official histories or standard statistical tables for the number that died.

75. Timothy Harrison Place, *Military Training in the British Army: From Dunkirk to D-Day* (London, 2000), p. 19.

76. Ministry of Information/RAF Film Production Unit, *Journey Together* (1944), Imperial War Museum, film collection, APY 26. The trainees are posted to operational units on return from Canada, missing OTUs and HCUs.

77. Sir Arthur Harris, *Bomber Offensive* (London, 1990; first publ. 1947), pp. 263–5.

78. Harris, *Despatch on War Operations*, Appendix F, paras. 67–8.

79. He returned to Leyland after the war, and became one of the leading figures in the motor industry.

80. Ministry of Information, *Merchantmen at War: The Official Story of the Merchant Navy, 1939–1944* (London, 1944), p. 16.

81. H. M. D. Parker, *Manpower: A Study of War-Time Policy and Administration* (London, 1957), pp. 326–30. C. P. Snow claimed later that he, not Hankey, came up with the radio scheme and the state bursaries. He claimed that Hankey did not get out much and had no imagination: C. P. Snow to S. Roskill, 6 June 1974, The Papers of Lord Hankey of the Chart, HNKY 12/1, Churchill Archives Centre.

82. Parker, *Manpower*, p. 146.

83. Ibid., p. 299.

84. Australian universities expanded teaching and research during the war with medical and dental, and science and engineering, students exempt from conscription: Michelle Freeman, 'Australian Universities at War', in Roy MacLeod (ed.), *Science and the Pacific War: Science and Survival in the Pacific, 1939–1945* (Dordrecht, 2000), pp. 121–38. See also Jagdish N. Sinha, *Science, War and Imperialism: India in the Second World War* (Leiden, 2008), and Donald H. Avery, *The Science of War: Canadian Scientists and Allied Military Technology during the Second World War* (Toronto, 1998).

85. Male arts students were able to start studies when conscription began at twenty, but when it was reduced to eighteen and a half, a special one-year deferment was usually granted; but this stopped in 1942 except for cases of postponement when students had an exam less than nine months from their call-up date: Parker, *Manpower*, pp. 313–14.

86. Ibid., p. 328.

87. CSO, *Annual Abstract of Statistics*, No. 84, 1935–1946 (London, 1948), Table 120, p. 100.

88. Parker, *Manpower*, pp. 322–3.

89. Graham R. Fleming and David Phillips, 'George Porter KT OM, Lord Porter of Luddenham. 6 December 1920–31', *Biographical Memoirs of Fellows of the Royal Society* 50 (2004), pp. 257–83.

90. B. H. Liddell Hart, *History of the Second World War* (London, 1973; first publ. 1970), noted the problems in advancing in Normandy caused by the 'lavishness of the British and American scales of supply', 700 daily tons for a British division, against 200 for a German (p. 590). A. D. Harvey, *Collision of Empires: Britain in Three World Wars, 1793–1945* (London, 1994; first publ. 1992), also made much of the lavishness of supply for British forces (pp. 560–63).

91. Basil Liddell Hart, *The Tanks: The History of the Royal Tank Regiment and Its Predecessors, Heavy Branch Machine-Gun Corps, Tank Corps and Royal Tank Corps, 1914–1945*, Vol. 2 (London, 1959), pp. 92–8, 154–6; Liddell Hart, *History of the Second World War*, pp. 280ff.

92. Major L. F. Ellis, *Victory in the West*, Vol. 1: *The Battle of Normandy* (London, 1962), p. 229.

93. I. S. O. Playfair, *The Mediterranean and the Middle East*, Vol. 3: *British Fortunes Reach Their Lowest Ebb* (London, 1960), pp. 29–31.

94. Ibid., p. 430.

95. I. S. O. Playfair and C. J. C. Molony, *The Mediterranean and the Middle East*, Vol. 4: *The Destruction of the Axis Forces in Africa* (London, 1966), p. 9.

96. David Fletcher, *The Great Tank Scandal: British Armour in the Second World War*, Part I (London 1989); David Fletcher, *The Universal Tank: British Armour in the Second World War*, Part II (London, 1993); and Peter Beale, *Death by Design: British Tank Development in the Second World War* (Stroud, 1998), all point to serious deficiencies and multiplication of types of British tank, and to successful British tanks at the end of the war, but the lack of proper quantitative data on tanks makes constructing a picture of what tanks were available where and when impossible to do.

97. I. S. O. Playfair, *The Mediterranean and the Middle East*, Vol. 1: *The Early Successes against Italy* (London, 1954), p. 342.

98. Pier Paolo Battistelli, *Rommel's Afrika Korps: Tobruk to El Alamein* (Oxford, 2006), pp. 64–5. See also the detailed analysis of gun–tank relations at http://www.wwiiequipment.com/pencalc; accessed 13 April 2010.

99. I. S. O. Playfair, *The Mediterranean and the Middle East*, Vol. 2: *The Germans Come to the Help of Their Ally* (London, 1956), pp. 341–4.

100. Ibid., p. 175; Playfair and Molony, *Mediterranean and the Middle East*, Vol. 4: *Destruction of the Axis Forces*, pp. 8–9.

101. Playfair, *Mediterranean and the Middle East*, Vol. 3: *Their Lowest Ebb*, p. 430.

102. Ibid., Vol. 2: *The Germans Come*, p. 175; Playfair and Molony, *Mediterranean and the Middle East*, Vol. 4: *Destruction of the Axis Forces*, p. 9.

103. Playfair, *Mediterranean and the Middle East*, Vol. 2: *The Germans Come*, p. 175; Playfair and Molony, *Mediterranean and the Middle East*, Vol. 4: *Destruction of the Axis Forces*, pp. 8–9.

104. For the gun comparison, see Ellis, *Victory in the West*, Vol. 1: *The Battle of Normandy*, p. 549.

105. Playfair and Molony, *Mediterranean and the Middle East*, Vol. 4: *Destruction of the Axis Forces*, p. 8.

106. J. P. Harris, *Men, Ideas and Tanks: British Military Thought and Armoured Forces, 1903–1939* (Manchester, 1995).

107. John Ellis, *The World War II Databook: The Essential Facts and Figures for All the Combatants* (London, 1993), pp. 204–20.

108. Place, *Military Training in the British Army*, pp. 97–101.

109. John Buckley, *British Armour in the Normandy Campaign, 1944* (London, 2004), p. 120.

110. Ibid., pp. 118, 120.

111. Ellis, *The World War II Databook*, p. 125.

112. Stephen A. Hart, *Sherman Firefly vs Tiger, Normandy 1944* (Oxford, 2007), p. 5. It was superior to the long 88mm mounted on the King

Tiger as well as the other 88mm tank guns. The 128mm on the Jagd-Tiger tank destroyer, produced in tiny numbers, was the only superior German gun: Ellis, *Victory in the West*, Vol. 1: *The Battle of Normandy*, p. 549.

113. Hart, *Sherman Firefly vs Tiger*, p. 25.

114. Ibid., p. 24.

115. Ellis, *Victory in the West*, Vol. 1: *The Battle of Normandy*, p. 547.

116. Peter Chamberlain and Chris Ellis, *British and American Tanks of World War Two: The Complete Illustrated History of British, American and Commonwealth Tanks, 1939–45* (London, 2000), p. 207.

117. Ibid., pp. 44, 67, 115.

118. Ibid., pp. 202–3.

119. See Patrick Delaforce, *Churchill's Secret Weapons: The Story of Hobart's Funnies* (London, 1998).

120. See Buckley, *British Armour in the Normandy Campaign*, for a detailed discussion of the arguments of historians and a rebuttal in detail. The argument was continuous from the end of the war. A particularly interesting case was the military historian Cyril Falls, whose *The Second World War: A Short History* (London, 1948), unusually notes that British tank production was well ahead of German especially given the relative size of the armies, but he insists on the lower quality of British tanks and also underplays British quantitative superiority in North Africa (see pp. 120, 127).

121. For the standard view, see David French, *Raising Churchill's Army: The British Army and the War against Germany 1919–1945* (Oxford, 2000), Chapter 3.

122. Webster and Frankland, *Strategic Air Offensive*, Vol. IV: *Annexes and Appendices*, Appendix 49, Table iii(a).

123. CSO, *Statistical Digest of the War*, Table 122; Webster and Frankland, *Strategic Air Offensive*, Vol. IV: *Annexes and Appendices*, Appendix 49, Table iii(a).

124. Fredric Boyce and Douglas Everett, *SOE: The Scientific Secrets* (Stroud, 2003), p. 276.

125. See the test at http://www.youtube.com/watch?v=V591FhD_y30, which suggests that the MP40 was only 'marginally better' than the Sten; accessed 14 April 2010.

126. Mark Harrison, *Medicine and Victory: British Military Medicine in the Second World War* (Oxford, 2004), p. 24.

127. Ibid., p. 35.

128. *Story of DDT: Dichloro Diphenyl Trichloroethane* (1944). Film: CVN 238, Imperial War Museum.

129. Norman Lewis, *Naples '44: An Intelligence Officer in the Italian Laby-rinth* (London, 2002; first publ. 1978), p. 175. The reference is probably to the shipyard Navalmeccanica in Castellammare di Stabia.

130. Calculated from Valerie Holman, *Print for Victory: Book Publishing in England 1939–1945* (London, 2009), Appendix 8, comparing 1944 with 1939.

131. Ibid., Appendix 10.

132. One economist compared Washington to an imperial posting. The weather was like that of Bombay, and the ubiquitous 'negro servants' were an 'inferior race of bearers and carriers' to those of New Delhi: letter from Austin Robinson, 21/10/44, in Prue Kerr, 'An Account of the Washington Lend-Lease Negotiations 1944 by E. A. G. Robinson', *Contributions to Political Economy* 22 (2003), pp. 63–78. The collection of letters gives a vivid picture of British officials in Washington.

133. Philip Ziegler, *Mountbatten: The Official Biography* (London, 1985), pp. 279, 280.

134. F. R. Banks, *I Kept No Diary: 60 years with Marine Diesels, Automobile and Aero Engines* (Shrewsbury, 1978). The second volume of Commander Sir Stephen King-Hall's unpublished memoirs contains a vivid chapter on his time at the MAP under Beaverbrook. King-Hall was then a Member of Parliament and worked as director of the Factory Defence Section. I am grateful to the late Ann King-Hall.

135. Anthony Furse, *Wilfrid Freeman: The Genius behind Allied Survival and Air Supremacy* (Staplehurst, 2000).

136. Harold Macmillan, *The Blast of War: 1939–1945* (London, 1967), pp. 89–90.

137. Ibid., p. 98.

138. J. M. Reid, *James Lithgow, Master of Work* (London, 1964), p. 212.

139. Sir Harold Emmerson, *The Ministry of Works* (London, 1956), p. 20. Despite this, Sir Harold listed all his predecessors and successors as permanent secretary, but not the Directors-General.

140. The only cases I have come across are the appointment of J. W. Bowen to the Air Ministry panel before the war, and a G. Thompson who was a member of the Tank Board: see M. M. Postan, D. Hay and J. D. Scott, *Design and Development of Weapons: Studies in Government and Industrial Organisation* (London, 1964), pp. 331–3. Both roles were essentially advisory.

141. In 1950 P. Sargant Florence was to use the term 'realistic economics' to denote an economics based on observations of economic life and whose method was research. This he contrasted with the economics practised

in Oxford and Cambridge, taught by people without research degrees, and not based on economic research: P. Sargant Florence, 'Patterns in Recent Social Research', *The British Journal of Sociology* 1 (September 1950), pp. 221–39, particularly p. 223; Ely Devons, *Planning in Practice: Essays in Aircraft Planning in Wartime* (Cambridge, 1950).

142. They had set up the National Institute of Economic and Social Research in London, like the Manchester unit a pioneer of applied economic research, and took the NIESR team into the ministry in 1939.

143. Angus Calder, *The People's War: Britain 1939–1945* (London, 1969), p. 18.

8. BOFFINS

1. Report of meeting, 'Use of Scientific Personnel: discussion at a meeting of the Parliamentary and Scientific Committee, House of Commons, 24 March 1942, The Papers of A. V. Hill, AVHL I 2/2, Churchill Archives Centre.

2. Ibid.

3. The most obvious thing that used to be thought about the place of science in British culture was that it was subordinate to the literary and socially inferior. Although in this as in much else he was being original, the great propagator of this myth was C. P. Snow. A former research scientist and don and novelist, his own career refuted his arguments. Not only did he (and many others) bridge his own 'two cultures' (he is joined in the list by his own assistant, Harry Hoff, better known as the novelist William Cooper), but he made his career in government as a scientist and expert on science. During the war he was at the Ministry of Labour, dealing with scientific and technical personnel. He was senior enough to get a CBE.

4. 'Research and Development of Aircraft and Aero-Engines', memorandum by the Minister of Aircraft Production, 3 March 1941, War Cabinet, WP (41) 49, CAB 66/15/22, The National Archives.

5. M. M. Postan, D. Hay and J. D. Scott, *Design and Development of Weapons: Studies in Government and Industrial Organisation* (London, 1964), pp. 6–9. But see G. P. Bulman, *An Account of Partnership – Industry, Government and the Aero Engine: The Memoirs of George Purvis Bulman*, Rolls-Royce Heritage Trust, Historical Series No. 31 (Derby, 2002), pp. 269–73, 366.

6. Details in specification dated 19 July 1940, Vickers Papers, 780, Cambridge University Library.

7. Wallis to Alexander Dunbar, 31 July 1940, ibid.
8. Ibid., 21 July 1940.
9. Alexander Dunbar to Lord Beaverbrook, 1 November 1940, ibid.
10. Wallis to Alexander Dunbar, 31 July 1940, ibid.
11. Sir Charles Kingsley Webster and Noble Frankland, *The Strategic Air Offensive against Germany 1939–1945*, Vol. IV: *Annexes and Appendices* (London, 1961), p. 33.
12. http://www.bomberhistory.co.uk/Viaduct; accessed 14 April 2010. The bypass is not mentioned in the official history.
13. Postan, Hay and Scott, *Design and Development of Weapons*, p. 194.
14. Ibid., p. 197.
15. Ibid., pp. 199–203.
16. Ibid., p. 100.
17. Ibid., p. 213.
18. Andrew Nahum, *Frank Whittle: Invention of the Jet* (Thriplow, 2004).
19. H. E. Wimperis, 'Research and Development in Aeronautics', 31st Thomas Hawksley Lecture, November 1944, *Proceedings of the Institution of Mechanical Engineers* 152 (1945), pp. 353–61, particularly p. 358.
20. Engineering apprentices, typically from local grammar schools, had to pass an exam in Maths, English, Sciences and Latin, French or German; while apprentices they were expected to take courses leading to HNC qualifications at technical colleges. They worked first in the tool room and drawing office, and later on the shopfloor. There were also a very few student apprentices, who were graduates in engineering: Graham Birchmore and Roy Burges, *The Lads of Enfield Lock: 172 Years of Apprentice Training at the Royal Small Arms Factory, Enfield, Middlesex, England, 1816–1988* (Enfield, 2005), p. 28. A case in point was David Luxton, who became an engineering apprentice at Enfield aged seventeen. He spent two days a week in college studying and taking evening classes in French and English three nights a week. He later took a higher qualification at evening classes at Regent Street Polytechnic in central London, and after a year on the shopfloor qualified as a member of the Institution of Mechanical Engineers. At Enfield he was responsible for making drawings with his team of ten for the tools and gauges for the Bren machine gun. In 1937 he went as a technical assistant to the new ordnance factory at Nottingham, which was to make the 3.7in AA gun; in 1940 he went as an assistant manager to a new gun factory in Newport; in 1941 he went to Fazakerley to sort out Sten production and later became manager there. He went on to become assistant director of Woolwich, and director of Nottingham, retiring with a CBE in 1973: ibid., pp. 65–7.

21. This was, it seems, common: Eustace Smith, who was to inherit the ship-builder Smiths Dock of Teesside, studied engineering at Leeds in the early 1920s, while Denis Rebbeck, son of Sir Frederick Rebbeck of Harland and Wolff, Belfast, studied engineering at Cambridge in the early 1930s. The son was to be the first graduate on the board. Furthermore many Oxbridge graduates started joining the emergent corporations like Unilever, Shell, ICI and so on as employees.

22. Peter Elphick, *Liberty: The Ships That Won the War* (London, 2001), pp. 78–9.

23. Imperial Chemical Industries, *Pharmaceutical Research in ICI, 1936–1957* (London, 1957), p. 3.

24. David Greenwood, 'Conflicts of Interest: The Genesis of Synthetic Anti-malarial Agents in Peace and War', *Journal of Antimicrobial Chemotherapy* 36 (1995), pp. 857–72. Chloroquine, another drug still used today, also emerged from a wartime research effort just after the war. It was done in the USA, though it was noted that German scientists had made it but had not fully understood its effectiveness.

25. Norman Wooding, 'Christopher Frank Kearton Baron Kearton, of Whitchurch, Bucks, Kt, O.B.E. 17 February 1911–2 July 1992', *Biographical Memoirs of Fellows of the Royal Society* 41 (November 1995), pp. 221–41.

26. See Christopher Hinton, 'Memoirs', TS 1970, The Papers of Lord Hinton of Bankside, HINT 4/2, Churchill Archives Centre.

27. Postan, Hay and Scott, *Design and Development of Weapons*, pp. 428–30.

28. *GEC Journal* XIV, No. 1 (1946), p. 14.

29. Robert Clayton and Joan Algar, *The GEC Laboratories 1919–1984* (London, 1989), p. 395. See also many entries in Clifford Paterson (eds. R. J. Clayton and J. Algar), *A Scientist's War: The War Diary of Sir Clifford Paterson 1939–45* (London, 1991).

30. Roger Hayward, 'British Air-Dropped Depth Charges and Anti-Ship Torpedoes', *Royal Air Force Historical Society Journal* 45 (2009), pp. 121–36. I am grateful to Stephen Marsh, who tells me that some sources give much longer ranges.

31. Postan, Hay and Scott, *Design and Development of Weapons*, p. 430. Robert Charles Alexander, *The Inventor of Stereo: The Life and Works of Alan Dower Blumlein* (Oxford, 2000).

32. J. V. Field, 'British Cryptanalysis: The Breaking of "Fish" Traffic', in Ad Maas and Hans Hooijmaijers (eds.), *Scientific Research in World War II: What Scientists Did in the War* (London, 2009), pp. 213–31.

33. G. D. Speake, 'The Marconi Research Centre – A Historical Perspective', *Physics in Technology* 16 (1985), pp. 275–81.

34. Mark Frankland, *Radio Man: The Remarkable Rise and Fall of C. O. Stanley* (London, 2002), pp. 112–22.

35. Ibid., p. 89.

36. Ibid., p. 91.

37. 'Scientists and Technologists in the War Program', *The Scientific Monthly* 55 (November 1942), p. 470.

38. Interview with N. F. Mott, *Bristol Evening World* (15 June 1945).

39. 'The Royal Society and the Central Register', *Notes and Records of the Royal Society of London* 2, No. 2. (November 1939), pp. 176–8.

40. For a very clear post-war description, which clearly regarded that lack of understanding of the distinction as very problematic, see Paul Freedman, *The Principles of Scientific Research* (London, 1949). Freedman was an industrial researcher, the head of lamp research with Crompton Parkinson.

41. Charles Goodeve, 'Uniform or plain clothes for officers concerned with naval development', submitted to the Admiralty Advisory Panel on Scientific Research, 12 December 1941, The Papers of Sir Charles Goodeve, GOEV 3/1, Churchill Archives Centre.

42. Admiral Sir Francis Pridham, TS memoir, Vol. 2, pp. 187, 189, 194, 195, The Papers of Vice-Admiral Sir (Arthur) Francis Pridham, PRID 2, Churchill Archives Centre.

43. Ibid., p.188.

44. E. C. Bullard, draft letter to Prof. Fowler, 9/6/41, The Papers of Sir Edward Bullard, BLRD, E.3, Churchill Archives Centre.

45. Ibid.

46. See R. D. Keynes, 'Papers re resignation from HM A/SEE Fairlie in 1942', The Papers of Richard Keynes, KEYN 1, Churchill Archives Centre.

47. John Lennard-Jones, 'Some Wartime Experiences', lecture to the University Chemical Society, Cambridge, 12 March 1947, The Papers of Sir John Lennard-Jones, LENJO 46, Churchill Archives Centre.

48. Pridham, TS memoir, Vol. 2, p. 197, Pridham Papers, PRID 2, Churchill Archives Centre.

49. Ibid., p. 190.

50. G. F. Whitby, 'Sir Frank Ewart Smith. 31 May 1897–14 June 1995', *Biographical Memoirs of Fellows of the Royal Society* 42 (November 1996), pp. 421–31.

51. Lennard-Jones, 'Some Wartime Experiences'.

52. Sykes, Garner, Curtis and Mott: Lennard-Jones, 'Some Wartime Experiences'.

53. John Lennard-Jones, daily journal, Notebook 58, Lennard-Jones Papers, LENJO 26, Churchill Archives Centre.

54. Ibid., Notebook 33, Lennard-Jones Papers, LENJO 25, Churchill Archives Centre.

55. Meeting with Captain Sinclair, DCS, RD, 13 October 1942, Lennard-Jones's daily journal, Notebook 1, Lennard-Jones Papers, LENJO 24, Churchill Archives Centre.

56. 'Advice' following entry for 18 November 1942, in ibid. See also 5 December entry, Notebook 2, which shows he raised it with Sir William Douglas of the Ministry of Supply.

57. Short description by Lennard-Jones of his work, written 1945 or 1946, Lennard-Jones Papers, LENJO 8, Churchill Archives Centre.

58. Ministry of Supply, 'How We Defeated the Heavy German Tanks', papers for press conference, 18 July 1946, Lennard-Jones Papers, LENJO 46, Churchill Archives Centre.

59. Winston S. Churchill, *The Second World War*, Vol. V: *Closing the Ring* (London, 1952), pp. 463–4.

60. Sir Arthur T. Harris, *Despatch on War Operations: 23rd February 1942 to 8th May 1945* (London, 1995), Appendix C, p. 94.

61. D. P. MacDougall and N. M. Newark, *Explosives and Terminal Ballistics: A Report Prepared for the AAF Science Advisory Group* (Wright Field, Dayton, O., 1946), p. 82.

62. Joel A. Vilensky, *Dew of Death: The Story of Lewisite, America's World War I Weapon of Mass Destruction* (Bloomington, Ind., 2005), p. 66.

63. See Adrian Fort, *Prof: The Life of Frederick Lindemann* (London, 2004), p. 171; R. Berman, 'Lindemann in Physics', *Notes and Records of the Royal Society of London* 41 (June 1987), pp. 186–7.

64. Fort, *Prof*, p. 172.

65. A. R. Todd to Vice-Chancellor, 7 March 1940, VCA 7/358/6, University of Manchester Archives.

66. Patrick Blackett to Vice-Chancellor, 22 June 1939, VCA 7/239/3, University of Manchester Archives; Patrick Blackett, 'Manchester University Physics Department Research School', 2nd draft, April 1944, VCA 7/71/3, University of Manchester Archives.

67. J. M. Nuttal to Norman Smith (Registrar), 16 April 1943, in staff commitment correspondence, box A-L, VCA 7/561, University of Manchester Archives.

68. Prof. A. M. Tyndall, FRS, 'A history of the Department of Physics in Bristol, 1876–1948, with personal reminiscences', TS August 1956, DM639, University of Bristol Archive.

69. Vera Brittain, *England's Hour* (London, 2005; first publ. 1941), pp. 34–9.

70. University of Bristol Calendar, 1939/40, Bristol University Archive.

71. University of Bristol minutes of Senate meeting of 16 October 1939, printed and bound copies in Bristol University Archive.

72. In economics John Hicks and T. S. Ashton remained (but were to leave after the war), and in chemistry both Alexander Todd and Michael Polanyi were very much resident for most of the war. Michael Polanyi's naturalization as a British citizen came through, late, on the eve of the war – he would otherwise, like a few Manchester colleagues, have been interned. Polanyi hoped to join the LMS railway laboratory in Derby and later did some research for ICI on synthetic rubber. Polanyi had a quiet war, unlike his juniors in the Berlin-Hungarian group, Szilard, Teller and Wigner, who went to the USA: William Taussig Scott and Martin X. Moleski, *Michael Polanyi: Scientist and Philosopher* (Oxford, 2005), pp. 180, 190–92. Thanks to Ralph Desmarais.

73. A. R. Todd to Bursar, 17 November 1939, VCA 7/358/6, University of Manchester Archives.

74. George Kington, *Bombs and Rockets: The University of Bristol Chemistry Department in World War II: A View from the Laboratory Bench* (Truro, 2004), pp. 6–14.

75. Annual report of the physics department for 1944–5 session, University of Manchester Archives.

76. Short description by Lennard-Jones of his work, written 1945 or 1946, Lennard-Jones Papers, LENJO 8, Churchill Archives Centre.

77. Prof. Mark Oliphant, preface to J. T. Randall, H. A. H. Boot and S. M. Duke, 'Magnetron Development in the University of Birmingham', March 1941, The Papers of Sir John Randall, RNDL 1/2/2, Churchill Archives Centre.

78. Robert Bud, *Penicillin: Triumph and Tragedy* (Oxford, 2007), pp. 29–36.

79. Ibid., p. 49.

80. Vilensky, *Dew of Death*, pp. 78–82, 99.

81. See John Brooks, 'The Midshipman and the Secret Gadget', in Peter Hore (ed.), *Patrick Blackett: Sailor, Scientist and Socialist* (London, 2003), pp. 72–96. He also shows that Blackett had co-invented a related device while a midshipman on *Barham* and was concerned with servo mechanisms in the 1930s.

82. Fredric Boyce and Douglas Everett, *SOE: The Scientific Secrets* (Stroud, 2003). See also Hannah Gay, *The History of Imperial College London, 1907–2007* (London, 2007), pp. 249–50.

83. *School for Secrets*, written and produced by Peter Ustinov, Two Cities Films (1946). Thanks to Maximilian Stadler.

84. This also happened in Australia, D. P. Mellor, *Australia in the War of 1939–1945. Series 4 – Civil – Vol. V: The Role of Science and Industry* (Canberra, 1958), p. 641.

85. Ronald W. Clark, *Sir Edward Appleton, GBE, KCB, FRS* (Oxford, 1971), p. 113.

86. Anon., 'Brain Children', *Journal of the Royal Naval Scientific Service* 1 (November 1945).

87. December 1940 in *Flight*. See http://airminded.org; accessed 14 April 2010.

88. Hayward, 'British Air-Dropped Depth Charges and Anti-Ship Torpedoes'.

89. *The Burney Toraplane*, Imperial War Museum film collection, MTE 601, two reels.

90. Chris Henry and Brian Delf, *British Anti-Tank Artillery 1939–45* (London, 2004), pp. 41–2.

91. For his influence, see Harold Macmillan Papers, c. 279, folios 1–53, Bodleian Library, University of Oxford.

92. Colonel R. Stuart Macrae, *Winston Churchill's Toyshop* (Kineton, 1971), p. 5. Colonel Macrae was a member from beginning to end. After the war Jefferis went to India and between 1950 and 1953 was chief super-intendent of the Military Engineering Experimental Establishment.

93. Winston S. Churchill, *The Second World War*, Vol. II: *Their Finest Hour* (London, 1949), pp. 148–9.

94. Ibid., pp. 148–50.

95. Churchill to General Ismay, 6 May 1942, in Winston S. Churchill, *The Second World War*, Vol. IV: *The Hinge of Fate* (London, 1951), for example, p. 766; p. 806 reproduces the minute to Ismay of 13 November 1942, 'I saw the Jefferis gun last week', asking how many had been ordered, etc. Another of 10 March 1943 to Ismay and Mountbatten, about piers on beaches, asked among other things whether Brigadier Jefferis had been consulted.

96. Lord Cherwell took personal control: see Lord Birkenhead, *The Prof in Two Worlds: The Official Life of Professor F. A. Lindemann, Viscount Cherwell* (London, 1961), p. 216.

97. Copy of letter sent to Mr Rootes, Ministry of Supply, by Lt-Col. Jefferis, 8 September 1941, The Papers of Colonel R. S. Macrae, MCRA 2/2, Churchill Archives Centre.

98. List of devices developed by MD1 (3) 27 May 1942, Macrae Papers, MCRA 2/3, Churchill Archives Centre.

99. List of weapons developed by Directorate MD1 accepted for service use, 17 October 1945, Macrae Papers, MCRA 2/7, Churchill Archives Centre.

100. Postan, Hay and Scott, *Design and Development of Weapons*, p. 269.

101. Ibid., p. 175.

102. See the figures in Guy Hartcup, *The Effect of Science on the Second World War* (Houndmills, 2003), Table 4.1.

103. Barnaby Blacker (ed.), *The Adventures and Inventions of Stewart Blacker, Soldier, Aviator, Weapons Inventor* (Barnsley, 2006), pp. 173–96.

104. Jefferis to Sir William Douglas, Ministry of Supply, 21 December 1944, Macrae Papers, MCRA 3/24, Churchill Archives Centre.

105. Colonel Macrae to Jefferis, 20 October 1944, Macrae Papers, MCRA 3/24, Churchill Archives Centre.

106. Churchill, *Second World War*, Vol. IV: *The Hinge of Fate*, pp. 826–9.

107. Ibid., p. 843.

108. Gerard Pawle, *The Secret War, 1939–45* (London, 1956), is a history by a member of the department.

109. Edward Terrell, *Admiralty Brief* (London, 1958), p. 13.

110. His full name was Nevil Shute Norway.

111. Terrell, *Admiralty Brief*, pp. 45–61.

112. Ibid., pp. 227–33.

113. Dr J. P. Laurie, Royal Naval Scientific Service, 'The Triumph of Plastic Armour', *Shipping World* 29 (August 1945), quoted in Terrell, *Admiralty Brief*, p. 228.

114. Terrell, *Admiralty Brief*, p. 161.

115. Ibid., pp. 201–7.

116. Ibid., Chapters 15 and 16.

117. *Fourth and Final Report of the Royal Commission on Awards to Inventors (Awards to Inventors) 1955–56*, Cmd 9744 (London, 1956), para 50.

118. Sir Henry Tizard, speech at the University of Manchester, reported in the *Manchester Guardian*, Saturday, 10 November 1945.

119. Wimperis, 'Research and Development in Aeronautics', p. 353.

120. *The Economist* (7 January 1939), p. 11

121. *The Times* in 1942 carried thirty-six uses of the term in this very narrow sense; many were references to the Imperial College of Science and Technology, one was a broad reference and three were ambiguous, including an advert for Chubb locks which states: 'Improvements in the technology of Locks and Safes are at all times stimulated by the applied intelligence of the burglar', *The Times*, 18 February 1942.

122. *The Economist* (11 February 1939), pp. 287–9.

123. My observations have been inspired by Eric Schatzberg, 'Technik Comes to America: Changing Meanings of Technology before 1930', *Technology and Culture* 47 (2006), pp. 486–512.

124. Sir Edward Appleton, 'The Scientist in Wartime', 32nd Thomas Hawksley Lecture, November 1945, *Proceedings of the Institution of Mechanical Engineers* 154 (1946), pp. 303–16, particularly p. 316.

125. Clark, *Sir Edward Appleton*, pp. 110, 115, 121.

126. J. G. Crowther and Richard Whiddington, *Science at War* (London, 1947).

127. Richard Whiddington Papers, Brotherton Library, University of Leeds.

128. Professor J. D. Bernal, 'Science in Architecture: A Paper Read to an Informal Meeting on Tuesday, 12 February 1946 . . .', *Journal of the Royal Institute of British Architects* (March 1946); Sir Frank E. Smith, vote of thanks and discussion of Bernal, ibid.

9. MACHINES AND MODERNITIES

1. On the latter, see Sabine Clarke, 'A Technocratic Imperial State? The Colonial Office and Scientific Research, 1940–1960', *Twentieth Century British History* 18 (2007), pp. 1–28.

2. *C. R. Attlee Speaks to Britain*, 25 June 1945, Gaumont British Newsreel BGU409280012 at www.Itnsource.com. Tellingly, Winston Churchill's funeral in 1965 was a very national affair, surprisingly so for so great a believer in Empire and the alliance with the United States: see Wendy Webster, *Englishness and Empire 1939–1965* (Oxford, 2005).

3. Ministry of Information, *What Britain Has Done 1939–1945: A Selection of Outstanding Facts and Figures* (London, 2007; first publ. 1945).

4. Central Statistical Office, *Statistical Digest of the War* (London, 1951).

5. For example, Stephen Broadberry and Peter Howlett, 'The United Kingdom: "Victory at all costs"', in Mark Harrison (ed.), *The Economics of World War II: Six Powers in International Comparison* (Cambridge, 1998), pp. 43–80.

6. Unsupported statement in Ministry of Information, *What Britain Has Done*, p. 8.

7. Winston S. Churchill, *The Second World War*, Vol. III: *The Grand Alliance* (London, 1950), p. 454.

8. W. K. Hancock and M. M. Gowing, *British War Economy* (London, 1949), pp. 369–70.

9. Angus Calder, *The People's War: Britain 1939–1945* (London, 1969), p. 321.

10. A recent instance is David Marquand, *Britain Since 1918: The Strange Career of British Democracy* (London, 2008), pp. 103–4: 'a level of mobilisation unsurpassed by any other belligerent' (p. 104).

11. A. J. P. Taylor, *English History, 1914–1945* (Oxford, 1965; Harmondsworth, 1975), pp. 616–17.

12. Calder, *The People's War*, p. 331.

13. Adam Tooze, *The Wages of Destruction: The Making and Breaking of the Nazi Economy* (London, 2006), pp. 358–9.

14. Sir Charles Kingsley Webster and Noble Frankland, *The Strategic Air Offensive against Germany, 1939–1945*, Vol. IV: *Annexes and Appendices* (London, 1961), Appendix 49, Table iv, p. 472.

15. CSO, *Statistical Digest of the War*, Table 9, p. 8.

16. Combined Production and Resources Board, *The Impact of the War on Civilian Consumption in the United Kingdom, the United States and Canada: A Report to the Combined Production and Resources Board from a Special Combined Committee on Nonfood Consumption Levels* (London, 1945), pp. 10–12, 16.

17. Ibid.

18. In a well-known paper (Mark Harrison, 'Resource Mobilization for World War II: The U.S.A., U.K., U.S.S.R., and Germany, 1938–1945', *Economic History Review* 41 (1988), pp. 171–92), Mark Harrison suggests that 56 per cent of national income was spent on the war in Britain in 1944, but that this includes Lend-Lease, etc., and that therefore one needs to consider a different measure, that of resources domestically produced devoted to war, which he estimates at 47 per cent, considerably less than the USA at 54 per cent (see Table 3, p. 184). However, Lend-Lease is not included in the UK accounts he bases his calculations on, so his 56 per cent in fact captures domestic war output only, and thus the total resources from whatever source devoted to war as a proportion of national income is considerably higher than this. Thus by any measure the UK was more mobilized than the US.

19. R. G. D. Allen, 'Mutual Aid between the US and the British Empire, 1941–45', *Journal of the Royal Statistical Society* 109 (1946), pp. 243–77.

20. David Syrett (ed.), *The Battle of the Atlantic and Signals Intelligence: U-Boat Tracking Papers, 1941–1945* (Aldershot, 2007), p. 9.

21. Allen, 'Mutual Aid between the US and the British Empire', p. 220.

22. Hugh Weeks, 'Anglo-American Supply Relationships', in D. N. Chester (ed.), *Lessons of the British War Economy* (Cambridge, 1951), pp. 69–82, particularly p. 71.

23. Roberto Cortés Conde, *The Political Economy of Argentina in the Twentieth Century* (Cambridge, 2009), figures 4.1, 4.2, 4.3, 4.4 and 4.7.

24. M. H. J. Finch, *A Political Economy of Uruguay since 1870* (London, 1981), Tables 5.3 and 5.5.

25. Taylor, *English History*, pp. 662–3.

26. Cmd 6707 and Cmd 7099 quoted in H. A. Shannon, 'The Sterling Balances of the Sterling Area, 1939–49', *The Economic Journal* 60 (September 1950), pp. 531–51.

27. Hancock and Gowing, *British War Economy*, p. 373, recognize but do not emphasize enough, I think, the element of conscious choice here.

28. Webster and Frankland, *Strategic Air Offensive*, Vol. IV: *Annexes and Appendices*, Appendix 49, Table iii.

29. For the failures of Germans as imperialists in Europe, see Mark Mazower, *Hitler's Empire: Nazi Rule in Occupied Europe* (London, 2008).

30. Simon Ball, *The Bitter Sea: The Struggle for Mastery in the Mediterranean 1935–1949* (London, 2009).

31. On this, see Tooze, *Wages of Destruction, passim.*

32. Vera Brittain, *Testament of Experience: An Autobiographical Story of the Years 1925–1950* (London, 1957), pp. 331–6.

33. Vera Brittain, *Seed of Chaos: What Mass Bombing Really Means* (London, 1941), p. 29.

34. 'Bombing and Policy', *New Statesman and Nation* (23 January 1943), p. 51.

35. Squadron Leader John Strachey, 'The Age of Air Power', broadcast on BBC Home Service; printed in *The Listener* (20 January 1944).

36. J. M. Spaight, *Air Power Can Disarm* (London, 1948), p. 169.

37. J. M. Spaight, *Bombing Vindicated* (London, 1944), p. 143.

38. Ministry of Information, *What Britain Has Done*, p. 66.

39. It is highly likely that the British killed many more German civilians than German soldiers. At the level of prisoners of war (as significant militarily as killed), British forces trounced both the Germans and the Italians, even before Normandy. By the end of the North Africa campaign, mainly British and imperial forces had captured nearly 300,000 Germans, more than the total number of British and Commonwealth forces held by the Germans in 1945.

40. 'Merchant Navy, 27,000 British, excluding lascars, due directly and indirectly to enemy action – deaths and permanently damaged lives': C. B. A. Behrens, *Merchant Shipping and the Demands of War* (London, 1958), p. 181.

41. Sir Arthur Harris, *Bomber Offensive* (London, 1990; first publ. 1947), pp. 263–5.

42. Ibid., p. 269.

43. Ibid., p. 276.

44. Sir Arthur T. Harris, *Despatch on War Operations: 23rd February 1942 to 8th May 1945* (London, 1995), p. 40.

45. Ibid., p. 30.

46. Tooze, *Wages of Destruction*, Chapters 17 and 18.

47. See Harris, *Despatch on War Operations*, Table 9 and data abstracted from Table 6.

48. Ibid., paragraph 170.

49. Sebastian Cox (ed.), *The Strategic Air War against Germany, 1939–1945: Report of the British Bombing Survey Unit* (London, 1998).

50. United States Strategic Bombing Survey, *The Effects of Strategic Bombing on the German War Economy* (Washington DC, 1945).

51. Webster and Frankland, *Strategic Air Offensive*, Vols. I–IV.

52. Noble Frankland, *The Bombing Offensive against Germany: Outlines and Perspectives* (London, 1965).

53. Noble Frankland, *History at War: The Campaigns of an Historian* (London, 1998), p. 70. Frankland also published a short summary of the official history: Noble Frankland, *Bomber Offensive: The Devastation of Europe* (London, 1970).

54. Or rather the tonnage of bombs increased by these amounts; their explosive power almost certainly increased a great deal more, as did accuracy of delivery. We might also note that the bombing raids of early 1943 might have been perhaps a third more destructive had they used aluminized explosive.

55. Tooze, *Wages of Destruction*, pp. 597, 598, 604.

56. Ibid., pp. 598, 602.

57. Albert Speer, *Inside the Third Reich* (London, 1995), p. 728.

58. Flak did not divert resources from other aspects of the war effort to the extent generally thought, and flak's effectiveness against Allied bombers has been grossly underestimated: Edward B. Westermann, *Flak: German Anti-Aircraft Defenses, 1914–1945* (Lawrence, Kan., 2001), pp. 3–4.

59. Webster and Frankland, *Strategic Air Offensive*, Vol. IV: *Annexes and Appendices*, Appendix 49, Table iii.

60. Harris, *Despatch on War Operations*, para. 205. In his interrogation the German armaments minister, Speer, claimed that in 1944 30 per cent of output of guns were for anti-aircraft use, as were 20 per cent of heavy shells, 50–55 per cent of the 'electrotechnical industry' and 30 per cent of the optical industry: Speer interrogation in Webster and Frankland, *Strategic Air Offensive*, Vol. IV: *Annexes and Appendices*, p. 383.

61. Webster and Frankland, *Strategic Air Offensive*, Vol. IV: *Annexes and Appendices*, p. 297.

62. Ibid.

63. As convincingly done by Richard Overy, *The Air War, 1939–1945* (London, 1980); Richard Overy, *Why the Allies Won* (London, 1995); Richard Overy, *Bomber Command, 1939–1945* (London, 1997).

64. The 7 per cent figure comes from the Report of the British Bombing Survey Unit, and is generally accepted, though it is not always clear what this was a percentage of. In the survey it is a percentage of people-hours engaged in warlike activity (including arms production). This almost certainly underestimates the effort. A recent thesis has argued that 5.6 per cent of GDP, 10 per cent of government spending and 12 per cent of defence spending went on Bomber Command: John Fahey, 'Britain 1939–1945: The Economic Cost of Strategic Bombing', unpublished Ph.D. thesis, University of Sydney (2006).

65. Cyril Falls, *Ordeal by Battle* (London, 1943), pp. 159–60.

66. 'The Campaign in 1943', in Churchill, *Second World War*, Vol. III: *The Grand Alliance*, pp. 582–4.

67. Peter Calvocoressi, Guy Wint and John Pritchard, *The Penguin History of the Second World War* (London, 1999), p. 357. This was originally published as Peter Calvocoressi and Guy Wint, *Total War: Causes and Courses of the Second World War* (London, 1972).

68. J. D. Bernal and Maurice Cornforth, *Science for Peace and Socialism* (London, 1949), p. 25.

69. P. M. S. Blackett, *Military and Political Consequences of Atomic Energy* (London, 1948).

70. Tami Davis Biddle, *Rhetoric and Reality in Air Warfare: The Evolution of British and American Ideas about Strategic Bombing, 1914–1945* (Princeton, NJ, 2002); W. W. Rostow, *Pre-Invasion Bombing Strategy: General Eisenhower's Decision of March 25, 1944* (Austin, Tex., 1981), pp. 20–21.

71. Rostow, *Pre-Invasion Bombing Strategy*.

72. Randall T. Wakelam, *The Science of Bombing: Operational Research in RAF Bomber Command* (Toronto, 2009), p. 193.

73. See David Edgerton, 'British Scientific Intellectuals and the Relations of Science, Technology, and War', in Paul Forman and José Manuel Sanchez-Ron (eds.), *National Military Establishments and the Advancement of Science and Technology* (Dordrecht, 1996), pp. 1–35; and especially Wakelam, *The Science of Bombing*, pp. 150, 152.

74. For example, J. G. Crowther and Richard Whiddington, *Science at War* (London, 1947); J. G. Crowther, *Science in Modern Society* (London,

1967); H. Rose and S. Rose, *Science and Society* (Harmondsworth, 1969); R. W. Clark, *The Rise of the Boffins* (London, 1962); Calder, *The People's War*, Chapter 8. Generally see David Edgerton, *Warfare State: Britain, 1920–1970* (Cambridge, 2005), pp. 319–23. In 1945 Ritchie Calder noted: 'it is commonly estimated that our total muster of "scientists" is 45,000 ... Of the scientists whose contributions saved us in the war, the elite corps consisted of a few hundreds.' The only names he gave were Bernal, Blackett, Zuckerman and Sir Henry Dale: Ritchie Calder, 'Science and the State', *New Statesman & Nation*, 8 October 1945, p. 384.

75. Sir Henry Tizard, address to the annual conference of the IPCS, July 1946, reprinted in *State Service* (July 1946), copy in Tizard Papers, HTT 596, Imperial War Museum.

76. Professor J. D. Bernal, letter to *The Times*, 11 March 1947.

77. Professor J. D. Bernal, 'Science in Architecture: A Paper Read to an Informal Meeting on Tuesday, 12 February 1946 ...', *Journal of the Royal Institute of British Architects* (March 1946), p. 155.

78. Sir Henry Tizard, speech at the University of Manchester, reported in the *Manchester Guardian*, Saturday, 10 November 1945.

79. Winston S. Churchill, *The Second World War*, Vol. II: *Their Finest Hour* (London, 1949), p. 337.

80. Robert Bud, 'Penicillin and the New Elizabethans', *British Journal for the History of Science* 31 (1998), pp. 305–33.

81. 'Fog Dispersal', *Manchester Guardian*, Friday, 1 June 1945.

82. Attlee, the Prime Minister, released a statement by Churchill on the British bomb project which made very clear that the US had been overwhelmingly responsible for it: *Manchester Guardian*, 7 August 1945.

83. Labour Party, *Let Us Face the Future: A Declaration of Labour Policy for the Consideration of the Nation* (London, 1945). Also in Iain Dale (ed.), *Labour Party General Election Manifestos, 1900–1997* (London, 2000), p. 55.

84. See Gaumont British newsreel, *War Exhibit in Prague* (1946) BGU 410140034, www.ItnSource.com.

85. David Pritchard, *The Radar War: Germany's Pioneering Achievement 1904–45* (Wellingborough, 1989), p. 9.

86. Imperial War Museum film collection BTF 270 is a record of what is perhaps the whole motorized column.

87. In Max Arthur, *Dambusters: A Landmark Oral History* (London, 2008), p. 50.

88. A commonplace on the internet, unsourced, attributed to Churchill.

89. The claim is a commonplace, and is attributed to F. H. Hinsley, in

F. H. Hinsley and Alan Stripp (eds.), *Codebreakers: The Inside Story of Bletchley Park* (Oxford, 2001), pp. 10–13; yet it is based on an interesting speculation, with many intricate steps, that had Ultra not been available to win the Battle of the Atlantic in 1943, Overlord would have been delayed until 1946. That is not the same as shortening the war by two years, for such a suggestion relies on the idea that the Normandy landings would have been as crucial in 1946 as in 1944.

90. Indeed, 'we have assertions, images and impressions of technological decisiveness in war, but we have no detailed measurement, analysis or consensus': George Raudzens, 'War-Winning Weapons: The Measurement of Technological Determinism in Military History', *The Journal of Military History* 54 (October 1990), pp. 403–34, particularly p. 432.

91. Taylor, *English History*, pp. 726–7.

92. Calder, *The People's War*, p. 17.

93. For a critique in general and milk in schools in particular, see J. Macnicol, 'The Effect of the Evacuation of Schoolchildren on Official Attitudes to State Intervention', in Harold L. Smith (ed.), *War and Social Change: British Society in the Second World War* (Manchester, 1986), pp. 3–31.

94. Richard Titmuss, *Problems of Social Policy* (London, 1950), p. 501.

95. Ibid., p. 486.

96. Ibid., p. 489.

97. Ibid., p. 530.

98. Ibid., p. 531. Titmuss and many others have stressed the improving health and life expectancy of the British people, attributing this in one way or another to features taken to be specific to the war. Yet contrary to what is usually implied health improved at no greater rate than before the war, with the crucial exception of the years 1940 and 1941, where leaving aside such things as bombing and accidents, health outcomes got notably worse. Susannah Peng, 'Did the Second World War improve the health of the British people? A New Look', History of Medicine BSc Mini-Project, Imperial College London, 2011.

99. Combined Production and Resources Board, *The Impact of the War*, pp. 62–3.

100. *Statement of Services to be Provided for in Votes of Credit, 1942–43*, Cmd 6363 (London, 1943), shows in some detail that votes of credit were used for additional war expenditure, such as MoF, EMS, etc., but not for routine civil requirements like general health or education.

101. Jim Tomlinson, *Democratic Socialism and Economic Policy: The Attlee Years, 1945–1951* (Cambridge, 1997), p. 249.

102. See Hugh Rockoff, 'The United States: From Ploughshares to Swords',

in Harrison (ed.), *The Economics of World War II*, pp. 81–121, particularly Table 3.11, p. 101.

103. Hancock and Gowing, *British War Economy*, pp. 366–7.

104. Andrew Roberts, *The Storm of War: A New History of the Second World War* (London, 2009), p. 112.

105. Broadberry and Howlett, 'The United Kingdom', pp. 68–9.

106. Alan S. Milward, *The Economic Effects of the Two World Wars on Britain*, 2nd edn (London, 1984), pp. 65, 68. Of this only some £200m came from the USA, with the vast majority coming from the sterling area: see *Statistical Material Presented during the Washington Negotiations*, Cmd 6707 (London, 1945).

107. Ibid., Table 4. Total sales in the USA accounted for one quarter of receipts.

108. Ibid., p. 69.

109. £860m destruction on land plus £380m shipping and cargo plus internal disinvestment of £612m minus internal investment in factories and plant of £513m. This gives £1,339m, which is 6.9 per cent of the pre-war physical capital stock of £19,520m. Source: Broadberry and Howlett, 'The United Kingdom', Table 2.20, p. 69.

110. A small element of the loan was intended to pay for Lend-Lease goods already in transit. All in all it was a loan which replaced Lend-Lease.

111. Alan P. Dobson, *US Wartime Aid to Britain 1940–1946* (London, 1986). Richard Toye, 'Keynes, the Labour Movement, and "How to Pay for the War"', *Twentieth Century British History* 10 (1999), pp. 255–81.

112. Jim Tomlinson, 'Balanced Accounts? Constructing the Balance of Payments Problem in Post-War Britain', *The English Historical Review* CXXIV (2009), pp. 863–84; 'The Attlee Government and the Balance of Payments, 1945–1951', *Twentieth Century British History* 2 (1991), pp. 47–66.

113. Cyril Falls, *The Second World War: A Short History* (London, 1948), p. 295.

114. See the discussion in Edgerton, *Warfare State*, Chapters 2 and 7.

115. In *Warfare State* I discuss this also in relation to historiography in the 1960s, noting the importance of critiques of liberalism from the left and right, and the common insistence, which persists to this day, that post-war Britain was never national or nationalistic enough: Edgerton, *Warfare State*, Chapter 7.

116. Similarly wartime and post-war planning had really no connection to the in any case very limited British economic or political thinking about planning.

117. This is the key argument of Overy in *Why the Allies Won*.

Bibliography

ARCHIVAL COLLECTIONS

Bodleian Library, University of Oxford
Harold Macmillan Papers
R. R. Stokes Papers

Bristol University Library
A. R. Collar Papers
Alfred Pugsley Papers
Cecil Burch Papers
Cecil Powell Papers
Charles Frank Papers
Robin Ralph Jamison Papers
University of Bristol Archive

Brotherton Library, University of Leeds
Liddle Collection 1939–1945
Richard Whiddington Papers
Robert Spence Papers

Cambridge University Library
Vickers plc Archive

Churchill Archives Centre, Churchill College, Cambridge
A. V. Hill Papers
Austen Albu Papers
Charles Goodeve Papers
Christopher Hinton Papers
Donald MacDougall Papers
Edward Bullard Papers

Eric Megaw Papers
Harold Hartley, including Trimpell Papers
Hugh Clausen Papers
John Baker Papers
John Lennard-Jones Papers
John Randall Papers
Leo Amery Papers
Maurice Hankey Papers
Naval Radar Trust Papers
R. V. Jones Papers
Ralph Alger Bagnold Papers
Richard Keynes Papers
Robert Stuart Macrae Papers
Stephen King-Hall Papers
Vice-Admiral Francis Pridham Papers
William Douglas Weir Papers

Imperial War Museum
Imperial War Museum film collection
Sir Henry Tizard Papers

Nuffield College, University of Oxford
Cherwell Papers

Royal Society of London
Patrick Blackett Papers

The National Archives
Cabinet and War Cabinet Papers, Official Historians' Papers (CAB 102),
    Security Service Papers (KV), Prime Ministers' Papers (PREM), etc.

University of Manchester
University of Manchester Archives
Vice-Chancellor's Papers
John Rylands University Library of Manchester, Labour Party Newspaper
    cuttings collection,

University of Sheffield
Hans Krebs Papers
Sorby Research Institute Collection

University of Sheffield Radar Archive
Vice-Chancellor's Papers

WEBSITES

http://www.itnsource.com (newsreels)
http://www.britishpathe.com (newsreels)
http://www.iwmcollections.org.uk (includes films)
http://airminded.org
http://gsn.ncl.ac.uk/
http://www.uboat.net/
http://www.wwiiequipment.com/
http://www.ibiblio.org/hyperwar/

PERIODICALS AND NEWSPAPERS

Aircraft Production: The Journal of the Aircraft Manufacturing Industry
British Machine Tool Engineering
British Science News
British Union Quarterly
Endeavour
Ford Times
GEC Journal
King-Hall News-Letter
*Manchester Guardian**
Monthly Science News
New Statesman and Nation
*The Economist**
The Engineer
The Listener
*The Times**

*Available online

PARLIAMENTARY AND GOVERNMENT PAPERS

House of Commons Debates and House of Lords Debates (Hansard), most
    now available online at http://hansard.millbanksystems.com/
Some notable command papers (available at http://parlipapers.chadwyck.co.uk):

DSIR, *Report for the Year 1947–48, with a Review of the Years 1938–48*, Cmd 7761 (1949)

Report of the Medical Research Council for the Years 1939–45, Cmd 7335 (1947–48)

First Report of the Royal Commission on Awards to Inventors 1949, Cmd 7586 (1949)

Second Report of the Royal Commission on Awards to Inventors, Cmd 7832 (1952)

Third Report of the Royal Commission on Awards to Inventors (Awards to Inventors) 1953, Cmd 8743 (1953)

Fourth and Final Report of the Royal Commission on Awards to Inventors (Awards to Inventors) 1955–56, Cmd 9744 (1956)

Statement of Services to be Provided for in Votes of Credit, 1942–43, Cmd 6363 (1943)

Statistical Abstract for the British Commonwealth for Each of the Ten Years 1936 to 1945 (Trade and Commerce Section), Sixty-ninth number, Cmd 7224 (1947)

Statistical Material Presented during the Washington Negotiations, Cmd 6707 (1945)

Statistics Relating to the War Effort of the United Kingdom 1944, Cmd 6564 (1944)

Wartime Tank Production, Cmd 6865 (1946)

Other

Central Statistical Office (CSO), *Annual Abstract of Statistics*, No. 85, 1937–1947 (London, 1948)

Combined Production and Resources Board, *The Impact of the War on Civilian Consumption in the United Kingdom, the United States and Canada: A Report to the Combined Production and Resources Board from a Special Combined Committee on Nonfood Consumption Levels* (London, 1945)

Ministry of Information, *What Britain Has Done 1939–1945: A Selection of Outstanding Facts and Figures* (London, 2007; first publ. 1945)

BIOGRAPHICAL SOURCES

Oxford Dictionary of National Biography: http://www.oxforddnb.com
Who Was Who: http://ukwhoswho.com
Biographical Memoirs of Fellows of the Royal Society (jstor, etc.)

PUBLISHED SOURCES
Official histories

Behrens, C. B. A., *Merchant Shipping and the Demands of War* (London, 1958)

Butler, J. R. M., and others, *Grand Strategy*, 6 vols. (London, 1956–1970)

Butlin, S. J. and C. B. Schedvin, *Australia in the War of 1939–1945. Series 4 – Civil – Vol. IV: War Economy, 1942–1945* (Canberra, 1977)

Central Statistical Office (CSO), *Statistical Digest of the War* (London, 1951)

Ellis, L. F., *The War in France and Flanders, 1939–1940* (London, 1953)

—, *Victory in the West*, 2 vols. (London, 1962, 1968)

Frey, John W. and H. Chandler Ide (eds.), *A History of the Petroleum Administration for War 1941–1945* (Washington DC, 2005; first publ. 1946)

Gowing, Margaret, *Britain and Atomic Energy 1939–1945* (London, 1964)

Hall, H. Duncan, *North American Supply* (London, 1955)

Hall, H. Duncan and C. C. Wrigley (with a chapter by J. D. Scott), *Studies of Overseas Supply* (London, 1956)

Hammond, R. J., *Food*, 3 vols. (London, 1951–62)

Hancock, W. K. and M. M. Gowing, *British War Economy* (London, 1949)

Hornby, W., *Factories and Plant* (London, 1958)

Hurstfield, Joel, *The Control of Raw Materials* (London, 1953)

Lloyd, E. M. H., *Food and Inflation in the Middle East 1940–1945* (Stanford, 1956)

Mellor, D. P., *Australia in the War of 1939–1945. Series 4 – Civil – Vol. V: The Role of Science and Industry* (Canberra, 1958)

Murray, Keith M., *Agriculture* (London, 1955)

Parker, H. M. D., *Manpower: A Study of War-Time Policy and Administration* (London, 1957)

Payton-Smith, D. J., *Oil: A Study of War-Time Policy and Administration* (London, 1971)

Playfair, I. S. O. (and C. J. C. Molony), *The Mediterranean and the Middle East*, 4 vols. (London, 1954–66)

Postan, M. M., *British War Production* (London, 1952)

Postan, M. M., D. Hay and J. D. Scott, *Design and Development of Weapons: Studies in Government and Industrial Organisation* (London, 1964)

Roskill, S., *The War at Sea*, 3 vols. (London, 1954–64)

Royal Army Service Corps, *The Story of the Royal Army Service Corps, 1939–1945* (London, 1955)

Ruppenthal, Roland G., *United States Army in World War II. European*

Theater of Operations: Logistical Support of the Armies, Vol. I: *May 1941–September 1944* (Washington DC, 1953)

Scott, J. D. and R. Hughes, *The Administration of War Production* (London, 1955)

Titmuss, Richard, *Problems of Social Policy* (London, 1950)

Webster, Sir Charles Kingsley and Noble Frankland, *The Strategic Air Offensive against Germany 1939–1945*, 4 vols. (London, 1961)

Other books and articles

Acworth, Bernard, 'Sea Power Surrendered', *English Review* LX (1935), pp. 195–200

Addison, Paul, 'By-Elections of the Second World War', in Chris Cook and John Ramsden (eds.), *By-Elections in British Politics* (London, 1973), pp. 165–190

—, *The Road to 1945: British Politics and the Second World War* (London, 1975)

Aldgate, Anthony and Jeffrey Richards, *Britain Can Take It: The British Cinema in the Second World War*, 2nd edn (London, 2007)

Alexander, Martin S. and William J. Philpott (eds.), *Anglo-French Defence Relations between the Wars* (London, 2002)

Alexander, Robert Charles, *The Inventor of Stereo: The Life and Works of Alan Dower Blumlein* (Oxford, 2000)

Allen, R. G. D., 'Mutual Aid between the US and the British Empire, 1941–45', *Journal of the Royal Statistical Society* 109 (1946), pp. 243–77

Allibone, T. E. and Guy Hartcup, *Cockcroft and the Atom* (Bristol, 1984)

Anker, Peder, *Imperial Ecology: Environmental Order in the British Empire, 1895–1945* (Cambridge, Mass., 2001)

Annan, Noel, *Changing Enemies: The Defeat and Regeneration of Germany* (London, 1996)

Anon., 'Brain Children', *Journal of the Royal Naval Scientific Service* 1 (November 1945)

Anon., *Science in War* (Harmondsworth, 1940)

Appleton, Edward, 'The Scientist in Wartime', 32nd Thomas Hawksley Lecture, November 1945, *Proceedings of the Institution of Mechanical Engineers* 154 (1946), pp. 303–16

Argonaut, *Give Us the Tools: A Study of the Hindrances to Full War Production and How to End Them* (London, 1942)

Arnold, Ken, et al., *War and Medicine* (London, 2008)

Arthur, Max, *Dambusters: A Landmark Oral History* (London, 2008)

Association of Scientific Workers, *Science and the Nation* (Harmondsworth, 1947)

Austin, Brian, *Schonland: Scientist and Scholar* (London, 2001)

Avery, Donald H., *The Science of War: Canadian Scientists and Allied Military Technology during the Second World War* (Toronto, 1998)

Bailey, Gavin, 'The Narrow Margin of Criticality: The Question of the Supply of 100-Octane Fuel in the Battle of Britain', *English Historical Review* CXXIII (April 2008), pp. 394–411

Bailey, J. B. A., *Field Artillery and Fire Power* (Annapolis, 2004)

Baker, John Fleetwood, *Enterprise Versus Bureaucracy: The Development of Structural Air-Raid Precautions during the 2nd World War* (Oxford, 1978)

Ball, Simon, 'The German Octopus: The British Metal Corporation and the Next War, 1914–1939', *Enterprise & Society* 5 (2004), pp. 451–89

—, *The Bitter Sea: The Struggle for Mastery in the Mediterranean 1935–1949* (London, 2009)

Ballard, J. G., *Miracles of Life: Shangai to Shepperton: An Autobiography* (London, 2008)

Balmer, Brian, *Britain and Biological Warfare: Expert Advice and Science Policy, 1930–65* (Basingstoke, 2001)

Banks, F. R. (Rod), *I Kept No Diary: 60 years with Marine Diesels, Automobile and Aero Engines* (Shrewsbury, 1978)

Barnett, Correlli, *Britain and Her Army: A Military, Political and Social Survey* (London, 1974; first publ. 1970)

—, *The Collapse of British Power* (London, 1972)

—, *The Audit of War: The Illusion and Reality of Britain as a Great Nation* (London, 1986)

Batey, Mavis, *Dilly: The Man Who Broke Enigma* (London, 2009)

Bayly, Christopher and Tim Harper, *Forgotten Armies: The Fall of British Asia 1941–1945* (London, 2004)

—, *Forgotten Wars: The End of Britain's Asian Empire* (London, 2007)

Beale, Peter, *Death by Design: British Tank Development in the Second World War* (Stroud, 1998)

Beardsley, E. H., 'Secrets Between Friends: Applied Science Exchange between the Western Allies and the Soviet Union during World War II', *Social Studies of Science* 7 (1977), pp. 447–73.

Beaton, Kendall, *Enterprise in Oil: A History of Shell in the United States* (New York, 1957)

Beckett, Francis, *Enemy Within: The Rise and Fall of the British Communist Party* (London, 1998)

Beckett, Ian F. W., and John Gooch (eds.), *Politicians and Defence: Studies in the Formulation of British Defence Policy* (Manchester, 1981)

Beesly, Patrick, *Very Special Intelligence: The Story of the Admiralty's Operational Intelligence Centre 1939–1945* (London, 1977)

Bennett, G. H. and R. Bennett, *Survivors: British Merchant Seamen in the Second World War* (London, 1999)

Bennett, Gill, *Churchill's Man of Mystery: Desmond Morton and the World of Intelligence* (London, 2007)

Bennett, Ralph, *Behind the Battle: Intelligence in the War with Germany 1939–1945* (London, 1999)

Berman, R., 'Lindemann in Physics', *Notes and Records of the Royal Society of London* 41 (June 1987), pp. 186–7

Bernal, J. D., *The Social Function of Science* (London, 1939)

Best, Geoffrey, *Humanity in Warfare: The Modern History of the International Law of Armed Conflicts* (London, 1983)

—, *Churchill: A Study in Greatness* (London, 2001)

—, *Churchill and War* (London, 2005)

Bevan, Aneurin, *Why Not Trust the Tories?* (London, 1944)

Biddle, Tami Davis, *Rhetoric and Reality in Air Warfare: The Evolution of British and American Ideas about Strategic Bombing, 1914–1945* (Princeton, NJ, 2002)

Birchmore, Graham and Roy Burges, *The Lads of Enfield Lock: 172 Years of Apprentice Training at the Royal Small Arms Factory, Enfield, Middlesex, England, 1816–1988* (Enfield, 2006)

Birkenhead, F. E. Smith, Earl of, *The Prof in Two Worlds: The Official Life of Professor F. A. Lindemann, Viscount Cherwell* (London, 1961)

Blacker, Barnaby (ed.), *The Adventures and Inventions of Stewart Blacker, Soldier, Aviator, Weapons Inventor* (Barnsley, 2006)

Blackett, P. M. S., 'A Note on Certain Aspects of the Methodology of Operational Research', *The Advancement of Science* 5 (1948), p. 31

—, *Military and Political Consequences of Atomic Energy* (London, 1948)

—, 'Scientists at the Operational Level', *The Advancement of Science* 5 (1948), pp. 27–9

—, *Studies of War: Nuclear and Conventional* (Edinburgh and London, 1962)

Blundel, Richard and Angela Tregear, 'From Artisans to "Factories": The Interpenetration of Craft and Industry in English Cheese-Making, 1650–1950', *Enterprise & Society* 7 (2006), pp. 728–31

Bondi, Hermann, *Science, Churchill and Me: The Autobiography of Hermann Bondi* (Oxford, 1990)

Bornstein, Sam and Al Richardson, *War and the International: A History of the Trotskyist Movement in Britain 1937–1949* (London, 1986)

Bowen, Frank C., *The Flag of the Southern Cross 1939–1945* (London, n.d. but 1947)

Boyce, Fredric and Douglas Everett, *SOE: The Scientific Secrets* (Stroud, 2003)

Boyce, Robert and Joseph A. Maiolo, *The Origins of World War Two: The Debate Continues* (Basingstoke, 2003)

Boycott, A. G., *The Elements of Imperial Defence: A Study of the Geographical Features, Material Resources, Communications, and Organization of the British Empire* (Aldershot, n.d. but probably 1939)

Bragg, Michael, *RDF: The Location of Aircraft by Radar Methods 1935–1945* (Paisley, 1988)

Brandt, Karl, *The Management of Agriculture and Food in the German-Occupied and Other Areas of Fortress Europe* (Stanford, 1954)

Branson, Noreen, *History of the Communist Party of Great Britain 1941–1951* (London, 1997)

Brassley, Paul, 'Wartime Productivity and Innovation, 1939–1945', in Brian Short, Charles Watkins and John Martin (eds.), *The Front Line of Freedom: British Farming in the Second World War* (The Agricultural History Review Supplement Series 4) (London, 2006), pp. 16–54

Brendon, Piers, *Winston Churchill: A Brief Life* (London, 1984)

Briggs, Asa, *The History of Broadcasting in the United Kingdom*. Vol. III: *The War of Words* (Oxford, 1970; rev. edn 1995)

Brittain, Vera, *England's Hour* (London, 2005; first publ. 1941)

—, *Seed of Chaos: What Mass Bombing Really Means* (London, 1944)

—, *Testament of Experience: An Autobiographical Story of the Years 1925–1950* (London, 1957)

Broadbent, Ewen, *The Military and Government: From Macmillan to Heseltine* (London, 1988)

Broadberry, Stephen and Peter Howlett, 'The United Kingdom: "Victory at all costs"', in Mark Harrison (ed.), *The Economics of World War II: Six Powers in International Comparison* (Cambridge, 1998), pp. 43–80

Brockway, Fenner and Frederic Mullally, *Death Pays a Dividend* (London, 1945)

Brooke, Alan, *War Diaries 1939–1945: Field Marshal Lord Alanbrooke* (eds. Alex Danchev and Dan Todman) (London, 2001)

Brooke, Stephen, *Labour's War: The Labour Party during the Second World War* (Oxford, 1992)

Brooks, John, 'The Midshipman and the Secret Gadget', in Peter Hore (ed.), *Patrick Blackett: Sailor, Scientist and Socialist* (London, 2003), pp. 72–96

Brown, A. J., *Applied Economics: Aspects of the World Economy in War and Peace* (London, 1949)

Brown, Andrew, *J. D. Bernal: The Sage of Science* (Oxford, 2005)

Brown, D. K., *A Century of Naval Construction: The History of the Royal Corps of Naval Constructors 1883–1983* (London, 1983)

—, *Nelson to Vanguard: Warship Design and Development 1923–1945* (London, 2000)

Bruley, Sue (ed.), *Working for Victory: A Diary of Life in a Second World War Factory* (Stroud, 2001)

Buckley, John, *British Armour in the Normandy Campaign, 1944* (London, 2004)

Bud, Robert, 'Penicillin and the New Elizabethans', *British Journal for the History of Science* 31 (1998), pp. 305–33

—, *Penicillin: Triumph and Tragedy* (Oxford, 2007)

—, 'Upheaval in the Moral Economy of Science? Patenting, Teamwork and the World War II Experience of Penicillin', *History and Technology* 24 (2008), pp. 173–90

Bulman, G. P., *An Account of Partnership – Industry, Government and the Aero Engine: The Memoirs of George Purvis Bulman*, Rolls-Royce Heritage Trust, Historical Series No. 31 (Derby, 2002)

Bungay, Stephen, *The Most Dangerous Enemy: A History of the Battle of Britain* (London, 2000)

—, *Alamein* (London, 2002)

Burn, Duncan (ed.), *The Structure of British Industry*, 2 vols. (Cambridge, 1958)

Burns, R. W., 'Early History of the Proximity Fuze (1937–1940)' *IEE Proceedings-A* 140 (May 1993), pp. 224–36

Buss, David, 'The Changing Household Diet', in J. M. Slater, *Fifty Years of the National Food Survey 1940–1990* (London, 1991), pp. 47–54

Cain, P. J. and A. G. Hopkins, *British Imperialism 1688–2000*, 2nd edn (Harlow, 2002)

Cairncross, Alec, *Living with the Century* (Fife, 1998)

Cairncross, Sir Alec and Nita Watts, *The Economic Section 1939–1961: A Study in Economic Advising* (London, 1989)

Calder, Angus, *The People's War: Britain 1939–1945* (London, 1969)

—, *The Myth of the Blitz* (London, 1991)

Calder, Ritchie, 'Bernal at War', in Brenda Swann and Francis Aprahamiam (eds.), *J. D. Bernal: A Life in Science and Politics* (London, 1999), pp. 160–90

Calvocoressi, Peter and Guy Wint, *Total War: Causes and Courses of the Second World War* (London, 1972); updated as Peter Calvocoressi, Guy Wint and John Pritchard, *The Penguin History of the Second World War* (London, 1999)

Campbell, Reginald, *Teak-Wallah* (London, 1946)

Canetti, Elias, *Party in the Blitz: The English Years*, trans. Michael Hofmann (London, 2005)

Carr, E. H., *Nationalism and After* (London, 1945)

Carter, G. B., *Porton Down: 75 Years of Chemical and Biological Research* (London, 1992)

Carver, Michael, *Out of Step: The Memoirs of Field Marshal Lord Carver* (London, 1989)

—, *Britain's Army in the 20th Century* (London, 1999)

Chamberlain, Peter and Chris Ellis, *British and American Tanks of World War Two: The Complete Illustrated History of British, American and Commonwealth Tanks, 1939–45* (London, 2000)

Chandler, David, *The Oxford Illustrated History of the British Army* (Oxford, 1994)

Charmley, John, *Churchill: The End of Glory, a Political Biography* (London, 1993)

Chatfield, A. E. M., *The Navy and Defence: The Autobiography of Admiral of the Fleet Lord Chatfield* (London, 1942)

Chermayeff, Serge, *The New Research Laboratories of Imperial Chemical Industries Limited: Dyestuffs Group, Blackley, Manchester* (Birmingham, c. 1939)

Chesneau, Roger, *Conway's All The World's Fighting Ships, 1922–1946* (London, 1980)

Chester, Daniel Norman (ed.), *Lessons of the British War Economy* (Cambridge, 1951)

Churchill, Winston, *Step by Step 1936–1939* (London, 1939)

—, *Blood, Sweat, and Tears* (New York, 1941)

—, *The Second World War*, Vol. I: *The Gathering Storm* (London, 1948)

—, *The Second World War*, Vol. II: *Their Finest Hour* (London, 1949)

—, *The Second World War*, Vol. III: *The Grand Alliance* (London, 1950)

—, *The Second World War*, Vol. IV: *The Hinge of Fate* (London, 1951)

—, *The Second World War*, Vol. V: *Closing the Ring* (London, 1952)

—, *The Second World War*, Vol. VI: *Triumph and Tragedy* (London, 1954)

Churchill, Winston and Robert Rhodes James (ed.), *Winston S. Churchill: His Complete Speeches, 1897–1963*, Vols. 6 and 7 (London, 1974)

Clarence-Smith, William G., 'The Battle for Rubber in the Second World War: Cooperation and Resistance' (2009), Commodities of Empire Project, Working Paper No. 14

Clark, Ronald W., *Tizard* (London, 1965)

—, *Sir Edward Appleton, GBE, KCB, FRS* (Oxford, 1971)

—, 'Science and Technology, 1919–1945', in R. Higham (ed.), *British Military History: A Guide to Sources* (London, 1972), pp. 542–65

—, *The Life of Bertrand Russell* (Harmondsworth, 1978)

Clarke, Peter, *The Cripps Version: The Life of Sir Stafford Cripps 1889–1952* (London, 2002)

Clarke, R. W. B., *The Economic Effort of War* (London, 1940)

Clarke, Sabine, 'A Technocratic Imperial State? The Colonial Office and Scientific Research, 1940–1960', *Twentieth Century British History* 18 (2007), pp. 1–28

Clay, Henry, 'The Economic Outlook of the United Kingdom', *The American Economic Review* 37 (May 1947), pp. 12–20

Clayton, Robert and Joan Algar, *The GEC Laboratories 1919–1984* (London, 1989)

Cocroft, Wayne D., *Dangerous Energy: The Archaeology of Gunpowder and Military Explosives Manufacture* (Swindon, 2000)

Cole, Major D. H., *Imperial Military Geography: General Characteristics of the Empire in Relation to Defence*, 7th edn (London, 1934)

Collingham, Lizzie, *The Taste of War: World War Two and the Battle for Food* (London, 2011)

Colls, Robert, *Identity of England* (Oxford, 2002)

Connell, John, *Wavell: Scholar and Soldier* (London, 1964)

Connelly, Mark, *We Can Take It! Britain and the Memory of the Second World War* (London, 2004)

Cook, Chris and John Ramsden (eds.), *By-Elections in British Politics* (London, 1973)

Cooter, Roger, 'The Rise and Decline of the Medical Member: Doctors and Parliament in Edwardian and Interwar Britain', *Bulletin of the History of Medicine* 78 (2004), pp. 59–107

Cooter, R., S. Sturdy and M. Harrison (eds.), *War, Medicine and Modernity* (Stroud, 1998)

Copeland, B. Jack, and others, *Colossus: The Secrets of Bletchley Park's Codebreaking Computers* (Oxford, 2006)

Corley, T. A. B., *A History of the Burmah Oil Company*, Vol. II: *1924–1966* (London, 1988)

Cortés Conde, Roberto, *The Political Economy of Argentina in the Twentieth Century* (Cambridge, 2009)

Courtauld, Simon, *As I was Going to St Ives: A Life of Derek Jackson* (Norwich, 2007)

Cox, Sebastian (ed.), *The Strategic Air War against Germany, 1939–1945: Report of the British Bombing Survey Unit* (London, 1998)

Cripps, Sir Stafford, address to the BAAS conference on 'Mineral Resources and the Atlantic Charter', 25 July 1942, *The Advancement of Science* 2 (October 1942), p. 240

Crook, Paul, 'Science and War: Radical Scientists and the Tizard–Cherwell Area Bombing Debate in Britain', *War & Society* 12 (1994), pp. 69–101

—, 'The Case against Area Bombing', in Peter Hore (ed.), *Patrick Blackett: Sailor, Scientist, Socialist* (London, 2003), pp. 167–86

Croucher, Richard, *Engineers at War 1939–1945* (London, 1982)

Crow, Alwyn, 'The Rocket as a Weapon of War in the British Forces', 34th Thomas Hawksley Lecture, November 1947, *Proceedings of the Institution of Mechanical Engineers* 158 (1948), pp. 15–21

Crowcroft, Robert, '"What is Happening in Europe?" Richard Stokes, Fascism, and the Anti-War Movement in the British Labour Party during the Second World War and After', *History* 93 (October 2008), pp. 514–30

Crowson, N. J., *Facing Fascism: The Conservative Party and the European Dictators 1935–1940* (London, 1997)

Crowther, Geoffrey, *Ways and Means of War* (Oxford, 1940)

Crowther, J. G., *Science in Modern Society* (London, 1967)

Crowther, J. G. and Richard Whiddington, *Science at War* (London, 1947)

Crowther, J. G., O. J. R. Howarth and D. P. Riley, *Science and World Order* (Harmondsworth, 1942)

Dainton, Fred, *Doubts and Certainties: A Personal Memoir of the Twentieth Century* (Sheffield, 2001)

Dale, Iain (ed.), *Conservative Party General Election Manifestos, 1900–1997* (London, 2000)

— (ed.), *Labour Party General Election Manifestos, 1900–1997* (London, 2000)

— (ed.), *Liberal Party General Election Manifestos, 1900–1997* (London, 2000)

Dalton, Hugh, reviewing W. N. Medlicott, *The Economic Blockade*, Vol. I: *1939–41* (London, 1952), in *The Economic Journal* 63 (September 1953), pp. 660–68

Danchev, Alex, *Alchemist of War: The Life of Basil Liddell Hart* (London, 1998)

Darwin, John, *The Empire Project: The Rise and Fall of the British World-System, 1830–1970* (Cambridge, 2009)

Daunton, M. J., *Wealth and Welfare: An Economic and Social History of Britain, 1851–1951* (Oxford, 2007)

Daunton, Martin, 'Britain and Globalisation since 1850: II. The Rise of

Insular Capitalism, 1914–1939', *Transactions of the Royal Historical Society* 17 (2007), pp. 1–33

Daunton, Martin and Bernhard Rieger (eds.), *Meanings of Modernity: Britain from the Late-Victorian Era to World War II* (Oxford, 2001)

Davenport, Nicholas, *Vested Interests or Common Pool* (London, 1942)

Davy, M. J. B., *Air Power and Civilization* (London, 1941)

de Kerbrech, Richard P., *Harland & Wolff's Empire Food Ships 1934–1948: A Link with the Southern Dominions* (Isle of Wight, 1998)

Dear, I. C. B. (ed.), *The Oxford Companion to the Second World War* (Oxford, 1995)

Delaforce, Patrick, *Churchill's Secret Weapons: The Story of Hobart's Funnies* (London, 2006)

D'Este, Carlo, *Warlord: A Life of Churchill at War, 1874–1945* (London, 2009)

Devons, Ely, *Planning in Practice: Essays in Aircraft Planning in Wartime* (Cambridge, 1950)

Dewey, Peter, 'The Supply of Tractors in Wartime', in Brian Short, Charles Watkins and John Martin (eds.), *The Front Line of Freedom* (London, 2006), pp. 89–100

—, *'Iron Harvests of the Field': The Making of Farm Machinery in Britain since 1800* (Lancaster, 2008)

Dobinson, Colin, *Fields of Deception: Britain's Bombing Decoys of World War II* (London, 2000)

—, *AA Command: Britain's Anti-Aircraft Defences of the Second World War* (London, 2001)

Dobson, Alan P., *US Wartime Aid to Britain 1940–1946* (New York, 1986)

Donnelly, Mark, *Britain in the Second World War* (London, 1999)

Downing, Taylor, *Churchill's War Lab: Code Breakers, Boffins and Innovators: The Mavericks Churchill Led to Victory* (London, 2010)

Dummelow, John, *1899–1949* (history of Metropolitan-Vickers) (Manchester, 1949)

Dyson, Freeman, *Disturbing the Universe* (New York, 1979)

Edgerton, David, *England and the Aeroplane: An Essay on a Militant and Technological Nation* (London, 1991)

—, 'Liberal Militarism and the British State', *New Left Review* 185 (Jan–Feb 1991), pp. 138–69.

—, 'The Prophet Militant and Industrial: The Peculiarities of Correlli Barnett', *Twentieth Century British History* 2 (1991), pp. 360–79

—, 'British Scientific Intellectuals and the Relations of Science, Technology, and War', in Paul Forman and José Manuel Sanchez-Ron (eds.), *National*

Military Establishments and the Advancement of Science and Technology (Dordrecht, 1996), pp. 1–35

—, *Science, Technology and the British Industrial 'Decline', 1870–1970* (Cambridge, 1996)

—, *Warfare State: Britain, 1920–1970* (Cambridge, 2005)

—, *The Shock of the Old: Technology and Global History since 1900* (London, 2007)

—, 'The Primacy of Foreign Policy? Britain in the Second World War', in William Mulligan and Brendan Simms (eds.), *The Primacy of Foreign Policy in British History, 1660–2000: How Strategic Concerns Shaped Modern Britain* (London, 2010).

—, 'War, Reconstruction and the Nationalisation of Britain, 1939–1951', in David Feldman, Mark Mazower and Jessica Reinisch (eds.), *Post-War Reconstruction in Europe: International Perspectives, 1945–1949* (Oxford, 2010)

Edgerton, D. E. H. and S. M. Horrocks, 'British Industrial Research and Development before 1945', *Economic History Review* 37 (1994), pp. 213–38

Egerer, Gerald, 'Protection and Imperial Preference in Britain: The Case of Wheat 1925–1960', *The Canadian Journal of Economics and Political Science/Revue Canadienne d'Economique et de Science Politique* 31 (August 1965), pp. 382–9

Eisler, Paul, *My Life with the Printed Circuit* (Bethlehem, 1989)

Ellis, John, *Brute Force: Allied Strategy and Tactics in the Second World War* (London, 1990)

—, *The World War II Databook: The Essential Facts and Figures for All the Combatants* (London, 1993)

Ellis, R. J., *He Walks Alone: The Public and Private Life of Captain Cunningham-Reid, DFC Member of Parliament, 1922–45* (London, 1945)

Elphick, Peter, *Liberty: The Ships That Won the War* (London, 2001)

Emmerson, Sir Harold, *The Ministry of Works* (London, 1956)

Emsley, Clive, Arthur Marwick and Wendy Simpson (eds.), *War, Peace and Social Change in Twentieth-Century Europe* (Milton Keynes, 1989)

Evans, Rob, *Gassed: British Chemical Warfare Experiments on Humans at Porton Down* (London, 2000)

Exton, Charles, *The Secret War Factory: Cowbridge Confidential* (Milton Keynes, 2007)

Falconer, Jonathan, *RAF Bomber Command in Fact, Film and Fiction* (Stroud, 1996)

Falls, Cyril, *Ordeal by Battle* (London, 1943)

—, *The Second World War: A Short History* (London, 1948)

Ferguson, Niall, *The Pity of War* (London, 1998)

—, *The War of the World: History's Age of Hatred* (London, 2006)

Fergusson, James, *The Vitamin Murders: Who Killed Healthy Eating in Britain?* (London, 2007)

Fernbach, David, 'Tom Wintringham and Socialist Defence Strategy', *History Workshop Journal* XIV (1982), pp. 63–91

Ferrier, R. W. and J. H. Bamberg, *The History of the British Petroleum Company*, Vol. 2 (Cambridge, 1994)

Ferry, Georgina, *Max Perutz and the Secret of Life* (London, 2007)

Feske, Victor, *From Belloc to Churchill: Private Scholars, Public Culture and the Crisis of British Liberalism 1900–1939* (Chapel Hill, 1996)

Field, J. V., 'British Cryptanalysis: The Breaking of "Fish" Traffic', in Ad Maas and Hans Hooijmaijers (eds.), *Scientific Research in World War II: What Scientists Did in the War* (London, 2009), pp. 213–31

Fielding, Steven, 'The Second World War and Popular Radicalism: The Significance of the "Movement Away from Party"', *History* 80 (1995), pp. 38–58

Finch, M. H. J., *A Political Economy of Uruguay since 1870* (London, 1981)

Fitzpatrick, Jim, *The Bicycle in Wartime: An Illustrated History* (Washington DC, 1998)

Fletcher, David, *The Great Tank Scandal: British Armour in the Second World War*, Part I (London, 1989)

—, *The Universal Tank: British Armour in the Second World War*, Part II (London, 1993)

Foot, Michael, *Aneurin Bevan: A Biography*, Vol. 1: *1897–1945* (London, 1967)

Foot, William, *Beaches, Fields, Streets, and Hills: The Anti-Invasion Landscapes of England, 1940* (York, 2006)

Forbes, R. J. and D. R. O'Beirne, *The Technical Development of the Royal Dutch/Shell* (Leiden, 1957)

Fort, Adrian, *Prof: The Life of Frederick Lindemann* (London, 2004)

Foss, Christopher F. and Peter McKenzie, *The Vickers Tanks* (Wellingborough, 1988)

Fox, Robert and Graeme Gooday (eds.), *Physics in Oxford 1839–1939* (Oxford, 2005)

Francis, Martin, *The Flyer: British Culture and the Royal Air Force, 1939–1945* (Oxford, 2008).

Frankland, Mark, *Radio Man: The Remarkable Rise and Fall of C. O. Stanley* (London, 2002)

Frankland, Noble, *The Bombing Offensive against Germany: Outlines and Perspectives* (London, 1965)

—, *Bomber Offensive: The Devastation of Europe* (London, 1970)

—, *History at War: The Campaigns of an Historian* (London, 1998)

Franklin, George, *Britain's Anti-Submarine Capability, 1919–1939* (London, 2003)

Franks, Norman, *Claims to Fame: The Lancaster* (London, 1995)

Freedman, Paul, *The Principles of Scientific Research* (London, 1949)

French, David, *The British Way in Warfare 1688–2000* (London, 1990)

—, 'Colonel Blimp and the British Army: British Divisional Commanders in the War against Germany, 1939–1945', *English Historical Review* 111 (1996), pp. 1182–1201.

—, *Raising Churchill's Army: The British Army and the War against Germany 1919–1945* (Oxford, 2000)

Friedman, Norman, *British Carrier Aviation: The Evolution of Ships and Their Aircraft* (London, 1988)

Frieser, Karl-Heinz, *The Blitzkrieg Legend: The 1940 Campaign in the West* (Annapolis, 2005)

Fry, Geoffrey, *The Politics of Crisis* (London, 2001)

Fuller, J. F. C., *Armament and History: A Study of the Influence of Armament on History from the Dawn of Classical Warfare to the Second World War* (London, 1946)

—, *The Conduct of War 1789–1961* (London, 1961)

Furse, Anthony, *Wilfrid Freeman: The Genius behind Allied Survival and Air Supremacy* (Staplehurst, 2000)

Gardiner, Juliet, *Wartime: Britain 1939–1945* (London, 2004)

Garrett, Stephen A., *Ethics and Airpower in World War II* (London/New York, 1993)

Gat, Azar, *War in Human Civilization* (Oxford, 2006)

Gay, Hannah, *The History of Imperial College London, 1907–2007* (London, 2007)

Gerrard, F., *Meat Technology* (London, 1951)

Gilbert, Martin, *Road to Victory: Winston S. Churchill 1941–1945* (London, 1986)

— (ed.), *The Churchill War Papers*, Vol. 1: *At the Admiralty, September 1939–May 1940* (London, 1993)

Glancey, Jonathan, *Spitfire: The Illustrated Biography* (London, 2008)

Goldsmith, M., *Sage: A Life of J. D. Bernal* (London, 1980)

Golley, John, *Whittle: The True Story* (Washington, DC, 1987)

Gooch, John (ed.), *Airpower: Theory and Practice* (London, 1995)

Gordon, G. A. H., *British Seapower and Procurement between the Wars: A Re-Appraisal of Rearmament* (London, 1988)

Goulter, C. J. M., *A Forgotten Offensive: Royal Air Force Coastal Command's Anti-Shipping Campaign, 1940–1945* (London, 1995)

Grayson, Richard, 'Leo Amery's Imperialist Alternative to Appeasement in the 1930s', *Twentieth Century British History* 17 (2006), pp. 489–515

Green, E. H. H. and D. M. Tanner, *The Strange Survival of Liberal England: Political Leaders, Moral Values and the Reception of Economic Debate* (Cambridge, 2007)

Greenleaf, W. H., *The British Political Tradition*, 3 vols. (London, 1983–7)

Greenwood, David, 'Conflicts of Interest: The Genesis of Synthetic Anti-malarial Agents in Peace and War', *Journal of Antimicrobial Chemotherapy* 36 (1995), pp. 857–72

Griffiths, Richard, *Fellow Travellers of the Right: British Enthusiasts for Nazi Germany 1933–39* (Oxford, 1983)

Grove, Eric J., *The Royal Navy since 1815: A New Short History* (Houndmills, 2005)

Guderian, Heinz, *Panzer Leader* (New York, 2001)

Hackmann, Willem, *Seek and Strike: Sonar, Anti-Submarine Warfare and the Royal Navy 1914–54* (London, 1984)

Halévy, Elie, *The Era of Tyrannies: Essays on Socialism and War*, trans. R. Webb (London, 1967)

Hanbury-Brown, R., *Boffin* (Bristol, 1991)

Hannah, Leslie, 'Logistics, Market Size, and Giant Plants in the Early Twentieth Century: A Global View', *Journal of Economic History* 68 (March 2008), pp. 46–79

Harbottle, H. R., 'The Telecommunications Network for Defence, Part II: The Network of Telephone Circuits and the Defence Telecommunications Control', *Post Office Electrical Engineers' Journal* 38 (January 1946; Victory number)

Hardie, D. W. F., *A History of the Chemical Industry in Widnes* (London, 1950)

Harris, Sir Arthur T., *Despatch on War Operations: 23rd February 1942 to 8th May 1945* (London, 1995)

Harris, Sir Arthur, *Bomber Offensive* (London, 1990; first publ. 1947)

Harris, J. P., *Men, Ideas and Tanks: British Military Thought and Armoured Forces, 1903–1939* (Manchester, 1995)

Harris, José, 'Society and the State in Twentieth-Century Britain', in F. M. L. Thompson, *Cambridge Social History of Britain*, Vol. 3: *Social Agencies and Institutions* (Cambridge, 1990), pp. 63–118

Harrison, Mark, 'Resource Mobilization for World War II: The U.S.A., U.K., U.S.S.R., and Germany, 1938–1945', *Economic History Review* 41 (1988), pp. 171–92

— (ed.), *The Economics of World War II: Six Great Powers in International Comparison* (Cambridge, 1998)

Harrison, Mark, 'The Medicalisation of War, the Militarisation of Medicine', *Social History of Medicine* 9 (1996), 267–76

—, 'Medicine and the Management of Modern Warfare', *History of Science* 34 (1996), pp. 379–410

—, *Medicine and Victory: British Military Medicine in the Second World War* (Oxford, 2004)

Harrison, Michael, *Mulberry: The Return in Triumph* (London, 1965)

Harrod, Roy, *The Prof: A Personal Memoir of Lord Cherwell* (London, 1959)

Hart, Stephen A., *Sherman Firefly vs Tiger, Normandy 1944* (Oxford, 2007)

Hartcup, Guy, *Challenge of War: Scientific and Engineering Contributions to World War Two* (London, 1970)

—, *Code Name Mulberry: The Planning, Building and Operation of the Normandy Harbours* (Newton Abbot, 1977)

—, *The War of Invention: Scientific Developments, 1914–18* (London, 1988)

—, *The Silent Revolution: The Development of Conventional Weapons 1945–85* (London, 1993)

—, *The Effect of Science on the Second World War* (Houndmills, 2003)

Harvey, A. D., *Collision of Empires: Britain in Three World Wars, 1793–1945* (London, 1994; first publ. 1992)

Harvey, W. J. and R. J. Solly, *BP Tankers: A Group Fleet History* (London, 2005)

Hastings, Max, *Finest Years: Churchill as Warlord 1940–1945* (London, 2009)

Hayward, Roger, 'British Air-Dropped Depth Charges and Anti-Ship Torpedoes', *Royal Air Force Historical Society Journal* 45 (2009), pp. 121–36

Helmore, W., *Air Commentary* (London, 1942)

Hennessy, Peter, *Whitehall* (London, 1989)

Henry, Chris and Brian Delf, *British Anti-Tank Artillery 1939–45* (London, 2004)

Hill, A. V., *The Ethical Dilemma of Science* (London, 1960)

Hill, Alexander, 'British Lend-Lease Aid and the Soviet War Effort, June 1941–June 1942', *Journal of Military History* 71 (July 2007), pp. 773–808

Hinsley, F. H. and Alan Stripp (eds.), *Codebreakers: The Inside Story of Bletchley Park* (Oxford, 2001)

Hinton, James, 'Coventry Communism: A Study of Factory Politics in the Second World War', *History Workshop* 10 (Autumn 1980), pp. 90–118

—, *Protests and Visions: Peace Politics in 20th-Century Britain* (London, 1989)

Hodge, Joseph Morgan, *The Triumph of the Expert: Agrarian Doctrines of Development and the Legacies of British Colonialism* (Athens, Ohio, 2007)

Holden-Reid, Brian, *J. F. C. Fuller: Military Thinker* (Basingstoke, 1987)

Holley, Jr., I. B., *Ideas and Weapons: Exploitation of the Aerial Weapon by the United States During World War II* (Hamden, 1971)

Holman, Valerie, *Print for Victory: Book Publishing in England 1939–1945* (London, 2009)

Hooker, Sir S., *Not Much of an Engineer: An Autobiography* (Shrewsbury, 1984)

Hooper, Bill, *Pilot Officer Prune's Picture Parade!* (London, 1991)

Hore, Peter (ed.), *Patrick Blackett: Sailor, Scientist, Socialist* (London, 2003)

Horrocks, Sally, 'Nutrition Science and the Food and Pharmaceutical Industries in Inter-War Britain', in David F. Smith (ed.), *Nutrition in Britain: Science, Scientists and Politics in the Twentieth Century* (London, 1997), pp. 53–74

House, Frank H., *Timber at War: An Account of the Organisation and Activities of the Timber Control 1939–1945* (London, 1965)

Howard, Michael, *The Continental Commitment: The Dilemma of British Defence Policy in the Era of the Two World Wars* (London, 1972)

—, *Captain Professor: A Life in War and Peace* (London, 2006)

Howe, Anthony, *Free Trade and Liberal England 1846–1946* (Oxford, 1997)

Howe, Stephen, 'Empire and Ideology', in Sarah Stockwell (ed.), *The British Empire: Themes and Perspectives* (Oxford, 2008), pp. 157–76

Hucker, Daniel, 'Franco-British Relations and the Question of Conscription in Britain, 1938–1939', *Contemporary European History* 17 (2008), pp. 437–56

Hughes, Matthew and William J. Philpott (eds.), *Palgrave Advances in Modern Military History* (Basingstoke, 2006)

Imlay, Talbot C., *Facing the Second World War: Strategy, Politics, and Economics in Britain and France 1938–1940* (Oxford, 2003)

—, 'A Reassessment of Anglo-French Strategy during the Phony War, 1939–1940', *English Historical Review* CXIX (2004), pp. 333–72

Imlay, Talbot C. and Monica Duffy Toft (eds.), *The Fog of Peace and War Planning: Military and Strategic Planning under Uncertainty* (London, 2006)

Imperial Chemical Industries, Ltd, *Imperial Chemical Industries, Ltd: A Short Account of the Activities of the Company* (London, 1929)

Imperial Chemical Industries, *Pharmaceutical Research in ICI, 1936–1957* (London, 1957)

Isserlis, L., 'Tramp Shipping Cargoes, and Freights', *Journal of the Royal Statistical Society* 101 (1938), pp. 53–146

Jackson, Ashley, *The British Empire and the Second World War* (London, 2006)

James, John, *The Paladins: A Social History of the RAF up to the Outbreak of World War II* (London, 1990)

James, Lawrence, *Warrior Race: A History of the British at War* (London, 2001)

Jeffery, Keith, 'The Second World War', in Judith M. Brown and Wm Roger

Louis (eds.), *The Oxford History of the British Empire: The Twentieth Century* (Oxford, 1999), pp. 306–28

Jefferys, Kevin, *The Churchill Coalition and Wartime Politics, 1940–1945* (Manchester, 1991)

Jenkins, Roy, *Churchill* (London, 2001)

Jewkes, J., *Ordeal by Planning* (London, 1948)

Johnman, Lewis and Hugh Murphy, *British Shipbuilding and the State since 1918: A Political Economy of Decline* (Exeter, 2002)

Johnson, Brian, *The Secret War* (London, 1978)

Johnson, S. W., 'Fray Bentos, Uruguay – 1920–1940', *The Bulletin* LII (May 2009), p. 13 (published Buenos Aires)

Jones, R. and Oliver Marriott, *Anatomy of a Merger* (London, 1970)

Jones, R. V., *Most Secret War* (London, 1978)

—, *Reflections on Intelligence* (London, 1989)

Jones, Thomas, *A Diary with Letters, 1931–1950* (London, 1954)

Jordan, Roger W., *The World's Merchant Fleets, 1939: The Particulars and Wartime Fates of 6000 Ships* (Annapolis, 1999)

Kaldor, Mary, *The Baroque Arsenal* (London, 1982)

Keegan, John, *The Price of Admiralty: War at Sea from Man-of-War to Submarine* (London, 1988)

—, *Churchill* (London, 2002)

Kemp, Norman, *The Devices of War* (London, 1956)

Kendall, David and Kenneth Post, 'The British 3-Inch Anti-Aircraft Rocket. Part One: Dive-Bombers', *Notes and Records of the Royal Society of London* 50 (July 1996), pp. 229–39

—, 'The British 3-Inch Anti-Aircraft Rocket. Part Two: High-Flying Bombers', *Notes and Records of the Royal Society of London* 51 (January 1997), pp. 133–40

Kennedy, Carol, *ICI: The Company That Changed Our Lives*, 2nd edn (London, 1993)

Kennedy, Paul, *The Rise and Fall of British Naval Mastery* (London, 1983)

Kerr, Prue, 'An Account of the Washington Lend-Lease Negotiations 1944 by E. A. G. Robinson', *Contributions to Political Economy* 22 (2003), pp. 63–78

Kidd, Alan and David Nicolls (eds.), *The Making of the British Middle Class? Studies of Regional and Cultural Diversity since the Eighteenth Century* (Stroud, 1998)

Kier, Elizabeth, *Imagining War: French and British Military Doctrines between the Wars* (Princeton, 1997)

Kiesling, Eugenia C., 'Illuminating Strange Defeat and Pyrrhic Victory: The

Historian Robert A. Doughty', *Journal of Military History* 71 (July 2007), pp. 875–88

Kindersley, R. M., 'British Overseas Investments, 1937', *The Economic Journal* 48 (December 1938), pp. 609–34

King-Hall, Stephen, *Total Victory* (London, 1941)

—, *My Naval Life 1906–1929* (London, 1952)

Kington, George, *Bombs and Rockets: The University of Bristol Chemistry Department in World War II: A View from the Laboratory Bench* (Truro, 2004)

Kinsey, G., *Orfordness, Secret Site: A History of the Establishment, 1915–1980* (Lavenham, 1981)

Kirby, Maurice W., *Operational Research in War and Peace: The British Experience from the 1930s to 1970* (London, 2003)

Kitchen, Martin, *Rommel's Desert War* (Cambridge, 2009)

Koch, H. W., 'The Strategic Air Offensive against Germany: The Early Phase, May–September 1940', *The Historical Journal* 34 (March 1991), pp. 117–41

Kraus, Jerome, 'The British Electron-Tube and Semi-Conductor Industry 1935–1962', *Technology and Culture*, 11 (1968), pp. 544–61.

Krebs, Hans, *Reminiscences and Reflections* (Oxford, 1981)

Krige, John and Dominique Pestre (eds.), *Science in the Twentieth Century* (Amsterdam, 1997)

Lampe, David, *Pyke, the Unknown Genius* (London, 1959)

Lane, Tony, *The Merchant Seamen's War* (Liverpool, 1990)

Latham, Colin and Anne Stobbs, *Radar: A Wartime Miracle* (Stroud, 1996)

Laurie, J. P., Royal Naval Scientific Service, 'The Triumph of Plastic Armour', *Shipping World* 29 (August 1945)

Le Bailly, Louis, *From Fisher to the Falklands* (London, 1991)

Lee, J. M., *Reviewing the Machinery of Government, 1942–1952: An Essay on the Anderson Committee and Its Successors* (1977)

Lee, Sabine, *Sir Rudolf Peierls: Selected Private and Scientific Correspondence*, Vol. 1 (London, 2007)

Lesch, John E., *The First Miracle Drugs: How the Sulfa Drugs Transformed Medicine* (Oxford, 2007)

Lewis, Roy and Angus Maude, *The English Middle Classes* (London, 1949)

Liddell Hart, Basil, *The Tanks: The History of the Royal Tank Regiment and Its Predecessors, Heavy Branch Machine-Gun Corps, Tank Corps and Royal Tank Corps, 1914–1945*, Vol. 2 (London, 1959)

—, *History of the Second World War* (London, 1970)

Lipman, M. J., *Memoirs of a Socialist Business Man* (London, 1980)

Llewellyn-Jones, Malcolm, 'A Clash of Cultures: The Case for Large Convoys', in Peter Hore (ed.), *Patrick Blackett: Sailor, Scientist and Socialist* (London, 2003), pp. 138–66

—, *The Royal Navy and Anti-Submarine Warfare, 1917–49* (London, 2005)

Long, David, *Towards a New Liberal Internationalism: The International Theory of J. A. Hobson* (Cambridge, 1996)

Long, David and Peter Wilson (eds.), *Thinkers of the Twenty Years' Crisis: Inter-War Idealism Reassessed* (Oxford, 1995)

Lovell, Bernard, 'Patrick Maynard Stuart Blackett', *Biographical Memoirs of Fellows of the Royal Society* (London, 1975)

Ludvigsen, Karl, *Battle for the Beetle* (London, 2000)

Lukacs, John, *Five Days in London, May 1940* (London, 1999)

Lund, Erik, 'The Industrial History of Strategy: Reevaluating the Wartime Record of the British Aviation Industry in Comparative Perspective, 1919–1945', *Journal of Military History* 66 (January 1998), pp. 75–99

Maas, Ad and Hans Hooijmaijers (eds.), *Scientific Research in World War II: What Scientists Did in the War* (London, 2009)

MacDougall, Donald, *Don and Mandarin: Memoirs of an Economist* (London, 1987)

MacDougall, G. D. A., 'The Prime Minister's Statistical Section', in D. N. Chester (ed.), *Lessons of the British War Economy* (Cambridge, 1951), pp. 58–68

Mackenzie, S. P., *The Battle of Britain on Screen: 'The Few' in British Film and Television Drama* (Edinburgh, 2007)

MacLeod, Roy, '"All for Each and Each for All": Reflections on Anglo-American and Commonwealth Scientific Cooperation, 1940–1945', *Albion* 26 (Spring, 1994), pp. 79–112

—, 'Science for Imperial Efficiency and Social Change: Reflections on the British Science Guild, 1905–1936', *Public Understanding of Science* (1994), pp. 155–93

— (ed.), *Science and the Pacific War: Science and Survival in the Pacific, 1939–1945* (Dordrecht, 2000)

Macmillan, Harold, *The Blast of War: 1939–1945* (London, 1967)

Macnicol, J., 'The Evacuation of Schoolchildren', in Harold L. Smith (ed.), *War and Social Change: British Society in the Second World War* (Manchester, 1986), pp. 3–31

Macrae, R. Stuart, *Winston Churchill's Toyshop* (Kineton, 1971)

Maddox, Brenda, *Rosalind Franklin: The Dark Lady of DNA* (London, 2003)

Maiolo, Joseph A., *The Royal Navy and Nazi Germany, 1933–39: A Study in Appeasement and the Origins of the Second World War* (Basingstoke, 1998)

—, *Cry Havoc: The Arms Race and the Second World War, 1931–1941* (London, 2010)

Mandler, Peter, *The English National Character: the history of an idea from Edmund Burke to Tony Blair* (London, 2006)

Marks, H. F., *A Hundred Years of British Food & Farming: A Statistical Survey* (London, 1989)

Marlowe, John, *Late Victorian: The Life of Sir Arnold Talbot Wilson* (London, 1967)

Marquand, David, *Britain Since 1918: The Strange Career of British Democracy* (London, 2008)

Marr, Andrew, *The Making of Modern Britain: From Queen Victoria to V.E. Day* (London, 2009)

Marwick, Arthur, *Britain in the Century of Total War: War, Peace and Social Change 1900–1967* (Harmondsworth, 1970)

—, *The Explosion of British Society 1914–1970* (London, 1971)

Mass-Observation, *Enquiry into British War Production*, Part 1: *People in Production* (London, 1942)

—, *War Factory* (London 1987; first publ. 1943)

Matless, David, *Landscape and Englishness* (London, 1998)

Matthews, R. C. O., C. H. Feinstein and J. C. Odling-Smee, *British Economic Growth 1856–1973* (Oxford, 1982)

Mayer, A. K. and C. J. Lawrence (eds.), *Regenerating England: Science, Medicine and Culture in Interwar Britain* (Amsterdam, 2000),

Mayhew, E. R., *The Reconstruction of Warriors: Archibald McIndoe, the Royal Air Force, and the Guinea Pig Club* (London, 2004)

Mazower, Mark, *Hitler's Empire: Nazi Rule in Occupied Europe* (London, 2008)

McGucken, William, 'The Central Organisation of Scientific and Technical Advice in the United Kingdom during the Second World War', *Minerva* 17 (1979), pp. 33–69

—, *Scientists, Society and the State* (Columbus, O., 1984)

McKercher, B., 'The Greatest Power on Earth: Britain in the 1920s', *International History Review* 13 (1991), pp. 661–880

McKibbin, Ross, *Classes and Cultures: England 1918–1951* (Oxford, 1998)

—, 'Politics and the Medical Hero: A. J. Cronin's *The Citadel*', *English Historical Review* CXXIII (2008), pp. 651–77

McLaine, Ian, *Ministry of Morale: Home Front Morale and the Ministry of Information in World War II* (London, 1979)

McNeill, William H., *The Pursuit of Power: Technology, Armed Force, and Society since A.D. 1000* (Oxford, 1982)

Mearsheimer, John J., *Liddell Hart and the Weight of History* (Ithaca, NY, 1988)

—, *The Tragedy of Great Power Politics* (New York, 2003)

Mellanby, Kenneth, *Human Guinea Pigs* (London, 1945)

Mensforth, Eric, 'Airframe Production', *Proceedings of the Institution of Mechanical Engineers* 156 (1947), pp. 24–38

—, *Family Engineers* (London, 1981)

Middlemas, Keith, *Politics in Industrial Society: The Experience of the British System since 1911* (London, 1979)

—, *Power, Competition and the State*, 3 vols. (London, 1986–1991)

Milward, Alan S., *War, Economy and Society 1939–1945* (London, 1977)

—, *The Economic Effects of the Two World Wars on Britain*, 2nd edn (London, 1984)

Milward, Alan S. and George Brennan, *Britain's Place in the World: A Historical Enquiry into Import Controls, 1945–60* (London, 1996)

Mitchell, Timothy, *Rule of Experts: Egypt, Techno-Politics, and Modernity* (Berkeley, Calif., 2002)

Mitchell, W. H. and L. A. Sawyer, *Empire Ships of World War II* (Liverpool, 1965)

Montagu, Ivor, *The Traitor Class* (London, 1940)

Morgan, David and Mary Evans, *The Battle for Britain: Citizenship and Ideology in the Second World War* (London, 1993)

Morpurgo, J. E., *Barnes Wallis: A Biography* (London, 1972)

Morrell, Jack, *Science at Oxford 1914–1939: Transforming an Arts University* (Oxford, 1997)

Moss, Michael S. and John R. Hume, *Workshop of the British Empire: Engineering and Shipbuilding in the West of Scotland* (London, 1977)

—, *Shipbuilders of the World: 125 Years of Harland and Wolff, Belfast 1861–1986* (Belfast, 1986)

Mott, Sir Nevill, *A Life in Science* (London, 1986)

Mulligan, William and Brendan Simms (eds.), *The Primacy of Foreign Policy in British History, 1660–2000: How Strategic Concerns Shaped Modern Britain* (London, 2010)

Murphy, J. T., *Victory Production!* (London, 1942)

Murray, Williamson and Allan R. Millett, *A War to be Won: Fighting the Second World War, 1937–1945* (Cambridge, 2000)

Murrow, Edward R., *In Search of Light: The Broadcasts of Edward R. Murrow 1938–1961* (New York, 1967)

Nahum, Andrew, 'Two-Stroke or Turbine? The Aeronautical Research Committee and British Aero Engine Development in World War II', *Technology and Culture* 38 (April 1997), pp. 312–54

—, *Frank Whittle: Invention of the Jet* (Cambridge, 2004)

Nancarrow, F. E., 'The Telecommunications Network for Defence, Part I: The

Defence Teleprinter Network', *Post Office Electrical Engineers' Journal* 38 (January 1946; Victory number)

National Army Museum, *'Against All Odds': The British Army of 1939–40* (London, 1990)

Nevell, Michael, John Roberts and Jack Smith, *A History of the Royal Ordnance Factory Chorley* (Lancaster, 1999)

Newberry, Charles Allen, *Wartime St Pancras: A London Borough Defends Itself* (London, 2006)

Newman, Michael, *John Strachey* (Manchester, 1989)

Nicholas, Siân, *The Echo of War: Home Front Propaganda and the Wartime BBC, 1939–45* (Manchester, 1996)

Nicolson, Harold, *Public Faces: A Novel* (London, 1932)

—, *Diaries and Letters, 1939–1945* (London, 1967)

Norris, Robert S., *Racing for the Bomb: General Leslie R. Groves, the Manhattan Project's Indispensable Man* (South Royalton, Vt, 2002)

Nye, Mary Jo, *Blackett: Physics, War and Politics in the Twentieth Century* (Cambridge, Mass., 2004)

O'Connell, Robert L., *Of Arms and Men: A History of War, Weapons, and Aggression* (New York, 1989)

O'Dell, T. H., *Inventions and Official Secrecy: A History of Secret Patents in the United Kingdom* (Oxford, 1994)

Ollerenshaw, Philip, 'War, Industrial Mobilisation and Society in Northern Ireland, 1939–1945', *Contemporary European History* 16 (2007), pp. 169–97

Olson, Lynne, *Troublesome Young Men: The Churchill Conspiracy of 1940* (London, 2008)

Olson, Jr., Mancur, *The Economics of the Wartime Shortage: A History of British Food Supplies in the Napoleonic War and in World Wars I and II* (Durham, NC, 1963)

Omissi, David E., *Air Power and Colonial Control* (Manchester, 1990)

Ord, Lewis C., *Secrets of Industry* (London, 1944)

Orwell, George, *The Lion and the Unicorn: Socialism and British Genius* (London, 1941)

Overy, Richard, *The Air War, 1939–1945* (London, 1980)

—, *The Road to War* (London, 1989)

—, *Why the Allies Won* (London, 1995)

—, *Bomber Command, 1939–1945* (London, 1997)

Pam, David, *The Royal Small Arms Factory, Enfield and Its Workers* (Enfield, 1998)

Paris, Michael, *Warrior Nation: Images of War in British Popular Culture, 1850–2000* (London, 2000).

Parker, R. A. C., *The Second World War: A Short History*, rev. edn (Oxford, 1997)

Pawle, Gerard, *The Secret War, 1939–45* (London, 1956)

Pearton, Maurice, *The Knowledgeable State: Diplomacy, War and Technology since 1830* (London, 1980)

Peden, George, *British Rearmament and the Treasury 1932–1939* (Edinburgh, 1979)

—, *Arms, Economics and British Strategy: From Dreadnoughts to Hydrogen Bombs* (Cambridge, 2007)

Peebles, Hugh B., *Warshipbuilding on the Clyde: Naval Orders and the Prosperity of the Clyde Shipbuilding Industry, 1889–1939* (Edinburgh, 1987)

Peierls, Rudolf, *Bird of Passage: Recollections of a Physicist* (Princeton, 1985)

Pelling, Henry, *Britain and the Second World War* (London, 1970)

Penney, John, *Bristol at War* (Derby, 2002)

— (ed.), *German Air Operations against the Bristol Area, 1939–1945*, Vol. 4: *November 25th to December 11th 1940* (Bristol, 1991)

Perkin, Harold, *The Rise of Professional Society: England since 1880*, 2nd edn (London, 2002; first publ. 1989)

Perry, F. W., *The Commonwealth Armies* (Manchester, 1988)

Place, Timothy Harrison, *Military Training in the British Army: From Dunkirk to D-Day* (London, 2000)

Pollard, Sidney, *The Development of the British Economy 1914–1980*, 3rd edn (London, 1983)

Poole, J. B. and Kay Andrews (eds.), *The Government of Science in Britain* (London, 1972)

Price, Alfred, 'The Rocket-Firing Typhoons in Normandy', *Royal Air Force Historical Society Journal* 45 (2009), pp. 109–20

Priestley, J. B., *Postscripts* (London, 1940)

Pritchard, David, *The Radar War: Germany's Pioneering Achievement 1904–45* (Wellingborough, 1989)

Purcell, Hugh, *The Last English Revolutionary: Tom Wintringham, 1898–1949* (Stroud, 2004)

Rankin, Nicholas, *Churchill's Wizards: The British Genius for Deception 1914–1945* (London, 2008)

Rau, Erik P., 'Technological Systems, Expertise and Policy Making: The British Origins of Operational Research', in Michael Allen and Gabrielle Hecht (eds.), *Technologies of Power: Essays in Honor of Thomas Parke Hughes and Agatha Chipley Hughes* (Cambridge, Mass., 2001), pp. 215–52

Raudzens, George, 'War-Winning Weapons: The Measurement of Technological Determinism in Military History', *The Journal of Military History* 54 (October 1990), pp. 403–34

Reader, W. J., *Imperial Chemical Industries: A History*, Vol. II: *The First Quarter Century, 1926–1952* (London, 1975)

Redford, Duncan, 'The March 1943 Crisis in the Battle of the Atlantic: Myth and Reality', *History* 92 (2007), pp. 64–83

Reid, J. M., *James Lithgow, Master of Work* (London, 1964)

Reimann, Guenter, *Patents for Hitler* (London, 1945)

Rex, Millicent B., 'The University Constituencies in the Recent British Election', *The Journal of Politics* 8 (May 1946), pp. 201–11

Reynolds, David, *Rich Relations: The American Occupation of Britain 1942–1945* (London, 1996)

—, *In Command of History: Churchill Fighting and Writing the Second World War* (London, 2004)

—, *From World War to Cold War: Churchill, Roosevelt, and the International History of the 1940s* (Oxford, 2006)

Reynolds, Jaime and Ian Hunter, 'Tom Horabin: Liberal Class Warrior', *Journal of Liberal Democrat History* 28 (Autumn 2000), pp. 17–21

Richards, Jeffrey and Anthony Aldgate, *Best of British: Cinema and Society from 1930 to the Present* (London, 1999)

Richardson, Dick, *The Evolution of British Disarmament Policy in the 1920s* (London, 1989)

Ritchie, Sebastian, 'A New Audit of War: The Productivity of Britain's Wartime Aircraft Industry Reconsidered', *War & Society* 12 (1994), pp. 125–47

—, *Industry and Air Power: The Expansion of British Aircraft Production, 1935–1941* (London, 1997)

Roberts, Andrew, *Masters and Commanders* (London, 2008)

—, *The Storm of War: A New History of the Second World War* (London, 2009)

Roberts, Edwin A., *The Anglo-Marxists: A Study in Ideology and Culture* (London, 1997)

Roberts, Gerrylyn and Anna E. Simmons, 'British Chemists Abroad 1887–1971: The Dynamics of Chemists' Careers', *Annals of Science* 66 (2009), pp. 103–28

Roberts, R. O., 'Comparative Shipping and Shipbuilding Costs', *Economica* 14 (November 1947), pp. 296–309

Robertson, Dennis H., *Britain in the World Economy* (London, 1954)

Robinson, Austin, 'The National Institute: The Early Years', *National Institute Economic Review* 124 (1988), pp. 63–6

Rooth, Tim, *British Protectionism and the International Economy: Overseas Commercial Policy in the 1930s* (Cambridge, 1993)

Rose, Hilary and Steven Rose, *Science and Society* (Harmondsworth, 1969)

Rose, Kenneth, *Elusive Rothschild: The Life of Victor, Third Baron* (London, 2003)

Rose, Sonya O., *Which People's War? National Identity and Citizenship in Wartime Britain 1939–1945* (Oxford, 2003)

Rosenhead, Jonathan, 'Operational Research at the Cross-Roads: Cecil Gordon and the Development of Post-war OR', *Journal of the Operational Research Society* 40 (1989), pp. 3–28

Rostow, W. W., *Pre-Invasion Bombing Strategy: General Eisenhower's Decision of March 25, 1944* (Austin, Tex., 1981)

Royal Society of London, 'A Discussion of the Effects of the Two World Wars on the Organisation and Development of Science in the United Kingdom', *Proceedings of the Royal Society of London*, Pt A, 342 (1975), pp. 441–586

Rubinstein, W. D., 'The Secret of Leopold Amery (An Important Figure in the British Conservative Party with a Concealed Jewish Background)', *Historical Research* 73 (2000), pp. 175–96

Russell, Archibald, *Span of Wings* (Shrewsbury, 1992)

Ryan, Alan, *Bertrand Russell: A Political Life* (London, 1988)

Samasuwo, Nhamo, 'Food Production and War Supplies: Rhodesia's Beef Industry during the Second World War, 1939–1945', *Journal of Southern African Studies* 29 (June 2003), pp. 487–502

Samuel, Raphael (ed.), *Patriotism: The Making and Unmaking of British National Identity*, Vol. II: *Minorities and Outsiders* (London, 1989)

Sanderson, Michael, *The Universities and British Industry* (London, 1972)

Saunders, C. T., 'Consumption of Raw Materials in the United Kingdom: 1851–1950', *Journal of the Royal Statistical Society. Series A (General)* 115 (1952), pp. 313–53

Scarth, Richard N., *Echoes from the Sky: A Story of Acoustic Defence* (Hythe, 1999)

Schandl, Heinz and Niels Schulz, *Using Material Flow Accounting to Operationalize the Concept of Society's Metabolism. A Preliminary MFA for the United Kingdom for the Period of 1937–1997*, ISER Working Papers, Number 2000–3 (2000), Institute for Social and Economic Research, University of Essex, UK

Schatzberg, Eric, *Wings of Wood, Wings of Metal: Culture and Technical Choice in American Airplane Materials 1914–1945* (Princeton, 1999)

—, 'Technik Comes to America: Changing Meanings of Technology before 1930', *Technology and Culture* 47 (2006), pp. 486–512

Scott, J. D., *Vickers: A History* (London, 1962)

Scott, Peter, 'Path Dependence and Britain's "Coal Wagon Problem"', *Explorations in Economic History* 38 (2001), pp. 366–85

—, *Triumph of the South: A Regional Economic History of Britain during the Early Twentieth Century* (Aldershot, 2007)

413

Scott, William Taussig and Martin X. Moleski, *Michael Polanyi: Scientist and Philosopher* (Oxford, 2005)

Searle, Adrian, *Pluto: Pipe-Line under the Ocean* (Newport, 1995)

Searle, G. R., *Country Before Party: Coalition and the Idea of 'National Government' in Modern Britain, 1885–1987* (London, 1995)

Sebag-Montefiore, Hugh, *Dunkirk: Fight to the Last Man* (London, 2006)

Self, Robert, *Neville Chamberlain: A Biography* (Aldershot, 2006)

Shannon, H. A., 'The Sterling Balances of the Sterling Area, 1939–49', *The Economic Journal* 60 (September 1950), pp. 531–51

Shephard, Ben, *A War of Nerves: Soldiers and Psychiatrists in the Twentieth Century* (London, 2000)

Short, Brian, et al., *The National Farm Survey* (Oxford, 2000)

Shute, Nevil, *Slide Rule: The Autobiography of an Engineer* (London, 1956)

Sinha, Jagdish N., *Science, War and Imperialism: India in the Second World War* (Leiden, 2008)

Skennerton, Ian, *The Lee-Enfield: A Century of Lee-Metford & Lee-Enfield Rifles & Carbines* (Labrador, Qld, 2007)

Skidelsky, Robert, *John Maynard Keynes*, Vol. III: *Fighting for Britain, 1937–1946* (London, 2000)

Slim, William, *Unofficial History* (London, 1959)

Small, James S., *The Analogue Alternative: The Electronic Analogue Computer in Britain and the USA, 1930–1975* (London, 2001)

Smith, Adrian, *Mountbatten: Apprentice War Lord 1900–1943* (London, 2010)

Smith, David F. (ed.), *Nutrition in Britain: Science, Scientists and Politics in the Twentieth Century* (London, 1997)

Smith, Harold L. (ed.) *War and Social Change: Britain in the Second World War* (Manchester, 1986)

Smith, Kevin, *Conflict Over Convoys: Anglo-American Logistics Diplomacy in the Second World War* (Cambridge, 1996)

Smith, Malcolm, *Britain and 1940: History, Myth and Popular Memory* (London, 2000)

Smith, Paul (ed.), *Government and the Armed Forces in Britain* (London, 1996)

Smithies, Edward, *War in the Air: Men and Women Who Built, Serviced and Flew War Planes Remember the Second World War* (London, 1990)

Spaight, J. M., *Bombing Vindicated* (London, 1944)

—, *Air Power Can Disarm* (London, 1948)

Speake, G. D., 'The Marconi Research Centre – A Historical Perspective', *Physics in Technology* 16 (1985), pp. 275–81

Speer, Albert, *Inside the Third Reich* (London, 1995)

Spiekermann, Uwe, 'Brown Bread for Victory: German and British Whole-meal Politics in the Inter-War Period', in Frank Trentmann and Flemming Just (eds.), *Food and Conflict in Europe in the Age of the Two World Wars* (London, 2006), pp. 143–71

Stanier, Sir William, 'The Influence of War', *The Engineer* (1956; Centenary number)

Stapleton, Julia, *Political Intellectuals and Public Identities in Britain since 1850* (Manchester, 2001)

Steel, Robert W., 'The Trade of the United Kingdom with Europe', *The Geographical Journal* 87 (June 1936), pp. 525–33

Stewart, Andrew, *Empire Lost: Britain, the Dominions, and the Second World War* (London, 2008)

Stockwell, Sarah E. (ed.), *The British Empire: Themes and Perspectives* (Oxford, 2008)

Stoney, Barbara, *Twentieth Century Maverick: The Life of Noel Pemberton Billing* (East Grinstead, 2004)

Strachan, Hew, 'The British Way in Warfare Revisited', *Historical Journal*, 26 (1983), pp. 447–61

—, *The Politics of the British Army* (Oxford, 1997)

Stranges, Anthony S., 'From Birmingham to Billingham: Synthetic Fuels in Great Britain, 1910–1945', *Technology and Culture* 26 (1985), p. 726–57

Sturmey, S. G., *British Shipping and World Competition* (London, 1962)

Summerfield, Penny, *Women Workers in the Second World War: Production and Patriarchy in Conflict* (London, 1984)

— *Reconstructing Women's Wartime Lives: Discourse and Subjectivity in Oral Histories of the Second World War* (Manchester, 1998)

Sutherland, John, *Stephen Spender: The Authorized Biography* (London, 2002)

Swann, Brenda and Francis Aprahamian (eds.), *J. D. Bernal: A Life in Science and Politics* (London, 1999)

Sweetman, John (ed.), *Sword and Mace: Twentieth Century Civil–Military Relations in Britain* (London, 1966)

Syrett, David (ed.), *The Battle of the Atlantic and Signals Intelligence: U-Boat Tracking Papers, 1941–1945* (Aldershot, 2007)

Taylor, A. J. P., *The Origins of the Second World War* (Harmondsworth, 1963)

—, *English History, 1914–1945* (Oxford, 1965; Harmondsworth, 1975)

Taylor, Bruce, *The Battlecruiser HMS Hood: An Illustrated Biography 1916–1941* (London, 2005)

Terraine, John, *The Right of the Line* (London, 1985)

Terrell, Edward, *Admiralty Brief* (London, 1958)

The Left & World War II: Selections from War Commentary 1939–1943 (London, 1989)

The Making of a Ruling Class: Two Centuries of Capital Development on Tyneside (Newcastle Upon Tyne, 1978)

Thomas, William, 'The Heuristics of War: Scientific Method and the Founders of Operations Research', *The British Journal for the History of Science* 40 (2007), pp. 251–74

Thomson, David, *England in the Twentieth Century* (Harmondsworth, 1965)

Thomson, George Malcolm, *Vote of Censure* (London, 1968)

Thorpe, Andrew, *Parties at War: Political Organization in Second World War Britain* (Oxford, 2009)

Tiratsoo, Nick (ed.), *From Blitz to Blair: A New History of Britain since 1939* (London, 1997)

Tomlinson, Jim, 'The Attlee Government and the Balance of Payments, 1945–1951', *Twentieth Century British History* 2 (1991), pp. 47–66

—, 'Welfare and the Economy: The Economic Impact of the Welfare State, 1945–1951', *Twentieth Century British History* 6 (1995), pp. 194–219

—, *Democratic Socialism and Economic Policy: The Attlee Years, 1945–1951* (Cambridge, 1997)

—, *The Politics of Decline: Understanding Post-War Britain* (London, 2000)

—, 'Balanced Accounts? Constructing the Balance of Payments Problem in Post-War Britain', *The English Historical Review* CXXIV (2009), pp. 863–84

Tooze, Adam, *The Wages of Destruction: The Making and Breaking of the Nazi Economy* (London, 2006)

Toye, Richard, 'Keynes, the Labour Movement, and "How to Pay for the War"', *Twentieth Century British History* 10 (1999), pp. 255–81

—, 'The Labour Party's External Economic Policy in the 1940s', *The Historical Journal* 43 (2000), pp. 189–215

—, 'Churchill and Britain's "Financial Dunkirk"', *Twentieth Century British History* 15 (2004), pp. 329–60

—, *Lloyd George and Churchill: Rivals for Greatness* (London, 2008)

Tracy, Nicholas (ed.), *The Collective Naval Defence of the Empire 1900–1940* (Aldershot, 1997)

Trentmann, Frank, *Free Trade Nation: Commerce, Consumption, and Civil Society in Modern Britain* (Oxford, 2008)

Trentmann, Frank and Flemming Just (eds.), *Food and Conflict in Europe in the Age of the Two World Wars* (London, 2006)

Turner, John T., *'Nellie': The History of Churchill's Lincoln-Built Trenching Machine*, Occasional Papers in Lincolnshire History and Archaeology, No. 7 (Society for Lincolnshire History and Archaeology, 1988)

Tweedale, Geoffrey, *Steel City: Entrepreneurship, Strategy and Technology in Sheffield, 1743–1993* (Oxford, 1995)

van Creveld, Martin, *Supplying War: Logistics from Wallenstein to Patton* (Cambridge, 1980)

—, *Technology and War: From 2000 B.C. to the Present* (London, 1991)

Vernon, James, *Hunger: A Modern History* (London, 2007)

Verulam, Frank, *Production for the People* (London, 1940)

Vilensky, Joel A., *Dew of Death: The Story of Lewisite, America's World War I Weapon of Mass Destruction* (Bloomington, Ind., 2005)

Vincent, David, *The Culture of Secrecy: Britain 1832–1998* (Oxford, 1998)

Waddington, C. H., *OR in World War 2: Operational Research against the U-Boat* (London, 1973)

Wakelam, Randall T., *The Science of Bombing: Operational Research in RAF Bomber Command* (Toronto, 2009)

Walkland, S. A., 'Science and Parliament: The Origins and Influence of the Parliamentary and Scientific Committee', *Parliamentary Affairs* XVII (1963), pp. 317–18

Waller, Maureen, *London 1945: A Life in the Debris of War* (London, 2005)

Warren, Kenneth, *Steel, Ships and Men: Cammell Laird, 1824–1993* (Liverpool, 1998)

Watson-Watt, Sir Robert, *Three Steps to Victory: A Personal Account by Radar's Greatest Pioneer* (London, 1957)

Watt, Donald Cameron, *How War Came: The Immediate Origins of the Second World War, 1938–1939* (London, 1989)

Webster, Charles, 'Government Policy on School Meals and Welfare Foods, 1939–70', in David F. Smith (ed.), *Nutrition in Britain: Science, Scientists and Politics in the Twentieth Century* (London, 1997), pp. 190–213

Webster, Mark, *Assembly: New Zealand Car Production 1921–98* (Birkenhead, Auckland, 2002)

Webster, Wendy, *Englishness and Empire 1939–1965* (Oxford, 2005)

Weeks, Hugh, 'Anglo-American Supply Relationships', in D. N. Chester (ed.), *Lessons of the British War Economy* (Cambridge, 1951), pp. 69–82

Weinberg, Gerhard L., *A World at Arms: A Global History of World War II* (Cambridge, 1994)

Weir, Erin, 'German Submarine Blockade, Overseas Imports, and British Military Production in World War II', *Journal of Military and Strategic Studies* (Spring–Summer 2003), article 1

Werskey, Gary, 'British Scientists and "Outsider" Politics, 1931–1945', *Science Studies* 1 (1971), pp. 67–83

—, *The Visible College: A Collective Biography of British Scientists and Socialists of the 1930s* (London, 1978)

—, 'The Visible College Revisited: Second Opinions on the Red Scientists of the 1930s', *Minerva* 45 (2007), pp. 305–19

Westermann, Edward B., *Flak: German Anti-Aircraft Defenses, 1914–1945* (Lawrence, Kan., 2001)

Wheeler-Bennett, Sir John W., *John Anderson, Viscount Waverley* (London, 1962)

White, George, *Tramlines to the Stars: George White of Bristol* (Bristol, 1995)

Williams, Robert J. P., et al. (eds.), *Chemistry at Oxford: A History 1600 to 2005* (Cambridge, 2009)

Williams-Ellis, Clough, *Britain and the Beast: A Survey by Twenty-Six Authors* (London, 1937)

Wilmot, Sarah, 'From "Public Service" to Artificial Insemination: Animal Breeding Science and Reproductive Research in Early Twentieth-Century Britain', *Studies in the History and Philosophy of the Biological & Biomedical Sciences* 38 (2007), pp. 411–41

Wilson, Peter, 'The New Europe Debate in Wartime Britain', in Philomena Murray and Paul Rich (eds.), *Visions of European Unity* (Boulder, Colo., 1996), pp. 39–62

Wilson, Thomas, *Churchill and the Prof* (London, 1995)

Wilt, Alan F., *Food for War: Agriculture and Rearmament in Britain before the Second World War* (Oxford, 2001)

Wimperis, H. E., 'Research and Development in Aeronautics', 31st Thomas Hawksley Lecture, November 1944, *Proceedings of the Institution of Mechanical Engineers* 152 (1945), pp. 353–61

Wolmar, Christian, *Fire and Steam: A New History of the Railways in Britain* (London, 2007)

Woods, Abigail, 'The Farm as Clinic: Veterinary Expertise and the Transformation of Dairy Farming, 1930–1950', *Studies in the History and Philosophy of the Biological & Biomedical Sciences* 38 (2007), pp. 462–87

Wright, Patrick, *On Living in an Old Country: The National Past in Contemporary Britain* (London, 1985)

—, *The Village That Died for England: The Strange Case of Tyneham* (London, 1995)

—, *Tank: The Progress of a Monstrous War Machine* (London, 2000)

Wylie, Neville, 'British Smuggling Operations from Switzerland, 1940–1944', *The Historical Journal* 48 (December 2005), pp. 1077–1102

Zaidi, Waqar H., 'The Janus-Face of Techno-nationalism: Barnes Wallis and the "Strength of England"', *Technology and Culture* 49 (2008), pp. 62–88

Zeitlin, Jonathan, 'Flexibility and Mass Production at War: Aircraft Manufacture in Britain, the United States, and Germany, 1939–1945', *Technology and Culture* 36 (January 1995), pp. 46–79

—, 'Rationalization Misconceived: Productive Alternatives and Public Choices in British Shipbuilding, 1940–1990', paper prepared for the annual meeting of the Society for the History of Technology (SHOT), Toronto, 17–20 October 2002

Zeitlin, Jonathan and Gary Herrigel, *Americanization and Its Limits: Reworking US Technology and Management in Post-War Europe and Japan* (Oxford, 2000)

Ziegler, Philip, *Mountbatten: The Official Biography* (London, 1985)

Zilliacus, K., *Can the Tories Win the Peace? And How They Lost the Last One* (London, 1945)

Zimmerman, David, *Top Secret Exchange: The Tizard Mission and the Scientific War* (Montreal, 1996)

—, *Britain's Shield Radar and the Defeat of the Luftwaffe* (Stroud, 2001)

Zimmern, Alfred, *Quo Vadimus?* (London, 1934)

Zuckerman, Lord Solly, *From Apes to Warlords 1904–46* (London, 1978)

—, *Six Men Out of the Ordinary* (London, 1992)

Zweiniger-Bargielowska, Ina, *Austerity in Britain: Rationing, Controls, and Consumption, 1939–1955* (Oxford, 2000)

UNPUBLISHED DISSERTATIONS

Chaston, Philip, 'Gentlemanly Professionals within the Civil Service: Scientists as Insiders during the Interwar Period', Ph.D. thesis, University of Kent at Canterbury (2009)

Crellin, J. C., 'Ford in Manchester', undergraduate dissertation, University of Manchester (1987)

David, Thomas Rhodri Vivian, 'British Scientists and Soldiers in the First World War (with Special Reference to Ballistics and Chemical Warfare)', Ph.D. dissertation, University of London (2009)

Desmarais, R. J., 'Science, Scientific Intellectuals, and British Culture in the Early Atomic Age, 1945–1956: A Case Study of George Orwell, Jacob Bronowski, J. G. Crowther and P. M. S. Blackett', Ph.D. thesis, Imperial College (2009)

Duff, Colin, 'British Armoury Practice: Technical Change and Small Arms Manufacture, 1850–1939', M.Sc. thesis, University of Manchester (1990)

Fahey, John, 'Britain 1939–1945: The Economic Cost of Strategic Bombing', Ph.D. dissertation, University of Sydney (2006)

Giffard, Hermione, 'The Development and Production of Turbojet Aero-Engines in Britain, Germany and the United States, 1936–1945', Ph.D. thesis, Imperial College London (2011)

Gloster, David, 'Architecture and the Air Raid: Shelter Technologies and the British Government, 1938–44', M.Sc. dissertation, Imperial College London (1997)

Houghton, Tim, 'Chemical Offensive: British Chemical Warfare Research, 1914–1945', M.Sc. dissertation, Imperial College (2004)

Judkins, Phillip Edward, 'Making Vision into Power: Britain's Acquisition of the World's First Radar-Based Integrated Air Defence System 1935–1941', Ph.D. dissertation, Cranfield University (2008)

Reinisch, Jessica, 'The Society for Freedom in Science, 1940–1963', M.Sc. dissertation, Imperial College (2000)

Stadler, Max, 'Assembling Life: Models, the Cell, and the Reformations of Biological Science, 1920–1960', Ph.D. dissertation, University of London (2010)

Weatherburn, Michael, 'Arnold T. Wilson, the New Victorians and the Forgotten Technocrats of Inter-War Britain', M.Sc. dissertation, Imperial College (2009)

Webber, Nick, 'Battling for the Future: A Critique of the Current Historiography Relations to War, Medicine and Modernity', M.Sc. dissertation, Imperial College, 2000

Zaidi, S. W. H., 'Technology and the Reconstruction of International Relations: Liberal Internationalist Proposals for the Internationalisation of Aviation and the International Control of Atomic Energy in Britain, USA and France, 1920–1950', Ph.D. dissertation, University of London (2009)

Acknowledgements

My first thanks are due to the Leverhulme Trust for granting me a Leverhulme Major Research Fellowship (2006–9), without which this book would not have been written. I am also grateful to Imperial College for granting me leave from most of my duties during the fellowship. My colleagues and graduate students in the Centre for the History of Science, Technology and Medicine at Imperial College have been hugely supportive in many ways. They read and commented freely and generously on some ill-digested draft chapters. My thanks to Bruno Cordovil, Ralph Desmarais, Hermione Giffard, Michael Kershaw, Andrew Mendelsohn, Alex Oikonomou, Andrew Rabeneck, Aparajith Ramnath, Kathleen Sherit, Maximilian Stadler, Michael Weatherburn, Abigail Woods and Waqar Zaidi. Waqar Zaidi has criticized more drafts than I care to remember, and helped in many ways large and small to improve the text again and again, and in many other ways too. Ralph Desmarais has also read more versions than I could have expected. I am particularly grateful to all those masters and doctoral students who have shared their expert knowledge of the Second World War with me. My old friend Jim Rennie commented on an early complete draft, which thoroughly reshaped the text; Chris Mitchell was also shrewd in his criticisms. Thanks also to Guy Ortolano and William Thomas.

I was fortunate in having the opportunity to present preliminary conclusions at seminars at the National Museum of Scotland; my own Centre; the Department of History, University of Sheffield; the Department of Economic and Social History, University of Glasgow; the British History 1815–1945 Seminar at the Institute of Historical Research, University of London; the Department of History, University of Sussex; the International Imperial and Global History Seminar, Centre for European Studies, St Anthony's College Oxford; the Birmingham – Sciences-Po Seminar 'Reconstructions, 1918–1939: France and Britain in Inter-War Europe', at Sciences-Po, Paris; the military history seminar at the Department of History, University of Kent; and at a conference on the history of Imperial Chemical Industries. I am most grateful to the organizers and audiences for their comments, and especially Simon Ball, Marc-Olivier Baruch, Clarisse Berthelezene, Valentino Cerretano, Patricia Clavin, Mark Connelly, Matthew Cragoe, Saul Dubow, Bill Griffiths, Roland Quinault, Klaus Staubermann, Mary Vincent and Benjamin Ziemann. An invitation from Sarah Dry focused my thoughts on British assessments of risk in the 1930s. My thinking was also stimulated by being asked to contribute to two collections of essays, one edited by William Mulligan and Brendan Simms, the other by David Feldman, Mark Mazower and Jessica Reinisch. A conference on the work of Keith Middlemas, organized with Saul Dubow, Peter Mandler and Pat Thane, was another stimulus.

Many scholars and others have taken the trouble to share thoughts and materials with me, among them Simon Ball, Leslie Hannah, Jose Harris, Mark Harrison, Kurt Jacobsen, the late S. W. Johnson, Malcolm Llewellyn-Jones, Dominique Pestre, Joanna Richardson, Nick Stevenson, Pat Thane, Dan Todman, Adam Tooze, Mark Webster and Abigail Woods. David Boyd very kindly allowed me to use his detailed production data on British tanks, which allowed for a like-for-like comparison with German tank production; Phil Judkins, whose remarkable book on radar will soon appear, shared his Ph.D. thesis and his remarkably illuminating costing of the early radar programme. For their hospitality when on visits to archives my thanks to Alicia Edgerton, Graeme Gooday, Paul Heywood, John Pickstone, Adam Tooze, Mary Vincent and Richard Whittington.

It is a pleasure to acknowledge and thank the owners of private papers for granting access to and permission to quote from collections in many archives: John Hull (R. R. Stokes papers), the Hon. Ann Keynes (Richard Keynes papers), Professor Lord Krebs (Hans Krebs papers), Tony and Peter Goodeve (Sir Charles Goodeve papers), Vickers Limited and the Syndics of Cambridge University Library (Vickers Archive), the Documents and Sound Section of the Imperial War

Museum and the Trustees of the Imperial War Museum (Sir Henry Tizard papers), John Lennard-Jones (Sir John Lennard-Jones papers), the Institution of Mechanical Engineers (Lord Hinton of Bankside papers), Lieutenant-Commander David Gould (Admiral Sir Francis Pridham papers), and the Churchill Archives Centre and the Hill family (A. V. Hill papers). It is a particular pleasure to note the significance of the remarkable collection of industrial, scientific and engineering papers at the Churchill Archives Centre at Churchill College, Cambridge, to this study: I have made much use of them and wish to express my gratitude to Allen Packwood and his staff for many kindnesses. I would also like to thank the special collections librarians and archivists at the Bodleian Library (University of Oxford), Bristol University Library, the Brotherton Library (University of Leeds), the John Rylands Library (University of Manchester) and the University of Sheffield Library, as well as the Bristol Record Office, Cambridge University Library, Nuffield College, Oxford, the Imperial War Museum, and the National Archives. I would like to thank the Science Museum Library for permission to browse their large collection of technical and industrial journals just before they were taken from London, and the volunteers of Project Liberty Ship for a memorable visit to the SS *John W. Brown* in Baltimore.

This book would not have materialized had it not been for the support and encouragement of David Cannadine, Martin Daunton and, above all, Adam Tooze. Simon Winder, has also been extraordinarily encouraging and enthusiastic even before this book was thought of. He is a prince among editors, a joy to work with and a great reader. My warmest thanks to them all and to my agent, Clare Alexander. Thanks also to Mark Handsley and Thi Dinh at Penguin. Authors' families know books as interlopers. My special thanks to Claire, and to our children Francesca, Lucía and Andrew, who have shared between a third and two thirds of their lives with this book.

Index

Page numbers in **bold** refer to maps and figures; page numbers in *italic* refer to tables.